International
REVIEW OF
Neurobiology
Volume 81

International
REVIEW OF
Neurobiology
Volume 81

SERIES EDITORS

The Neurobiology of Epilepsy and Aging

EDITED BY

R. EUGENE RAMSAY

International Center for Epilepsy, Department of Neurology, University of Miami
School of Medicine, Miami, Florida, USA; Department of Neurology
Miami VA Medical Center, Miami, Florida, USA

JAMES C. CLOYD

Department of Experimental and Clinical Pharmacology
College of Pharmacy, University of Minnesota, Minneapolis, Minnesota, USA

KEVIN M. KELLY

Department of Neurology, Allegheny General Hospital
Pittsburgh, Pennsylvania, USA

ILO E. LEPPIK

Epilepsy Research and Education Program, University of Minnesota
Minneapolis, Minnesota, USA

EMILIO PERUCCA

Clinical Pharmacology Unit, Department of Internal Medicine and Therapeutics
University of Pavia, Pavia, Italy; Institute of
Neurology IRCCS C. Mondino Foundation, Pavia, Italy

AMSTERDAM • BOSTON • HEIDELBERG • LONDON
NEW YORK • OXFORD • PARIS • SAN DIEGO
SAN FRANCISCO • SINGAPORE • SYDNEY • TOKYO

ELSEVIER   Academic Press is an imprint of Elsevier

Academic Press is an imprint of Elsevier
525 B Street, Suite 1900, San Diego, California 92101-4495, USA
84 Theobald's Road, London WC1X 8RR, UK

This book is printed on acid-free paper. ∞

For information on all Academic Press publications
visit our Web site at www.books.elsevier.com

ISBN-13: 978-0-12-374018-2
ISBN-10: 0-12-374018-5

PRINTED IN THE UNITED STATES OF AMERICA
07 08 09 10 9 8 7 6 5 4 3 2 1

DEDICATION

This book is dedicated to the life and memories of A. James Rowan, who lost his battle with cancer on August 27, 2006. Jim was a quiet man of modest physical stature, but he was a giant in the eyes of his family, friends, and colleagues. He always commanded our attention and gained our deep respect in everything he did. How will we remember him? A colleague and good friend, Jim Cloyd, said it well: "I will always remember Jim standing with his hands on his hips and his head slightly cocked, with an impish grin on his face."

Jim had many loves and passions. Beyond the satisfaction he derived from his roles as physician, teacher, and researcher, he adored his family, his friends, fine dining, reading, and writing. He was the consummate clinician with a tremendous devotion to his patients. His patients and their families continually commented on Dr. Rowan's caring and personable manner and knew him to be an outstanding physician.

Jim's strongest passion was the desire to teach physicians, particularly those in training. He loved to explore the question, "How do you teach the ART, and not just the science, of neurology?" He was an astute observer, which always served him well. He started each patient's examination as he entered a room: before he even talked to or touched the patient. His *Primer of EEG* has become a global standard, written as though he were talking to the reader in person. He put his heart and soul into writing—which for him was teaching—in his own clear and personalized style.

I was fortunate to partner with Jim during his pinnacle accomplishment: the planning, successful completion, and publication of the results of a large clinical trial in geriatric epilepsy. His eyes sparkled when we talked about the results and put together plans to educate physicians about our findings, knowing that this would substantially change the approach to treating geriatric patients with epilepsy. As a result, Jim has been instrumental in improving the lives of so many people.

The world is a better place because of Arbo James Rowan. We all will remember his passions for superb teaching, patient care, and finding answers to his own critical clinical questions. This book is but one of the many impressive achievements of his life.

R. Eugene Ramsay

CONTENTS

Epilepsy in the Elderly: Scope of the Problem
Ilo E. Leppik

Animal Models in Gerontology Research
Nancy L. Nadon

Animal Models of Geriatric Epilepsy
Lauren J. Murphree, Lynn M. Rundhaugen, and Kevin M. Kelly

Life and Death of Neurons in the Aging Cerebral Cortex

JOHN H. MORRISON AND PATRICK R. HOF

An *In Vitro* Model of Stroke-Induced Epilepsy: Elucidation of the Roles of Glutamate and Calcium in the Induction and Maintenance of Stroke-Induced Epileptogenesis

ROBERT J. DELORENZO, DAVID A. SUN, ROBERT E. BLAIR, AND SOMPONG SOMBATI

Mechanisms of Action of Antiepileptic Drugs

H. Steve White, Misty D. Smith, and Karen S. Wilcox

Epidemiology and Outcomes of Status Epilepticus in the Elderly

Alan R. Towne

Diagnosing Epilepsy in the Elderly

R. Eugene Ramsay, Flavia M. Macias, and A. James Rowan

Pharmacokinetics of Antiepileptic Drugs in Elderly Nursing Home Residents

ANGELA K. BIRNBAUM

The Impact of Epilepsy on Older Veterans

MARY JO V. PUGH, DAN R. BERLOWITZ, AND LEWIS KAZIS

Risk and Predictability of Drug Interactions in the Elderly

RENÉ H. LEVY AND CAROL COLLINS

Outcomes in Elderly Patients With Newly Diagnosed and Treated Epilepsy

MARTIN J. BRODIE AND LINDA J. STEPHEN

Recruitment and Retention in Clinical Trials of the Elderly

Flavia M. Macias, R. Eugene Ramsay, and A. James Rowan

Treatment of Convulsive Status Epilepticus

David M. Treiman

Treatment of Nonconvulsive Status Epilepticus

Matthew C. Walker

Antiepileptic Drug Formulation and Treatment in the Elderly: Biopharmaceutical Considerations

BARRY E. GIDAL

CONTRIBUTORS

Dan R. Berlowitz (153, 221), Center for Health Quality, Outcomes, and Economic Research, Bedford VA Medical Center, Bedford, Massachusetts 01730, USA; Department of Health Services, Boston University School of Public Health, Boston, Massachusetts 02118, USA

Angela K. Birnbaum (201, 211), Department of Experimental and Clinical Pharmacology, College of Pharmacy, University of Minnesota, Minneapolis, Minnesota 55455, USA

Robert E. Blair (59), Department of Neurology, Virginia Commonwealth University, Richmond, Virginia 23298, USA

Martin J. Brodie (253), Patient Services and Clinical Research Studies, Epilepsy Unit, Division of Cardiovascular and Medical Sciences, Western Infirmary, Glasgow, G11 6NT Scotland, United Kingdom

James C. Cloyd (201), Department of Experimental and Clinical Pharmacology, College of Pharmacy, University of Minnesota, Minneapolis, Minnesota 55455, USA

Carol Collins (235), Department of Pharmaceutics, University of Washington, Seattle, Washington 98195, USA

Robert J. DeLorenzo (59), Department of Pharmacology and Toxicology, Virginia Commonwealth University, Richmond, Virginia 23298, USA; Department of Neurology, Virginia Commonwealth University, Richmond, Virginia 23298, USA; and Department of Biochemistry and Molecular Biophysics, Virginia Commonwealth University, Richmond, Virginia 23298, USA

Lynn E. Eberly (165), Division of Biostatistics, School of Public Health, University of Minnesota, Minneapolis, Minnesota 55455, USA

Judith Garrard (165), Division of Health Policy and Management, School of Public Health, University of Minnesota, Minneapolis, Minnesota 55455, USA

Barry E. Gidal (299), School of Pharmacy and Department of Neurology, University of Wisconsin, Madison, Wisconsin 53705, USA

Susan L. Harms (165), Division of Health Policy and Management, School of Public Health, University of Minnesota, Minneapolis, Minnesota 55455, USA

Patrick R. Hof (41), Department of Neuroscience, Mount Sinai School of Medicine, New York, New York 10029, USA

Lewis Kazis (221), Center for Health Quality, Outcomes, and Economic Research, Bedford VA Medical Center, Bedford, Massachusetts 01730, USA; Department of Health Services, Boston University School of Public Health, Boston, Massachusetts 02118, USA

Kevin M. Kelly (29), Department of Neurology, Allegheny General Hospital, Pittsburgh, Pennsylvania 15212, USA

Ilo E. Leppik (1, 165), Epilepsy Research and Education Program, University of Minnesota, Minneapolis, Minnesota 55455, USA

René H. Levy (235), Department of Pharmaceutics, University of Washington, Seattle, Washington 98195, USA; Department of Neurological Surgery, University of Washington, Seattle, Washington 98195, USA

Flavia M. Macias (129, 265), Department of Neurology, Miami VA Medical Center, Miami, Florida 33125, USA

Susan Marino (201), Department of Pharmacy Practice, College of Pharmacy, University of Florida, Gainesville, Florida 32610, USA

John H. Morrison (41), Department of Neuroscience, Mount Sinai School of Medicine, New York, New York 10029, USA

Lauren J. Murphree (29), SRA International, Inc., Fairfax, Virginia 22033, USA; National Institute of Neurological Disorders and Stroke, National Institutes of Health, Bethesda, Maryland 20892, USA

Nancy L. Nadon (15), National Institute on Aging, Bethesda, Maryland 20892, USA

Emilio Perucca (183), Clinical Pharmacology Unit, Department of Internal Medicine and Therapeutics, University of Pavia, Pavia, Italy; Institute of Neurology IRCCS C. Mondino Foundation, Pavia, Italy

Mary Jo V. Pugh (153, 221), South Texas Veterans Health Care System (VERDICT), San Antonio, Texas 78229, USA; Department of Medicine, University of Texas Health Science Center at San Antonio, San Antonio, Texas 78229, USA

R. Eugene Ramsay (129, 265), International Center for Epilepsy, Department of Neurology, University of Miami School of Medicine, Miami, Florida 33136, USA; Department of Neurology, Miami VA Medical Center, Miami, Florida 33125, USA

A. James Rowan (129, 265), Department of Neurology, Bronx VA Medical Center, Bronx, New York 10468, USA; Mount Sinai School of Medicine, New York, New York 10029, USA

Lynn M. Rundhaugen (29), National Institute of Neurological Disorders and Stroke, National Institutes of Health, Bethesda, Maryland 20892, USA

Misty D. Smith (85), Anticonvulsant Drug Development Program, Department of Pharmacology and Toxicology, University of Utah, Salt Lake City, Utah 84108, USA

Sompong Sombati (59), Department of Neurology, Virginia Commonwealth University, Richmond, Virginia 23298, USA

Linda J. Stephen (253), Patient Services and Clinical Research Studies, Epilepsy Unit, Division of Cardiovascular and Medical Sciences, Western Infirmary, Glasgow, G11 6NT Scotland, United Kingdom

David A. Sun (59), Department of Neurological Surgery, Vanderbilt University Medical Center, Nashville, Tennessee 37232, USA

Alan R. Towne (111), Department of Neurology, Virginia Commonwealth University, Richmond, Virginia 23298, USA

David M. Treiman (273), Barrow Neurological Institute, Phoenix, Arizona 85013, USA

Matthew C. Walker (287), Department of Clinical and Experimental Epilepsy, Institute of Neurology, University College London, London WC1N 3BG, United Kingdom

H. Steve White (85), Anticonvulsant Drug Development Program, Department of Pharmacology and Toxicology, University of Utah, Salt Lake City, Utah 84108, USA

Karen S. Wilcox (85), Anticonvulsant Drug Development Program, Department of Pharmacology and Toxicology, University of Utah, Salt Lake City, Utah 84108, USA

ACKNOWLEDGMENT

Editorial support was provided by MedLogix Communications, LLC.

EPILEPSY IN THE ELDERLY: SCOPE OF THE PROBLEM

Ilo E. Leppik

Epilepsy Research and Education Program
University of Minnesota, Minneapolis, Minnesota 55455, USA

I. Introduction
II. Known Knowns
    A. Incidence and Prevalence of Epilepsy Is Higher in the Community-Dwelling Elderly Than in Younger Adults or Children
    B. Prevalence and Incidence of Epilepsy Is Higher Among Nursing Home Residents Than Community-Dwelling Elderly
    C. Retirement Age of 65 Is a Political Decision, Not Related to Medical Condition
    D. The Elderly Are Not a Homogeneous Group
    E. Patterns of AED Use Differ Markedly Between Countries and Between Community-Dwelling and Nursing Home Elderly
    F. In Nursing Homes, the Young-Old Are More Likely to Use an AED Than the Old-Old
    G. Causes of Epilepsy in the Elderly
    H. Many Seizures in the Elderly Are Not Epileptic
    I. AED Pharmacology Is Different in the Elderly
    J. Choosing AEDs for an Elderly Person Is Difficult
    K. Drug Interactions With Non-AEDs Are a Major Problem
III. Known Unknowns
IV. Unknown Unknowns
V. Unknown Knowns
    A. The "Therapeutic" Range of Total Phenytoin Is 10–20 mg/liter
    B. AED Levels Are Stable Over Time With Constant Doses
VI. Conclusion
    References

The knowledge base for treating elderly persons with epilepsy is limited. There are few known knowns, many known unknowns, and probably many unknown knowns, that is, the things we know that "ain't so." We know that the incidence and prevalence of epilepsy is higher in the elderly than any other age group, that the elderly are not a homogeneous group, that epilepsy is much more common in the nursing home population than in the community-dwelling elderly, and that antiepileptic drug (AED) use varies greatly among countries, but that in all, the older AEDs (phenytoin, phenobarbital, and carbamazepine) are the most commonly used. We also know that drugs that require hepatic metabolism for elimination are subject to pharmacokinetic changes with age and may be problematic because of

INTERNATIONAL REVIEW OF
NEUROBIOLOGY, VOL. 81
DOI: 10.1016/S0074-7742(06)81001-9

1

drug–drug interactions. There are many known unknowns in both the basic science of brain aging and the susceptibility to epilepsy, and many clinical issues remain unresolved. Some unknown knowns (i.e., misconceptions) are that the elderly need levels of AEDs similar to those for younger adults and that AED levels do not fluctuate widely. This book is designed to help the reader understand the issues and, hopefully, to stimulate research to provide answers for the known unknowns.

I. Introduction

The elderly are the most rapidly growing segment of the population, and demographic trends predict that by 2050 the proportion of the population older than 60 years will be more than 30% in many developed countries (United Nations, 2001). Epidemiological studies have shown that onset of epilepsy is higher in the elderly than in any other age group (Hauser *et al.*, 1996). With the combination of these two factors, elderly persons with epilepsy will represent an increasingly large group of patients needing expert care to maximize their quality of life and minimize health care costs. Because elderly persons with epilepsy may have different problems than younger adults, research which specifically identifies and addresses issues relevant to this population is needed. In a news briefing, knowledge was classified as "known knowns, known unknowns, and unknown unknowns" (Rumsfeld, 2002). However, the speaker left out perhaps the most critical class, the unknown knowns, or "things we know that ain't so" (Leppik, 2006b), to paraphrase Mark Twain. In medicine, this class consists of generally accepted facts that are found not to be true or applicable, after more information from research is available, and that may cause more harm than good. One example is the past practice of using aspirin to treat pain in children with hemophilia. Today, there may be many practices used in younger persons which may be inappropriate for the elderly. One cannot simplistically extrapolate knowledge gained from research done in younger patients to the elderly. Despite the magnitude of the problem, little basic and clinical research has been directed toward addressing crucial issues facing the elderly person with epilepsy (Leppik, 2006b). This chapter will review the scope of the problem and provide a context for the chapters included in this volume.

II. Known Knowns

A. INCIDENCE AND PREVALENCE OF EPILEPSY IS HIGHER IN THE COMMUNITY-DWELLING ELDERLY THAN IN YOUNGER ADULTS OR CHILDREN

Only recently have there been sufficient numbers of elderly in the general population to provide meaningful epidemiological data regarding the magnitude of the problem of epilepsy in the elderly. For a long time, epilepsy was considered to be primarily a disorder of children, with the incidence of epilepsy decreasing with age. However, studies have shown that the incidence of epilepsy is represented by a U-shaped curve, with the highest incidence being at both ends of the age spectrum (Cloyd et al., 2006; Hauser et al., 1996). One early study found that the incidence of seizures begins to increase after age 50 and rises to 127/100,000 person-years in those aged 60 and older (Hauser and Hesdorffer, 1990). A study reported the incidence of epilepsy to be 10.6/100,000 person-years for the ages of 45–59, 25.8/100,000 person-years for the ages of 60–74, and 101.1/100,000 person-years for the ages of 75–89 (Hussain et al., 2006). African-American subjects in this study had more than twice the incidence of epilepsy, 57.6/100,000 person-years compared to 26.1/100,000 person-years for Caucasians (Hussain et al., 2006). Among persons 75 years and older, the active epilepsy prevalence rate of persons living in the community is approximately 1.5%, about twice the rate of younger adults (Hauser et al., 1991). A study of data collected on 1,130,155 veterans aged 65 or older from 1997 to 1999 found that 20,558 of the veterans (1.8%) had a diagnosis of epilepsy (Perucca et al., 2006; Pugh et al., 2004). In the United States, approximately 181,000 persons developed epilepsy in 1995, and approximately 61,000 of these were over age 65 (Epilepsy Foundation of America, 1999). In Finland, the incidence of epilepsy decreased significantly in children and younger adults from 1986 to 2002, but increased significantly ($p < 0.0001$) in the elderly (Sillanpaa et al., 2006).

B. Prevalence and Incidence of Epilepsy Is Higher Among Nursing Home Residents Than Community-Dwelling Elderly

In the United States, all physicians' orders must carry an indication listed by an International Classification of Diseases, Ninth Revision (ICD-9), code. Using these codes, it has been possible to link specific antiepileptic drug (AED) use to diagnoses. This is very useful because some AEDs, such as valproate and carbamazepine, are often used for conditions other than epilepsy. However, the criteria or findings used to make a diagnosis of epilepsy are not well understood. By using these codes, studies have shown that approximately 9% of nursing home residents in the United States are classified as having epilepsy or seizures (Garrard et al., 2000). A report from a study of 2001 Italian nursing home residents stated that 5.3% of the men and 4.0% of the women were being treated with an AED, but only 3.5% of the men and 2.7% of the women had a history of epileptic seizures (Galimberti et al., 2006). A study of 565 residents of a nursing home in Germany found that 4.96% had an AED prescribed, but only 3.00% had a seizure-related diagnosis (Huying et al., 2006). Thus, the use of AEDs in nursing homes in the United States is much higher than in European countries, but interestingly, the proportion of AED prescriptions for epilepsy and other conditions is similar. Also, in all countries, the prevalence of epilepsy in nursing home residents is much higher than in the community-dwelling elderly (Garrard et al., 2000).

Only one study gives some indication of the incidence of epilepsy in nursing home residents; it is a study of 510 Beverly Enterprises nursing homes with 10,318 admissions over a 6-month period. On admission, 802 residents (7.8%) were using an AED. Of these, 57.7% had an epilepsy/seizure ICD-9 code, and most of the others had an indication suggesting behavioral issues. In the 3-month follow-up period, an additional 260 persons (2.7% of the admission cohort who were not using an AED) were placed on an AED, 20.9% for an epilepsy/seizure indication (Garrard et al., 2003). Thus, 54 of 9516 persons had an event resulting in a diagnosis of epilepsy/seizure within 3 months after admission. This is an incidence rate of approximately 571/100,000, or more than five times the incidence rate for the community-dwelling elderly. Another key finding of this study was that the odds ratio for being placed on an AED decreased with age. Setting 1.00 as the reference for the 65- to 74-year age group, the 75–84 cohort had an odds ratio of 0.68 ($p < 0.05$), and those over age 85 had an odds ratio of 0.47 ($p < 0.0001$) (Garrard et al., 2003). This is in marked contrast to the community-dwelling elderly, in whom the incidence of epilepsy increases with age (Hauser et al., 1996). Thus, the question arises as to whether the incidence of epilepsy actually decreases or the diagnosis is being missed in the older patient.

C. RETIREMENT AGE OF 65 IS A POLITICAL DECISION, NOT RELATED TO MEDICAL CONDITION

Although age 65 is generally accepted as the age for retirement under entitlement programs, this age is somewhat arbitrary. The concept of a government-sponsored retirement system came into being during the 1880s, at a time when German Chancellor Otto von Bismarck's political party was campaigning for reelection. For this first-ever government-sponsored entitlement program, the proposed retirement age was set at 70 years, when the average lifespan was only 48 years, but Bismarck's party nevertheless won the election. As it became apparent that very few would collect any benefits under Bismarck's scheme, the retirement age was reduced to 65 early in the twentieth century (http://www.ssa.gov/history/ottob.html; Social Security Administration history page). Since that time, "the elderly" have generally been defined as those 65 years or older, even though there is no medical evidence to support this. Aging is a gradual process, with many of the changes that manifest themselves in later years having their genesis much earlier, but remaining subclinical. Furthermore, there is a wide variation in the age at which persons begin to manifest these changes. Unfortunately, this demarcation between younger adults and the elderly has achieved wide acceptance, and until more accurate categories are developed and accepted, those over 65 will continue to be considered as elderly in this volume.

D. THE ELDERLY ARE NOT A HOMOGENEOUS GROUP

It is becoming clear that the elderly are not a homogeneous group. The elderly have been subdivided into the "young-old," those 65–74 years of age; the "middle-old" or "old," those 75–84 years of age; and the "old-old," those 85 years or older. However, as persons develop health issues at different times, further subdivisions have been proposed (Leppik, 2006a). These are the persons with only epilepsy, the persons with epilepsy and multiple medical problems, and the frail elderly. Thus, one must tailor the studies and interventions to nine categories (Table I) (Leppik, 2006a). In addition, there are major differences between the community-dwelling elderly (independent living) and those residing in nursing homes. In general, the healthy elderly are most likely community-dwelling, whereas the frail elderly reside in nursing homes.

Research in persons up to age 18 cannot be properly interpreted without using the subcategories of newborn, infant, and so on, and studies of elderly persons should also specify which population is being evaluated. Clearly, drug side effects, efficacy, absorption, and other factors may be markedly different between a 93-year-old healthy person and a 68-year-old frail person. Also, issues regarding health care delivery will likely differ between the community-dwelling elderly and the nursing home elderly. Studies should be designed with specific populations, and reports should specify the populations studied (Leppik, 2006a,b).

E. PATTERNS OF AED USE DIFFER MARKEDLY BETWEEN COUNTRIES AND BETWEEN COMMUNITY-DWELLING AND NURSING HOME ELDERLY

Use of AEDs in the elderly varies from country to country. In the United States, approximately 1.5% of the community-dwelling elderly are prescribed an AED, and phenytoin is the most commonly used AED (Pugh *et al.*, 2004). In a

TABLE I
FURTHER DIVISIONS OF THE ELDERLY POPULATION

Aged 65–74 years	Aged 75–84 years	Aged 85 years and older
Healthy young-old	Healthy middle-old	Healthy old-old
Young-old with multiple medical problems	Middle-old with multiple medical problems	Old-old with multiple medical problems
Frail young-old	Frail middle-old	Frail old-old

Adapted from Leppik (2006a), with permission from Elsevier, Copyright 2006.

US Veterans Affairs study of patients with epilepsy, phenytoin was used by 61% of the patients and carbamazepine by slightly more than 10%. However, phenobarbital was used in 18%, often in combination with phenytoin (Pugh et al., 2004). In a study of community use from a registry of 471,873 patients in Denmark, the eight most frequent AED regimens were all monotherapy: carbamazepine, oxcarbazepine, phenobarbital, valproic acid, lamotrigine, clonazepam, phenytoin, and primidone, in that order (Rochat et al., 2001). The estimated crude 1-year prevalence of AED use was greater in men, 0.83% of the population, than women, 0.77% ($p < 0.001$), and it increased with age for both genders. The prescription pattern revealed a surprisingly extensive use of phenobarbital and very low use of phenytoin (Rochat et al., 2001).

The prevalence of epilepsy, seizures, and AED use is much higher in nursing home residents. In a review of 45,405 people aged 65 years or older living throughout the United States in long-term care facilities, at least one AED was taken by 4573 (10.1%) of the residents (Cloyd et al., 1994). Additional studies have confirmed that the prevalence of AED use in the nursing home population varies between 10% and 11% (Garrard et al., 2000; Lackner et al., 1998; Schachter et al., 1998). Approximately 5% of all Americans reside in a nursing home at any given time (Hetzel and Smith, 2001), and 43% of persons older than age 65 are likely to enter one sometime before they die (Kemper and Murtaugh, 1991). A 1994 report indicated that approximately 1.4 million elderly people reside in nursing homes in the United States (Dey, 1996). Thus, as many as 150,000 elderly nursing home residents may be taking AEDs.

A closer examination of AED use in nursing homes revealed that approximately 7% of residents were receiving AEDs at the time of admission, but surprisingly, approximately 3% had an AED newly prescribed after admission (Garrard et al., 2003). Phenytoin is the most commonly used AED in nursing homes. In a 1-day, point-prevalence study of 21,551 nursing home residents in 24 states and the District of Columbia in 1995, 10.5% had an AED order (Garrard et al., 2000). Among the residents, 6.2% were using phenytoin, 1.8% carbamazepine, 0.9% valproic acid, 1.7% phenobarbital, and 1.2% other AEDs. In a subsequent study evaluation of AED initiation after admission, phenytoin was the most commonly initiated AED when the ICD-9 code was seizures or epilepsy, but valproate or carbamazepine was more common when a psychiatric or behavioral diagnosis was listed (Garrard et al., 2003). In a study of AED use in nursing homes in Germany, the most frequently prescribed AEDs were carbamazepine (37.1%), valproic acid (25.9%), and phenytoin (14.8%) (Huying et al., 2006). In a study of Italian nursing home residents, phenobarbitone was by far the most frequently prescribed AED (Galimberti et al., 2006). Thus, prescribing patterns differ greatly between countries.

F. In Nursing Homes, the Young-Old Are More Likely to Use an AED Than the Old-Old

Analysis of AED use in US nursing homes by age groups indicates that there are major differences in use by the designated age subdivisions. Paradoxically, in the nursing homes, far fewer of the old-old use AEDs, as compared to the young-old (Table II). This pattern has also been noted in Italy, where it was found that residents in early old age (60–74 years) were more likely than older residents to take an AED (Galimberti et al., 2006).

G. Causes of Epilepsy in the Elderly

The causes of epilepsy in the elderly are reasonably well known. The most common identifiable cause of epilepsy in the elderly is stroke, and it accounts for 30–40% of all cases (Hauser, 1997). Brain tumor, Alzheimer's disease, and head injury are other major causes. However, in approximately half of the cases, no specific etiology can be identified (cryptogenic). One study found that among "healthy" subjects, those without a history of stroke, head injury, or dementia, the cumulative risk of onset of epilepsy after 60 years of age was 1.1%, much lower than if the risk factors were present (Hussain et al., 2006). The 1.1% rate, however, is still higher than rates seen in younger adults. Persons with mild Alzheimer's disease had a cumulative incidence of unprovoked seizure risk of 8%, and the risk was highest (87-fold increase) in the 50–59 age group with early Alzheimer's disease (Amatniek et al., 2006).

A major concern, however, is that some seizures may not be epileptic, that is, they may originate from a medical problem outside the brain. In the elderly, a clear distinction must be made between the causes of seizures and the causes of epilepsy. Many conditions can cause seizures in the elderly: cardiac dysfunctions,

TABLE II
AED Use by Age Category

	AED users	Total	Percentage
Young-old	745	3143	23.7
Middle-old	950	7783	12.2
Old-old	616	10,625	5.8

AED indicates antiepileptic drug.
Data from Garrard et al. (2000).

metabolic derangements, respiratory compromise, alcohol abuse, and so on. These must be excluded before a diagnosis of an epileptic seizure can be made.

H. Many Seizures in the Elderly Are Not Epileptic

The most crucial differential diagnosis in the elderly population is that of convulsive syncope. In this condition, a seizure is provoked by lack of circulation to the brain. Cardiac causes are common, and in the absence of a known central nervous system disorder (stroke, tumor, or degeneration), syncope should be suspected early. An electroencephalogram with concomitant electrocardiogram (ECG) rhythm strip is essential.

Electrolyte imbalance, febrile illness, hypoglycemia, or hyperglycemia may also provoke seizures, but these conditions should be easily recognized by laboratory tests or physical examination, and they do not need to be treated with chronic AEDs. Psychogenic nonepileptic seizures (PNES) have been observed in the elderly and should not be overlooked (Behrouz et al., 2006).

I. AED Pharmacology Is Different in the Elderly

The age-related physiological changes that appear to have the greatest effect on AED pharmacokinetics involve protein binding and the reduction in liver volume and blood flow (Verbeeck et al., 1984; Wynne et al., 1989). By age 65, many individuals have low normal albumin concentrations or are frankly hypoalbuminemic (Verbeeck et al., 1984). Albumin concentration may be further reduced by conditions such as malnutrition, renal insufficiency, and rheumatoid arthritis. As serum albumin levels decline, the likelihood increases that drug binding will decrease, and measurement of total concentrations is misleading.

For low-clearance drugs, hepatic clearance of the older AEDs is primarily influenced by the extent of protein binding and the intrinsic metabolizing capacity (intrinsic clearance) of the liver. Because hepatic clearance affects steady-state drug concentrations, age-related alterations in protein binding or intrinsic clearance can affect serum drug concentrations.

Phase I reactions (oxidation, reduction, and hydroxylation) are thought to be affected to a greater extent than phase II reactions (glucuronidation, acetylation, and sulfonation). The decrease in oxidative metabolism (cytochrome P450) of some drugs in the elderly has been theorized to be due to changes in liver blood flow and liver mass as one ages (Woodhouse and Wynne, 1988). However, on the basis of very limited information, changes in the glucuronidation reactions appear to be spared with aging (Herd et al., 1991; Woodhouse and Herd, 1993).

Despite the theoretical effects of age-related physiological changes on drug disposition and the widespread use of AEDs in the elderly, few studies on AED pharmacokinetics in the elderly have been published.

J. Choosing AEDs for an Elderly Person Is Difficult

Assessment of AED treatment efficacy and toxicity in elderly patients is challenging because seizures are sometimes difficult to observe, signs and symptoms of toxicity can be attributed to other causes (Alzheimer's disease, stroke, and so on) or to comedications, and patients may not be able to accurately self-report problems. Treatment in the elderly carries more risks than in younger persons because they may experience more side effects and have a greater risk for drug interactions; in addition, the elderly may be less able to afford the costs of medications. Neither the benefits nor the risks of treatment have been investigated in this population; current practices are not based on evidence. As a cause of adverse reactions among the elderly, AEDs rank fifth among all drug categories (Moore and Jones, 1985).

At the present time, there is a great lack of data regarding the clinical use of AEDs in the elderly. The paucity of information makes it very difficult to recommend specific AEDs with any confidence that the outcomes will be optimal. Nevertheless, decisions need to be made, and indeed are being made. Consequently, the "comfort level" with some drugs may play a larger role in choosing them than actual experience or data.

A drug choice optimal for the elderly healthy except for epilepsy (EH) group may not be appropriate for the elderly with multiple medical problems (EMMP) group. Even in the EH group, there will be a decline in functioning of the various organ systems, leading to lower hepatic and renal clearance of AEDs, and a possible increase in sensitivity of the central nervous system to side effects.

Because AEDs may affect balance, falls and fractures may be related to inappropriate AED use, especially in nursing homes. In addition, the cost of AEDs may be an important factor for many patients. These issues are greatly complicated for those with multiple medical problems. The physician must thus be aware of the benefits and shortcomings of all AEDs and the state of physical and fiscal health of the patient.

The most commonly used AEDs in the elderly are phenytoin, carbamazepine, phenobarbital, and valproate. However, each of these has significant disadvantages for use in the elderly. Only a few studies focusing specifically on the elderly have been done. One study compared carbamazepine to lamotrigine in newly diagnosed elderly persons with epilepsy and found that lamotrigine use was associated with fewer adverse events (Brodie et al., 1999). A more recent study, which was conducted at US Veterans Affairs medical centers, compared carbamazepine to lamotrigine and gabapentin in 593 newly diagnosed patients with epilepsy. The primary end point was early termination from the trial, and by this criterion, 44.2% of lamotrigine-treated patients discontinued, compared to 51.0% for gabapentin and 64.5% for carbamazepine (Rowan *et al.*, 2005). The major issues leading to discontinuation are related to tolerability. An analysis of the tolerability of levetiracetam in older persons in studies of anxiety, cognitive disorders, and epilepsy found it to be well tolerated (Cramer *et al.*, 2003). Thus, it appears that tolerability rather than control of seizures may be the major factor when deciding on the best drug.

K. DRUG INTERACTIONS WITH NON-AEDs ARE A MAJOR PROBLEM

Concomitant medications taken by elderly patients can alter the absorption, distribution, and metabolism of AEDs, thereby increasing the risk of toxicity or therapeutic failure. Comedications are frequently used by patients in nursing homes receiving AEDs. Calcium-containing antacids and sucralfate reduce the absorption of phenytoin. The absorption of phenytoin, carbamazepine, and valproate may be reduced significantly by oral antineoplastic drugs that damage gastrointestinal cells. In addition, phenytoin concentrations may be lowered by intravenously administered antineoplastic agents. The use of folic acid for treatment of megaloblastic anemia may decrease serum concentrations of phenytoin, and enteral feedings can also lower serum concentrations in patients receiving orally administered phenytoin.

Many drugs displace AEDs from plasma proteins, an effect that is especially serious when the interacting drug also inhibits the metabolism of the displaced drug; this occurs when valproate interacts with phenytoin. Several drugs used on a short-term basis (including propoxyphene and erythromycin) or as maintenance therapy (such as cimetidine, diltiazem, fluoxetine, and verapamil) inhibit the metabolism of one or more AEDs that are metabolized by the P450 system. Certain agents can induce the P450 system or other enzymes, causing an increase in drug metabolism. The most commonly prescribed inducers of drug metabolism are phenytoin, phenobarbital, carbamazepine, and primidone. Ethanol, when used chronically, also induces drug metabolism. The interaction between antipsychotic drugs and AEDs is complex. Hepatic metabolism of certain antipsychotics, such as haloperidol, can be increased by carbamazepine, resulting in diminished psychotropic response. Antipsychotic medications, especially chlorpromazine, promazine, trifluoperazine, and perphenazine, can reduce the threshold for seizures. The risk of seizures has been thought to be related to the total number of psychotropic medications taken concurrently, their doses, abrupt increases in doses, and the presence of organic brain pathology. The epileptic patient taking antipsychotic drugs may need a higher dose of antiepileptic medication to control seizures. In contrast, central nervous system depressants are likely to lower the maximum dose of AEDs that can be administered before toxic symptoms occur.

III. Known Unknowns

Questions that exist in the field of epilepsy for which there are currently no answers include:

- Why do the nursing home young-old have a much higher rate of AED use than the old-old, in direct contrast with the community-dwelling elderly? This is especially puzzling in light of the known increase of epilepsy with advancing age.
- Does the aging brain become more prone to epilepsy? If so, what can be done to reduce the change in seizure threshold?
- Is kindling a good model in the aging brain?
- What are the basic mechanisms responsible for causing seizures after stroke?
- What is the role of the aging hippocampus?
- Why does status epilepticus appear to be more common in the elderly?
- What relationships are there between epilepsy and neuropsychiatric disorders and their treatment?
- What are the complications of chronic AED use in the elderly on bone health and other systems?
- Are the elderly more or less susceptible to idiosyncratic reactions?
- What are the major differences between young and old in pharmacokinetics?
- What are the problems of performing drug trials in the elderly? Are the clinical manifestations of seizures different in the elderly?
- What are the roles of the recently introduced AEDs in the treatment of the elderly?
- Is the cost/benefit calculation of treatment favorable?
- What is the best method to calculate the cost/benefit ratio? Are the factors to be considered in this calculation different in the elderly?
- How will our politicians solve the Social Security and Medicare crises without harming the medical and pharmaceutical systems? Is there another "Bismarckian" solution?

IV. Unknown Unknowns

There are likely very many unknown unknowns, but by their very nature, what they are is not known. However, there is a need to be prepared to be surprised.

V. Unknown Knowns

There are probably many things which are accepted knowledge or practice that will in the future be found to be not true. Like the unknown unknowns, it is not possible to predict what they will be. However, a few have recently been elucidated.

A. The "Therapeutic" Range of Total Phenytoin Is 10–20 mg/liter

This is probably not true for the elderly because they have decreased protein binding and probably have increased sensitivity to the side effects of this drug. In adults with normal binding, approximately 10% of phenytoin is free (Gidal *et al.*, 2002), so that the 10–20 mg/liter is approximately 1–2 mg/liter free. In the elderly, with altered binding, 15% or more may be free (Banh *et al.*, 2002), so that a total level of 20 mg/liter would be at least 3 mg/liter free. A more appropriate target range for the elderly may be 5–15 mg/liter of total phenytoin (Leppik, 2000).

B. AED Levels Are Stable Over Time With Constant Doses

A recently completed study of serial blood levels in elderly nursing home patients receiving stable doses of phenytoin revealed marked intrapatient variability (Birnbaum *et al.*, 2003). Blood levels of phenytoin in some varied by more than 100%. This will cause difficulties in interpreting blood level data and therapeutic decisions. Much of this variability may be caused by variable absorption in the frail elderly. This subject is more fully addressed in the chapter by Dr. Birnbaum in this volume.

VI. Conclusion

This short chapter, out of necessity for brevity, touched only briefly on the general problem of epilepsy in the elderly. The following chapters will provide much more detailed discussions of the issues outlined here.

References

Amatniek, J. C., Hauser, W. A., DelCastillo-Castaneda, C., Jacobs, D. M., Marder, K., Bell, K., Albert, M., Brandt, J., and Stern, Y. (2006). Incidence and predictors of seizures in patients with Alzheimer's disease. *Epilepsia* **47,** 867–872.

Banh, H. L., Burton, M. E., and Sperling, M. R. (2002). Interpatient and intrapatient variability in phenytoin protein binding. *Ther. Drug Monit.* **24,** 379–385.

Behrouz, R., Heriaud, L., and Benbadis, S. R. (2006). Late-onset psychogenic nonepileptic seizures. *Epilepsy Behav.* **8,** 649–650.

Birnbaum, A., Hardie, N. A., Leppik, I. E., Conway, J. M., Bowers, S. E., Lackner, T., and Graves, N. M. (2003). Variability of total phenytoin serum concentrations within elderly nursing home residents. *Neurology* **60,** 555–559.

Brodie, M. J., Overstall, P. W., and Giorgi, L. (1999). Multicentre, double-blind, randomised comparison between lamotrigine and carbamazepine in elderly patients with newly diagnosed epilepsy. *Epilepsy Res.* **37,** 81–87.

Cloyd, J. C., Lackner, T. E., and Leppik, I. E. (1994). Antiepileptics in the elderly. Pharmacoepidemiology and pharmacokinetics. *Arch. Fam. Med.* **3,** 589–598.

Cloyd, J., Hauser, W., Towne, A., Ramsay, R., Mattson, R., Gilliam, F., and Walczak, T. (2006). Epidemiological and medical aspects of epilepsy in the elderly. *Epilepsy Res.* **68**(Suppl. 1), S39–S48.

Cramer, J. A., Leppik, I. E., De Rue, K., Edrich, P., and Kramer, G. (2003). Tolerability of levetiracetam in elderly patients with CNS disorders. *Epilepsy Res.* **56,** 135–145.

Dey, A. N. (1996). Characteristics of elderly home health care users: Data from the 1994 National Home and Hospice Care Survey. *Adv. Data* **279,** 1–12.

Epilepsy Foundation of America (1999). "Epilepsy, a Report to the Nation." Epilepsy Foundation, Landover, MD.

Galimberti, C. A., Magri, F., Magnani, B., Arbasino, C., Cravello, L., Marchioni, E., and Tartara, A. (2006). Antiepileptic drug use and epileptic seizures in elderly nursing home residents: A survey in the province of Pavia, Northern Italy. *Epilepsy Res.* **68,** 1–8.

Garrard, J., Cloyd, J., Gross, C., Hardie, N., Thomas, L., Lackner, T., Graves, N., and Leppik, I. (2000). Factors associated with antiepileptic drug use among elderly nursing home residents. *J. Gerontol. A Biol. Sci. Med. Sci.* **55,** M384–M392.

Garrard, J., Harms, S., Hardie, N., Eberly, L. E., Nitz, N., Bland, P., Gross, C. R., and Leppik, I. E. (2003). Antiepileptic drug use in nursing home admissions. *Ann. Neurol.* **54,** 75–85.

Gidal, B. E., Garnett, W. R., and Graves, N. (2002). Epilepsy. *In* "Pharmacotherapy: A Pathophysiologic Approach" (J. T. DiPiro, R. L. Talbert, G. C. Yee, G. R. Matzke, B. G. Wells, and L. M. Posey, Eds.), pp. 1031–1060. McGraw-Hill, New York.

Hauser, W. A. (1997). Epidemiology of seizures in the elderly. *In* "Seizures and Epilepsy in the Elderly" (A. J. Rowan and R. E. Ramsay, Eds.), pp. 7–18. Butterworth-Heinemann, Boston.

Hauser, W. A., and Hesdorffer, D. C. (1990). "Epilepsy: Frequency, Causes, and Consequences." Demos, New York.

Hauser, W. A., Annegers, J. F., and Kurland, L. T. (1991). Prevalence of epilepsy in Rochester, Minnesota: 1940–1980. *Epilepsia* **32,** 429–445.

Hauser, W. A., Annegers, J. F., and Rocca, W. A. (1996). Descriptive epidemiology of epilepsy: Contributions of population-based studies from Rochester, Minnesota. *Mayo Clin. Proc.* **71,** 576–586.

Herd, B., Wynne, H., Wright, P., James, O., and Woodhouse, K. (1991). The effect of age on glucuronidation and sulphation of paracetamol by human liver fractions. *Br. J. Clin. Pharmacol.* **32,** 768–770.

Hetzel, L., and Smith, A. (2001). The 65 years and over population: 2000 [Census 2000 Brief] U.S. Census Bureau, Washington, DC.

Hussain, S. A., Haut, S. R., Lipton, R. B., Derby, C., Markowitz, S. Y., and Shinnar, S. (2006). Incidence of epilepsy in a racially diverse, community-dwelling, elderly cohort: Results from the Einstein aging study. *Epilepsy Res.* **71,** 195–205.

Huying, F., Klimpe, S., and Werhahn, K. J. (2006). Antiepileptic drug use in nursing home residents: A cross-sectional, regional study. *Seizure* **15,** 194–197.

Kemper, P., and Murtaugh, C. M. (1991). Lifetime use of nursing home care. *N. Engl. J. Med.* **324,** 595–600.

Lackner, T. E., Cloyd, J. C., Thomas, L. W., and Leppik, I. E. (1998). Antiepileptic drug use in nursing home residents: Effect of age, gender, and comedication on patterns of use. *Epilepsia* **39,** 1083–1087.

Leppik, I. E. (2000). "Contemporary Diagnosis and Management of the Patient With Epilepsy." Handbooks in Health Care, Newtown, PA.

Leppik, I. E. (2006a). Antiepileptic drug trials in the elderly. *Epilepsy Res.* **68,** 45–48.

Leppik, I. E. (2006b). Introduction to the International Geriatric Epilepsy Symposium (IGES). *Epilepsy Res.* **68**(Suppl. 1), 1–4.

Moore, S. R., and Jones, J. K. (1985). Adverse drug reaction surveillance in the geriatric population: A preliminary view. *In* "Geriatric Drug Use: Clinical & Social Perspectives" (S. R. Moore and T. W. Teal, Eds.), pp. 70–77. Pergamon Press, New York.

Perucca, E., Berlowitz, D., Birnbaum, A., Cloyd, J. C., Garrard, J., Hanlon, J. T., Levy, R. H., and Pugh, M. J. (2006). Pharmacological and clinical aspects of antiepileptic drug use in the elderly. *Epilepsy Res.* **68**(Suppl. 1), S49–S63.

Pugh, M. J., Cramer, J., Knoefel, J., Charbonneau, A., Mandell, A., Kazis, L., and Berlowitz, D. (2004). Potentially inappropriate antiepileptic drugs for elderly patients with epilepsy. *J. Am. Geriatr. Soc.* **52,** 417–422.

Rochat, P., Hallas, J., Gaist, D., and Friis, M. L. (2001). Antiepileptic drug utilization: A Danish prescription database analysis. *Acta Neurol. Scand.* **104,** 6–11.

Rowan, A. J., Ramsay, R. E., Collins, J. F., Pryor, F., Boardman, K. D., Uthman, B. M., Spitz, M., Frederick, T., Towne, A., Carter, G. S., Marks, W., Felicetta, J., *et al.* (2005). New onset geriatric epilepsy: A randomized study of gabapentin, lamotrigine, and carbamazepine. *Neurology* **64,** 1868–1873.

Rumsfeld, D. (2002). Department of Defense briefing. http://www.defenselink.mil/transcripts/2002/t02122002_t212sdv2.html.

Schachter, S. C., Cramer, G. W., Thompson, G. D., Chaponis, R. J., Mendelson, M. A., and Lawhorne, L. (1998). An evaluation of antiepileptic drug therapy in nursing facilities. *J. Am. Geriatr. Soc.* **46,** 1137–1141.

Sillanpaa, M., Kalviainen, R., Klaukka, T., Helenius, H., and Shinnar, S. (2006). Temporal changes in the incidence of epilepsy in Finland: Nationwide study. *Epilepsy Res.* **71,** 206–215.

United Nations (2001). "World Population Ageing: 1950–2050." United Nations Publications, New York.

Verbeeck, R. K., Cardinal, J. A., and Wallace, S. M. (1984). Effect of age and sex on the plasma binding of acidic and basic drugs. *Eur. J. Clin. Pharmacol.* **27,** 91–97.

Woodhouse, K., and Herd, B. (1993). The effect of age and gender on glucuronidation and sulphation in rat liver: A study using paracetamol as a model substrate. *Arch. Gerontol. Geriatr.* **16,** 111–115.

Woodhouse, K. W., and Wynne, H. A. (1988). Age-related changes in liver size and hepatic blood flow. The influence on drug metabolism in the elderly. *Clin. Pharmacokinet.* **15,** 287–294.

Wynne, H. A., Cope, L. H., Mutch, E., Rawlins, M. D., Woodhouse, K. W., and James, O. F. (1989). The effect of age upon liver volume and apparent liver blood flow in healthy man. *Hepatology* **9,** 297–301.

ANIMAL MODELS IN GERONTOLOGY RESEARCH

Nancy L. Nadon

National Institute on Aging, Bethesda, Maryland 20892, USA

Animal models have paved the way for the vast majority of advances in biomedical research. Studies on aged animals are essential for understanding the processes inherent in normal aging and the progression of age-related diseases. Animal models are used to identify physiological changes with age, to identify the genetic basis of normal aging and age-associated disease and degeneration, and to test potential therapeutic interventions. This chapter will focus on rodent models and will summarize important considerations for the use of animals in aging research in general and in modeling geriatric epilepsy.

I. Animal Models in Aging Research: Considerations for Experimental Design

Animal models provide potent tools for investigating the physiology, cell biology, behavioral biology, and genetics of normal aging and age-associated diseases. Rodents in particular have contributed enormously to the fast pace of advancement in biomedical research. The wealth of information available on the physiology and genetics of rodents provides a foundation on which to build studies of aging and geriatric disorders. Other animal models also play key roles in aging research. Nonhuman primates are essential to understanding normal brain aging and neurodegenerative processes in humans because of their similarity to humans in physiology, neurobiology, and behavior; and lower organisms, such as *Drosophila*, *Caenorhabditis elegans*, and even yeast, have contributed to the identification of genes involved in determining life span. Nevertheless, rats and

INTERNATIONAL REVIEW OF
NEUROBIOLOGY, VOL. 81
DOI: 10.1016/S0074-7742(06)81002-0

15

mice are the true workhorses of biomedical research, and effective use of these models will continue to fuel the discovery process. While not discounting the value of other animal models, this chapter will focus on the use of rodents in aging research.

Epilepsy is a complex phenotype with multiple etiologies. It has a polygenic basis, which makes identification of the contributing genetic factors a challenge. Many forms of epilepsy can be modeled in rodents, with seizures induced by electrical stimulation (Loscher and Schmidt, 1988), by chemical treatment such as kainate lesioning (Bouilleret *et al.*, 1999), or by routine stimuli such as handling. Spontaneous and induced mutations in rodents have created many models of specific aspects of epilepsy, as well as general models with which to study the physiology of epilepsy and potential therapeutic interventions. For example, El mice model the complex nature of epilepsy (Seyfried *et al.*, 1992). They provide a model for investigating the biochemical and physiological nature of seizure initiation and propagation and have been utilized to identify genes involved in susceptibility to epilepsy and seizure threshold (Legare *et al.*, 2000). Mice and rats will contribute greatly to our understanding of the origins and progression of epilepsy and to the development of therapeutic interventions. Modeling geriatric epilepsy adds one more dimension to the experimental design, and this chapter will discuss some of the issues to address in order to make the most of these valuable models.

II. The Age Factor

One of the most important variables to be considered in choosing a rodent model and designing the experimental protocol is the ages of the animals used. The strain of choice will impact the ages chosen, as discussed later, but there are some general principles to follow. Young control animals should be fully mature, not just sexually mature. Different physiological systems have different maturation rates, but in general, young control mice and rats should be 3–6 months old. At the other extreme, if the aged time points are too old, it becomes difficult to distinguish the effects of age from the effects of underlying diseases. Again, there is variation between different organ systems, but a rule of thumb is to use animals around the age where 50% of the population has died, to ensure that the animals are old enough to see age-related changes but not so old that all animals are compromised by pathologies common in extreme old age.

Most rodent strains have similarly shaped survival curves regardless of the actual mean life span of the strain (Fig. 1). Before choosing to use strains that have very short mean life spans or significantly skewed slopes in the mortality phase, the underlying diseases often responsible for the short life spans should be considered. AKR mice, for example, express an ecotropic retrovirus (AKV) from birth and as a

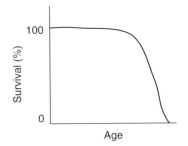

FIG. 1. Typical rodent survival curve. A composite survival curve typical of most rat and mouse strains, showing low mortality during most of the life span and a steep slope to the mortality phase. Most inbred and F1 hybrid strains show little variance from the general shape of the curve.

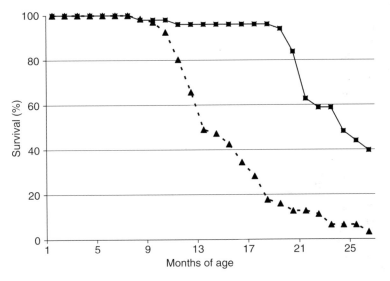

FIG. 2. Survival curve for BALB/cBy in an NIA colony. Survival curves for female (solid line with squares) and male (dotted line with triangles) BALB/cBy mice in the NIA colony ($n = 48$ for females, $n = 64$ for males) (N.L.N., unpublished data, 2002). The mice were fed ad libitum and housed individually in a barrier facility.

result have a very high incidence of leukemia and a median life span of only about 1 year (Festing and Blackmore, 1971). In BALB/cBy mice from a National Institute on Aging (NIA) colony in which the mice were individually housed to preclude losses from fighting, analysis showed that mortality began much earlier for males than for females, and median life span was significantly shorter for males than females (Fig. 2). The cause for this difference is not known.

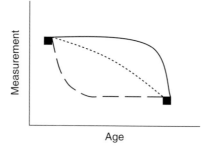

FIG. 3. Hypothetical patterns of change with age. This illustration shows different possible patterns of change in a variable as a function of age. Observing a difference between old and young animals does not necessarily indicate an age-related change; it could be a developmental change, as indicated by the dashed line.

The other major consideration relating to age is the number of age groups. Once the model system has been fully characterized, it may be acceptable to use only two time points (young and old), but the characterization should initially include some intermediate time points as well, because making measurements only at young and old ages could be very misleading. Figure 3 illustrates three possible changes in a parameter with age. The pattern shown by the dotted line suggests an age-associated change in function, whereas the change shown by the solid line may be age-related or may be influenced by diseases that increase in incidence in extreme old age. The change denoted by the dashed line is probably not an aging-associated change but rather a developmental change.

Finally, be sure to start with a sufficient number of animals to allow for mortality in the aged group. A good power analysis is a valuable first step in determining the minimum number of animals needed per age group. Factoring the cost of aged rodents in the power analysis can alter the experimental design and save money (Miller and Nadon, 2000).

III. Genetic Background

Genetic background is a key factor in animal selection. Because physiological variables may differ among various strains, genetics can define the types of studies that are possible with a certain strain. The use of inbred mice and rats provides genetic uniformity, which facilitates interpretation of results in biomedical research and also allows small sample sizes to represent the population. However, inbred rodents are subject to strain-specific pathologies, and there is great variation in pathologies and normal values among strains. Research findings in one strain may not hold for other strains.

F1 hybrid[1] rodents have far fewer strain-specific pathologies than their parental strains but retain the benefit that all individuals in the population are genetically identical. Hybrid rodents tend to be more robust and longer-lived than either parental strain. For example, approximate mean survival is 800 days for Fischer 344 (F344) rats and 900 days for Brown Norway (BN) rats, but for F344×BN F1 hybrid rats, mean survival is about 950 days (Turturro et al., 1999). Genetically heterogeneous (HET) rodents are mixtures of defined laboratory strains (usually four or eight strains). Because no individual animal is identical to any other, the population begins to resemble the mixed genetic makeup of humans but is still defined by the limited genetic variability of the parental strains. Larger sample sizes are needed because of the genetic diversity between individuals and the resultant greater phenotypic variability. Mice from a four-way cross, for example, would share approximately 50% of their genes with any other mouse from that cross, but the shared genes would vary between any two pairs in the population. Analysis of a four-way cross (CB6F1×C3D2F1) HET population determined that cause of death was from a wide range of pathologies with no single disease predominating (Miller et al., 1999). The mixing of defined genetic backgrounds allows HET mice to be used for gene mapping, gene discovery, and biomarker analysis (Miller et al., 1999).

Outbred rodents start with an undefined, genetically diverse population and maintain that diversity through random matings. There is extreme variability between individuals in the population, requiring larger sample sizes, and the undefined genetic background is not suitable for investigations into the genetic origin of normal changes or age-related disease. The genetic diversity of the population most closely reflects the diversity of the human population, but outbred populations can differ greatly from other populations of the same outbred strain as a result of the founder effect: the size of the starting population defines the diversity possible in each colony of outbred rodents. Even within one colony, the genetic constitution can change over time from differences in breeding success. As a result, outbred rodents are seldom the model of choice for aging research.

IV. Choice of Strain

Strain-specific characteristics that are relevant to the experimental design must be considered when using inbred strains. Strains can vary in the type and frequency of age-associated pathologies, some of which are summarized in Table I. A good

---

[1]The naming convention of first filial generation (F1) hybrids lists the standard strain abbreviation of the female strain first, followed by the standard strain abbreviation of the male strain; thus, B6C3 mice are the offspring of a C57BL/6×C3H/He cross, B6D2 are the offspring of a C57BL/6×DBA/2 cross, and CB6 are the offspring of a BALB/c×C57BL/6 cross.

TABLE I

Strain-Specific Prevalence of Common Pathologies in Aged Rats

	F344[a] (%)	BN[a] (%)	F344BNF1[b] (%)
Pituitary adenoma (males)	12.9	1.2	19
Pituitary adenoma (females)	33.9	16	23
Glomerulonephropathy (males)	56	0	34
Glomerulonephropathy (females)	22	0	19
Hydronephrosis (males)	0	62	45

[a]Lipman *et al.* (1999).
[b]Lipman *et al.* (1996).
Adapted from Kelly *et al.* (2006), with permission.

example is glomerulonephropathy, which has a very high incidence in aged F344 rats but is virtually absent in BN rats. Leukemia is also prevalent and a frequent cause of death in aged F344 rats (Lipman *et al.*, 1999). Extremely old F344 rats are likely to have at least one of these two pathologies, if not both. Pituitary adenomas are very common in aged rats and could influence experimental outcomes in studies of convulsive disorders. Strain-specific pathologies that are not age-related might also be an important consideration. BN rats have a high incidence of hydronephrosis, and the urinary blockage can affect the overall health of the rat even though it is not usually a life-threatening condition (Lipman *et al.*, 1999). C57BL/6 mice are prone to excoriative dermatitis, and although the cause is unknown, the incidence of this condition is influenced by diet and housing conditions (Turturro *et al.*, 2002). The development of open lesions can in turn influence aggressive behavior in cage mates, resulting in losses from the group that are independent of age or study design.

Something as basic as water intake could influence an experiment that relied on delivery of a therapeutic agent via drinking water. Analysis of water intake showed a wide variation among 28 mouse strains, with a twofold difference between the high and low extremes (Tordoff and Bachmanov, 2001). Another example is the phenotypic differentiation of stem cells in the adult brain, which can vary between mouse strains (Kempermann and Gage, 2002). These differences influence not only neurogenesis and plasticity in the adult brain, but birth of glia as well. There are also distinct strain differences in learning and in the aging of brain function. Ingram and Jucker (1999) investigated learning ability by using the Morris water maze to measure the ability of mice to use spatial cues to locate a submerged platform. Young and old C57BL/6J mice improved in similar fashion with each succeeding set of repeated trials. Young 129/SvJ mice also improved with successive trials, but old 129/SvJ mice showed no improvement, suggesting significant differences in the aging of brain function between these two strains.

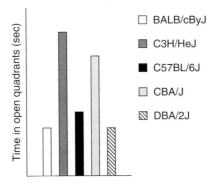

F<small>IG</small>. 4. Time spent in open quadrants. Data derived from a study in the Mouse Phenome Database showing the relative average amount of time spent in open areas for males of five mouse strains (Flaherty *et al.*, 2001). This measurement is used as an index of relative anxiety.

Behavioral characteristics vary greatly among rodent strains and can affect experiments in direct and indirect ways. BN rats, for example, are less anxiety-prone and more social than F344 rats and explore their environment more (Ramos *et al.*, 1997; Rex *et al.*, 1996). BN rats also perform better on learning tests (Spangler *et al.*, 1994), although the differences are minor. Inbred strains of mice also show significant variation in terms of anxiety. Figure 4 shows the result of a strain comparison for time spent in open areas, from a study in the Mouse Phenome Database (Flaherty *et al.*, 2001). The five strains chosen for this illustration include the four inbred strains available in the NIA aged rodent colony plus C3H/HeJ, one of the parental strains for B6C3F1 hybrid mice, also available in the NIA colony. There are considerable differences among the strains that could affect the outcome of behavioral and learning experiments requiring explorative behavior. The levels of anxiety evident may also indicate differences in stress hormones that could affect a variety of phenotypes.

Ferraro *et al.* (2002) compared maximal electroshock seizure threshold (MEST) in several strains of mice by measuring seizure susceptibility as a function of the level of current required to induce maximal seizure. The MEST varied greatly among the 10 strains tested, with a threefold difference between the lowest (DBA/2J mice) and the highest (C57BL/6J mice) MEST value. There are differences between C57BL/6 and DBA/2 mice in terms of susceptibility to audiogenic seizures as well. DBA/2 mice, especially those between 2 and 6 weeks of age, are extremely sensitive to audiogenic seizures; C57BL/6 mice are almost completely resistant to them (Seyfried *et al.*, 1986). Such strain differences can be exploited for gene discovery. To identify genes involved in specific aging functions, recombinant inbred (RI) strains are often generated, thereby transferring genomic regions that influence a phenotype to a genetic background that expresses a different phenotype. RI strains generated from C57BL/6×DBA/2 crosses have been used for over two decades

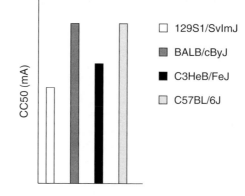

Fig. 5. CC50 for maximal hindlimb extension seizures. Data derived from a study in the Mouse Phenome Database showing the relative sensitivity to electrogenic seizures (Frankel and White, 2000). CC50 indicates the current magnitude at which 50% of the mice experienced a maximal hindlimb extension seizure. Adapted from Kelly *et al.* (2006), with permission.

to identify genes involved in convulsive disorders. In their 1986 review, Seyfried *et al.* (1986) provided a detailed synopsis of their use of RI strains to identify the *Ias* gene, which is involved in inhibiting the propagation of seizures.

Frankel and White (2000) addressed strain differences in sensitivity to electrogenic seizures in a study reported in the Mouse Phenome Database (http://aretha.jax.org/pub-cgi/phenome/mpdcgi?rtn=docs/home). Figure 5 illustrates the different intensities of stimulation required to induce hindlimb extension seizures in the mouse strains of that study. Of note is the difference in sensitivity between 129S1/SvImJ and C57BL/6J mice. Most knockout mice are generated by injection of strain 129 embryonic stem cells into C57BL/6 embryos, resulting in 129×C57BL/6 chimeric mice. Despite breeding the knocked out gene to homozygosity, the genetic background is a mixture of 129 and C57BL/6. For a knockout generated to study the role of a specific gene in seizures, the mixed background would be confounding, and it would be impossible to generate an appropriate control. These strain differences illustrate why it is important to cross knockout mice onto a uniform genetic background and how the choice of that background can influence results.

V. Environmental Influences

Husbandry and environmental conditions can influence the health and behavior of rodents and affect experimental outcomes. For rodents raised in specific pathogen-free (SPF) barriers, documentation of the SPF status provides a measure of assurance that the animals have no infectious diseases, which might influence experimental results. However, the specific pathogens excluded will vary to some degree, even in commercial colonies, and the SPF nature of the barrier should be determined by examining health reports. *Helicobacter* is a new genus of bacteria discovered in the 1990s, and it is still endemic in many commercial, barrier-raised colonies. Most *Helicobacter* species have no pathogenic effect on rodents, but some species do, such as *H. hepaticus*, which can migrate to the liver and cause hepatitis. BALB/c mice are particularly susceptible to hepatitis caused by *H. hepaticus* (Ward *et al.*, 1994), and in the portions of the NIA aged rodent colonies that are still housed in helicobacter-positive barriers, only BALB/c mice show a low incidence of hepatitis.

In conventionally housed colonies, the use of filter bonnets provides some protection from the spread of organisms between cages, but rodents in conventional colonies are routinely exposed to more pathogens than those protected by a barrier. If rodents are received from a barrier-protected colony and housed in a conventional facility for a lengthy period, exposure to new potential pathogens must be expected. Regardless of the type of housing, the most important consideration is to regularly monitor health status.

Most rodents are group-housed and thrive in that environment because they are social animals. For many rat strains, individuals that have been raised in different cages can be housed together, but this is not true for most mouse strains, and mice caged separately should not be recombined. Differences between mouse strains should be addressed when deciding on housing configurations. Male BALB/c and DBA/2 mice have a propensity to fight with cage mates, even when they have been caged together since weaning. C57BL/6 mice also tend to develop aggression between cage mates, with one mouse in the cage taking the position of dominance. Within the NIA colony, it was found that removing the dominant mouse did not alleviate the situation; another mouse simply assumed the position of dominance. When group-housing strains such as BALB/c, DBA/2, and C57BL/6 are caged for lengthy periods, some animal losses due to fighting must be anticipated.

Ad libitum (AL) feeding is the most common form of feeding rodents because of its ease and low cost. However, there are advantages to mild caloric restriction (CR), even for strains that do not routinely become obese. Mild (10%) CR is sufficient to keep rodents at a more stable weight throughout life. Higher levels of CR, such as the 40% CR used in the NIA Biomarkers Program, extend the life span of rodents and reduce the incidence and/or delay the onset of many age-related pathologies (Lipman *et al.*, 1996, 1999; Turturro *et al.*, 1999). The level of response to CR varies among strains, which again illustrates the importance of strain selection, and there are sometimes gender effects as well. C57Bl/6NNia mice show a robust response to 40% CR in both males and females, with about a 20–25% life span increase for both sexes (Turturro *et al.*, 1999); however, in

DBA/2JNia mice on a 40% CR protocol, females show about a 20% increase in median life span compared to only a <5% increase in males. Whereas most age-related pathologies are decreased in incidence and/or delayed in onset by CR, a few show different patterns. For example, the incidence of age-related pancreatic exocrine atrophy is decreased in CR compared to AL F344 rats, but increased in CR compared to AL BN rats (Lipman *et al.*, 1999).

CR may be a particularly useful paradigm for investigating physiological influences on geriatric epilepsy. Fasting provides a measure of protection against seizures, and raising seizure-prone El mutant mice under CR delayed the onset of seizures in response to handling (Greene *et al.*, 2001). These findings support the hypothesis that CR may lessen seizure incidence by reducing the energy sources available to the brain. A similar result can be seen in response to ketogenic diets (Mantis *et al.*, 2004), demonstrating that changes to energy sources in the brain may induce alterations in metabolic pathways that may influence seizure genesis. The rodent model is a potent tool to investigate these pathways because the diet and environment can be strictly controlled.

Environmental enrichment has not been widely used with rodents, but there is a body of literature documenting the benefits of enrichment, particularly on brain aging and plasticity. Enrichment protocols can employ social housing, exercise wheels, food enrichment, toys, activity stimuli, and even training. Aged mice raised in an enriched environment show increased neuroplasticity, and exercise appears as effective as enrichment for promoting increased neurogenesis (Kempermann *et al.*, 2002; van Praag *et al.*, 2000). Young *et al.* (1999) reported that environmental enrichment protected against kainate-induced seizure gene-sis in rats. They demonstrated that as little as 3 weeks of living in an enriched environment after weaning increased the volume of the granule cell layer in the hippocampus and suppressed seizures after lesioning with kainate. These reports not only illustrate important considerations for housing rodents, but also suggest optimism for the successful use of other approaches aimed at healing seizure-inducing brain damage.

VI. Genomic Manipulations

There is a plethora of genomic manipulations possible in rodents that cannot be done in larger mammals, and these manipulations provide valuable tools for discerning the genetic basis of normal aging and age-associated diseases. Genetic models include spontaneous and induced mutations, transgenic mice and rats, knockout and knockin mice, chimeric rodents, consomic and cogenic rodents, RI strains, and wild-derived strains. This discussion is not meant to review these methods, but one example will be given to demonstrate the power of these tools.

Chimeric animals have been used for decades and seem almost primitive when compared to today's advanced techniques for genomic engineering, but using chimeric animals to exploit strain differences can lead to scientific breakthroughs. Chimeric mice are generated by mixing cells from two embryos, generating a mouse with four parental contributions. They enable investigation of the influence of different genetic backgrounds on cell function, illuminate cell–cell interactions, and identify what cell types are affected by mutations. For example, chimeras have been used to determine whether the phenotype of a cell is influenced by the genotype of a neighboring cell (Mullen *et al.*, 1997).

An example unrelated to epilepsy but elegant in its design was published by Geiger and Van Zant (2002). They made chimeric C57BL/6:DBA/2 mice (median life spans of 27 and 23 months, respectively, for each strain) and analyzed the contribution of each strain to the peripheral blood compartment. The DBA/2 contribution steadily decreased over the life of the chimeric mice so that by very old age all blood cells were C57BL/6-derived. However, the DBA/2 hematopoietic stem cells (HSCs) were still present in the old chimeric mice; they could be reactivated by successive transplantation of bone marrow from the aged chimeric mice into young recipients. This suggested that the environment of the aged chimeric mice was more limiting to the DBA/2 HSCs than to the C57BL/6 HSCs.

VII. Resources

There are numerous resources available that can assist in the use of rodent models. Two publicly available online resources of note provide access to genotypic and phenotypic information on rat and mouse strains. The Rat Genome Database (http://rgd.mcw.edu) provides strain phenotypes, characteristics, genetics, and genomic resources for the rat model. The Mouse Phenome Database at the Jackson Laboratory Web site (http://aretha.jax.org/pub-cgi/phenome/mpdcgi?rtn=docs/home) provides phenotyping and links to expressed sequence tags for mouse strains.

The NIA supports resources for studies on aging and age-associated diseases, and the NIA Scientific Resources Web site (http://www.nia.nih.gov/ResearchInformation/ScientificResources) provides links to information on all NIA biological resources. The NIA aged rodent colony provides aged rats and mice from a limited number of strains: BN, F344, and F344BNF1 rats; and C57BL/6, BALB/cBy, CBA, DBA/2, B6C3F1, B6D2F1, and four-way cross mice. There is also a CR rat colony with rats raised under 40% CR from 4 months of age on, and AL controls are available for the CR rats. The NIA rodent colonies are barrier-raised under controlled environmental conditions.

NIA resources include a tissue bank from the aged rodent colonies, which provides flash frozen tissue and will soon include tissue arrays. To facilitate expression studies, the NIA Microarray Facility provides 17K mouse cDNA membrane arrays to the research community. The NIA Aging Cell Repository (http://ccr.coriell.org/nia) includes a large number of human and animal cell lines, and many of the human cell lines have phenotypic data available. The NIA Cell Repository contains cell lines from patients with different types of convulsive disorders, as does the Human Genetics Resource Center supported by the National Institute of Neurological Disorders and Stroke (http://ccr.coriell.org/ninds/genetic.html).

References

Bouilleret, V., Ridoux, V., Depaulis, A., Marescaux, C., Nehlig, A., and Le Gal La Salle, G. (1999). Recurrent seizures and hippocampal sclerosis following intrahippocampal kainate injection in adult mice: Electroencephalography, histopathology and synaptic reorganization similar to mesial temporal lobe epilepsy. *Neuroscience* **89,** 717–729.

Ferraro, T. N., Golden, G. T., Smith, G. G., DeMuth, D., Buono, R. J., and Berrettini, W. H. (2002). Mouse strain variation in maximal electroshock seizure threshold. *Brain Res.* **936,** 82–86.

Festing, M. F. W., and Blackmore, D. K. (1971). Life span of specified-pathogen-free (MRC category 4) mice and rats. *Lab. Anim.* **5,** 179–192.

Flaherty, L., Cook, M. N., and Williams, R. W. (2001). Anxiety-related behaviors in the elevated zero-maze. The Jackson Laboratory Web site. http://phenome.jax.org/pub-cgi/phenome/mpdcgi?rtn=projects/details&sym=Flaherty1.

Frankel, W. N., and White, H. S. (2000). Electroconvulsive thresholds. The Jackson Laboratory Web site. http://phenome.jax.org/pub-cgi/phenome/mpdcgi?rtn=projects/details&sym=Frankel1.

Geiger, H., and Van Zant, G. (2002). The aging of lympho-hematopoietic stem cells. *Nat. Immunol.* **3,** 329–333.

Greene, A. E., Todorova, M. T., McGowan, R., and Seyfried, T. N. (2001). Caloric restriction inhibits seizure susceptibility in epileptic EL mice by reducing blood glucose. *Epilepsia* **42,** 1371–1378.

Ingram, D. K., and Jucker, M. (1999). Developing mouse models of aging: A consideration of strain differences in age-related behavioral and neural parameters. *Neurobiol. Aging* **20,** 137–145.

Kelly, K. M., Nadon, N. L., Morrison, J. H., Thibault, O., Barnes, C. A., and Blalock, E. M. (2006). The neurobiology of aging. *Epilepsy Res.* **68**(Suppl. 1), S5–S20.

Kempermann, G., and Gage, F. H. (2002). Genetic influence on phenotypic differentiation in adult hippocampal neurogenesis. *Brain Res. Dev. Brain Res.* **134,** 1–12.

Kempermann, G., Gast, D., and Gage, F. H. (2002). Neuroplasticity in old age: Sustained fivefold induction of hippocampal neurogenesis by long-term environmental enrichment. *Ann. Neurol.* **52,** 135–143.

Legare, M. E., Bartlett, F. S., II, and Frankel, W. N. (2000). A major effect QTL determined by multiple genes in epileptic EL mice. *Genome Res.* **10,** 42–48.

Lipman, R. D., Chrisp, C. E., Hazzard, D. G., and Bronson, R. T. (1996). Pathologic characterization of Brown Norway, Brown Norway × Fischer 344, and Fischer 344 × Brown Norway rats with relation to age. *J. Gerontol. A Biol. Sci. Med. Sci.* **51A,** B54–B59.

Lipman, R. D., Dallal, G. E., and Bronson, R. T. (1999). Effects of genotype and diet on age-related lesions in ad libitum fed and calorie-restricted F344, BN, and BNF3F1 rats. *J. Gerontol. A Biol. Sci. Med. Sci.* **54A,** B478–B491.

Loscher, W., and Schmidt, D. (1988). Which animal models should be used in the search for new antiepileptic drugs? A proposal based on experimental and clinical considerations. *Epilepsy Res.* **2,** 145–181.

Mantis, J. G., Centeno, N. A., Todorova, M. T., McGowan, R., and Seyfried, T. N. (2004). Management of multifactorial idiopathic epilepsy in EL mice with caloric restriction and the ketogenic diet: Role of glucose and ketone bodies. *Nutr. Metab. (Lond.)* **1,** 1–11.

Miller, R. A., and Nadon, N. L. (2000). Principles of animal use for gerontological research. *J. Gerontol. A Biol. Sci. Med. Sci.* **55A,** B117–B123.

Miller, R. A., Burke, D., and Nadon, N. (1999). Announcement: Four-way cross mouse stocks: A new, genetically heterogeneous resource for aging research. *J. Gerontol. A Biol. Sci. Med. Sci.* **54A,** B358–B360.

Mullen, R. J., Hamre, K. M., and Goldowitz, D. (1997). Cerebellar mutant mice and chimeras revisited. *Perspect. Dev. Neurobiol.* **5,** 43–55.

Ramos, A., Berton, O., Mormede, P., and Chaouloff, F. (1997). A multiple-test study of anxiety-related behaviours in six inbred rat strains. *Behav. Brain Res.* **85,** 57–69.

Rex, A., Sondern, U., Voigt, J. P., Franck, S., and Fink, H. (1996). Strain differences in fear-motivated behavior of rats. *Pharmacol. Biochem. Behav.* **54,** 107–111.

Seyfried, T. N., Glaser, G. H., Yu, R. K., and Palayoor, S. T. (1986). Inherited convulsive disorders in mice. *Adv. Neurol.* **44,** 115–133.

Seyfried, T. N., Brigande, J. V., Flavin, H. J., Frankel, W. N., Rise, M. L., and Wieraszko, A. (1992). Genetic and biochemical correlates of epilepsy in the El mouse. *Neuroscience* **18**(Suppl. 2), 9–20.

Spangler, E. L., Waggie, K. S., Hengemihle, J., Roberts, D., Hess, B., and Ingram, D. K. (1994). Behavioral assessment of aging in male Fischer 344 and Brown Norway rat strains and their F1 hybrid. *Neurobiol. Aging* **15,** 319–328.

Tordoff, M. G., and Bachmanov, A. A. (2001). Food intakes, water intakes, and spout side preferences. The Jackson Laboratory Web site. http://phenome.jax.org/pub-cgi/phenome/mpdcgi?rtn=projects/details&sym=Tordoff2.

Turturro, A., Witt, W. W., Lewis, S., Hass, B. S., Lipman, R. D., and Hart, R. W. (1999). Growth curves and survival characteristics of the animals used in the Biomarkers of Aging Program. *J. Gerontol. A Biol. Sci. Med. Sci.* **54A,** B492–B501.

Turturro, A., Duffy, P., Hass, B., Kodell, R., and Hart, R. (2002). Survival characteristics and age-adjusted disease incidences in C57BL/6 mice fed a commonly used cereal-based diet modulated by dietary restriction. *J. Gerontol. A Biol. Sci. Med. Sci.* **57,** B379–B389.

van Praag, H., Kempermann, G., and Gage, F. H. (Praag 2000). Neural consequences of environmental enrichment. *Nat. Rev. Neurosci.* **1,** 191–198.

Ward, J. M., Anver, M. R., Haines, D. C., and Benveniste, R. E. (1994). Chronic active hepatitis in mice caused by Helicobacter hepaticus. *Am. J. Pathol.* **145,** 959–968.

Young, D., Lawlor, P. A., Leone, P., Dragunow, M., and During, M. J. (1999). Environmental enrichment inhibits spontaneous apoptosis, prevents seizures and is neuroprotective. *Nat. Med.* **5,** 448–453.

ANIMAL MODELS OF GERIATRIC EPILEPSY

Lauren J. Murphree,*,† Lynn M. Rundhaugen,† and Kevin M. Kelly‡

*SRA International, Inc., Fairfax, Virginia 22033, USA
†National Institute of Neurological Disorders and Stroke, National Institutes of Health
Bethesda, Maryland 20892, USA
‡Department of Neurology, Allegheny General Hospital
Pittsburgh, Pennsylvania 15212, USA

Geriatric epilepsy is a significant clinical problem that has not been studied adequately in animal models. This chapter will review the available literature with particular attention to models that have demonstrated how acute seizures and epilepsy in aged animals differ from those of younger animals. Studies include several strains of mice [e.g., El, DBA, senescence-accelerated mouse (SAM), Cacnb4 knockout] as well as acute seizure models in common strains of aged mice. Aged rats (including Fischer 344, Wistar, and Sprague–Dawley) have been used in acute seizure, lesion, and epilepsy models. This area of research remains largely unexplored and therefore provides numerous opportunities for new investigations.

INTERNATIONAL REVIEW OF
NEUROBIOLOGY, VOL. 81
DOI: 10.1016/S0074-7742(06)81003-2

29

I. Introduction

There are many unknowns concerning epilepsy in the elderly, including what differentiates seizures in the aged versus the young; this may result from insufficient animal studies of geriatric epilepsy. The lack of *in vivo* studies is due, in part, to the costs associated with older animals and the confounding comorbidities associated with aging. A PubMed search of articles from 1965 to 2006, combining the search terms "aged" or "aging" and "epilepsy" or "seizures," revealed that fewer than 50 papers were directly related to epilepsy or seizures in aged animals (Leppik *et al.*, 2006). In this chapter, studies using animal models that demonstrated how acute seizures and epilepsy in aged animals differ from those of younger animals will be reviewed. Because this area of research remains largely unexplored and because the proportion of the elderly continues to increase, the use of animal models to study geriatric epilepsy provides opportunities for new investigations. Invigorating the study of diseases such as geriatric epilepsy will provide enhanced benefits for public health.

II. Mouse Models

A. Senescence-Accelerated Mouse

The senescence-accelerated mouse (SAM), originally developed by Takeda in 1981 (Takeda *et al.*, 1981), has been widely used in geriatric research, with nearly 400 citations as of September 2006. However, its use in the field of geriatric epilepsy has been limited. The first publication of an investigation using these animals in the study of seizures was not until 1992, when Yamazaki *et al.* (1992) reported spontaneous seizures in a substrain of the SAM-resistant mouse (SAM-R/1/Eis). This substrain does not appear to have been used in subsequent epilepsy research. Two groups have used the SAM-prone strain (SAMP8) to investigate the effects of epilepsy in geriatric animals. Using a small sample, Kondziella *et al.* (2002, 2003) have shown differential effects of age on neurotransmitter metabolism following pentylenetetrazol (PTZ) kindling [the process of triggering seizures by small repeated stimulation of the brain (Goddard *et al.*, 1969)], but not on kindling rate or efficacy. Kim *et al.* (2002) demonstrated that aged SAMP8 mice were more prone to seizures and oxidative damage than age-matched senescence-resistant mice, following repeated kainic acid (KA) treatment. This chapter has been cited in the neurodegeneration literature, but not in works related to epilepsy.

B. Cacnb4 (lh) Mouse

Originally identified as the lethargic (*lh/lh*) mouse at Jackson Laboratory (Dickie, 1964), the Cacnb4 mouse has spontaneous absence-like seizures apparently caused by a mutation in the β subunit of voltage-gated calcium channels, which produce a truncated, less-abundant mRNA. These animals generated considerable interest when it was shown that they also had increased expression of gamma-aminobutyric acid (GABA)$_B$ receptors that contributed to the seizure phenotype (Hosford et al., 1992). A follow-up paper using aged animals showed that there was no significant difference in the amount of mRNA for cacnb4 or GABA$_B$ when comparing Cacnb4 mice to wild-type controls (Lin et al., 1999). Aside from this observation, there are no published studies investigating the absence phenotype in aged lethargic animals.

C. MTIII (−/−) MOUSE

Metallothionein-III (MTIII) is a zinc-binding protein that is abundant in the synapses of glutamatergic neurons that release zinc, which can modify neurotransmitter effects. Knocking out the MTIII gene in mice yields animals that are more susceptible to seizures in old age (Erickson et al., 1997). These animals have reduced zinc levels in several brain regions, including the hippocampus and cortex. The aged knockout mice were slightly more susceptible to severe KA-induced seizures than were wild-type controls, but were significantly more prone to injury of CA3 cells following seizure activity. Subsequent papers citing this study have not focused on geriatric aspects.

D. El MOUSE

The El mice are a mutant strain derived from ddY mice (Suzuki and Nakamoto, 1977). These animals develop convulsive seizures following repeated tossing or similar stimuli, and they are considered a model of complex partial seizures or temporal lobe epilepsy (Imaizumi and Nakano, 1964; Nagatomo et al., 2000). Only two published studies reported the use of these animals to study geriatric epilepsy. Tsuda et al. (1993) noted age- and seizure-dependent changes in neurotransmitter levels in stimulated animals when compared to age-matched ddY control mice; decreased noradrenaline levels in the striatum and the hippocampus of mice older than 12 weeks and decreased dopamine levels in the striatum of mice 16 weeks or older were observed. In a biochemical study, Nagatomo et al. (2000) observed a decrease in whole-brain nitric oxide concentrations over time for both stimulated and unstimulated mice, but aged stimulated animals displayed an increase in nicotinamide adenine dinucleotide phosphate diaphorase–positive neurons compared to age-matched, unstimulated controls.

E. DBA MOUSE

DBA mice are susceptible to audiogenic seizures (AGS) (Hall, 1947), with strains varying in age of maximum seizure risk (Fuller and Sjursen, 1967).

Engstrom *et al.* (1986) showed that aged DBA mice (26–115 days) were resistant to AGS but had a lower minimal electroshock threshold and required a greater dose of acetazolamide to prevent tonic extension in the maximal electroshock (MES) model than age-matched C57 mice (non-AGS susceptible). This study, which used mice with a maximal age of 115 days (~16 weeks), appears to be the only study published that used DBA mice of advanced age.

F. Acute Seizure Models in Common Strains

Several papers from the 1980s described the effect of aging on the seizure susceptibility of BDF1 mice. Results for these studies showed that the threshold for the tonic extensor component of MES seizures increased with age (out to 30 months; Nokubo *et al.*, 1986), as did the anticonvulsant effect of oxazepam in seizures caused by PTZ (Kitani *et al.*, 1989), although the LD_{50} (lethal dose in 50% of the animals) for PTZ actually decreased with age (Nokubo and Kitani, 1988). These aged animals were more susceptible to the neurotoxic effects of phenobarbital as demonstrated by the rotorod test (~50% decrease in minimal toxic concentration) and more responsive to its anticonvulsant effects (requiring ~10–20% of the young animal dose; Kitani *et al.*, 1988).

Two papers have focused on histological changes in aged mice following seizure. In C57BL/6J mice, Benkovic *et al.* (2006) showed that aged animals were more sensitive to the pathological effects of a single dose of KA, especially for markers of reactive gliosis (neuroinflammation). D'Costa *et al.* (1991) used a single electroconvulsive shock to induce seizures and observed c-fos expression in different brain regions of varying age groups. All ages exhibited a transient increase in c-fos immunoreactivity, but in aged animals the response was significantly less robust.

Many studies in the general epilepsy field have focused on gender differences in susceptibility to seizure. Manev *et al.* (1987) showed that in adult animals (3-month-old CBA mice), males were distinctly more susceptible to seizures induced by the GABA-blocking agents bicuculline and picrotoxin than were age-matched females. However, aged animals (24 months) did not display a gender difference in sensitivity to picrotoxin.

III. Gerbil Model

As described by Spangler *et al.* (1997), the Mongolian gerbil exhibits several age-related characteristics that make it an interesting model of aging for behavioral tests of psychomotor activity and learning. They noted the presence of a confounding seizure phenotype on exposure to novel situations and suggested that selective breeding may be able to reduce this characteristic. Cheal (1986) proposed that this gerbil could be a promising model of epilepsy, and Scotti *et al.* (1998) observed a kindling-like phenomenon in a subset of these animals, with the older animals in this subset being more refractory to seizures. These studies suggest that this species is worthy of additional study.

IV. Rat Models

There is substantially more published literature using rats than mice in the geriatric epilepsy area. However, there are fewer genetic models in rats because of the increased difficulty of genetic manipulations in this species. Models generally fall into three categories: acute, lesion, and chronic epilepsy.

A. Acute Seizure Models

In aged animals, the most widely used acute models employ a convulsant such as KA or PTZ. Strychnine, pilocarpine, nicotine, and electrical stimulation also induce acute seizures in animal models. The results from studies of epilepsy in aged animals have shown both increased and decreased seizure susceptibility with age, depending on the model.

1. *Kainic Acid*

Aged rats have an increased sensitivity to the convulsant effects of KA when compared to younger animals. Wozniak *et al.* (1991) showed that low doses of KA that were nontoxic in young Sprague–Dawley (SD) rats (5–6 months) caused status epilepticus, neuronal damage, and death in older animals (22–25 months). Middle-aged animals displayed intermediate effects. Expanding on these results, Dawson and Wallace (1992) showed that aged female Long–Evans rats (30 months) exhibited a higher rate of tonic-clonic seizures at lower doses than did their young counterparts (6 months). KA treatment caused a significant increase in release of aspartate, glutamate, and norepinephrine in the aged rats, which was not observed in the adult control animals.

In addition to increased sensitivity to the convulsant effects of KA, aged animals had altered responses to KA-induced status epilepticus. The electroencephalogram (EEG) responses in aged Fischer 344 (F344) rats (22–25 months) showed a reduction in some of the faster frequencies (12.5–35 Hz) when compared to the EEG responses of younger animals (7–8 months); aged rats also had a higher frequency of preseizure events and shorter latency to individual seizure benchmarks, for example class 5 seizures, according to a modification (Ben-Ari, 1985) of the progressive scale described by Racine (1972) (Darbin *et al.*, 2004). Other work has shown that the aged brain also regulates protein expression distinctly after KA treatment. Kelly *et al.* (2003) showed that post–status epilepticus brains of aged F344 animals (25–29 months) showed prominently decreased expression of α_{1A}, α_{1C}, and α_{1D} subunits of voltage-gated calcium channels when compared to post–status epilepticus brains of younger animals (5–14 months).

2. *Pentylenetetrazol*

Several studies in SD rats have investigated the effects of age on the brain's transcriptional response to acute PTZ-induced seizures. The earliest of such work focused on the immediate early gene c-fos, which has been associated with brain plasticity. Young animals given a convulsive dose of PTZ (50 mg/kg) transiently upregulated c-fos mRNA. In aged animals (30 months), this response was delayed (longer time to an observed increase) and blunted (smaller maximum upregulation), although the basic response remained intact (Retchkiman *et al.*, 1996). Using the same stimulus, the authors saw a similar pattern in the cortex and hippocampus with tissue plasminogen activator (tPA) mRNA. Interestingly, the cortical transcriptional response, while present in all layers in the young animals, was limited to layer V neurons in the older (18 months) animals (Popa-Wagner *et al.*, 2000). In contrast, transient increase in microtubule-associated protein 1B (MAP1B), which is also associated with brain plasticity, occurred at earlier times in older animals (although the responses were significantly reduced) than in their young counterparts (Popa-Wagner *et al.*, 1999). Also in this model, Schmoll *et al.* (2005) showed that old animals lose the ability to regulate axonal growth-associated protein 43 (GAP-43) in certain brain areas. Because GAP-43 is associated with repair and plasticity following acute neuronal injury (Bomze *et al.*, 2001), it could be an important marker of the brain's ability to recover from acute seizure activity.

3. *Other Acute Models*

Studies looking at the effects of age on acute seizures have also used a variety of other stimuli. Hunter *et al.* (1989) showed that senescent animals (24-month-old F344 rats) had an increased sensitivity to strychnine-induced seizures and mortality, which corresponded with a reduction in [^{3}H]strychnine and [^{3}H]GABA binding in the medulla and spinal cord. Aged F344 animals are frequently used in the study of the regulation of neurotransmitter systems in aging, but most of this work does not indicate that any seizure activity was observed.

The common convulsant pilocarpine also shows differential effects in older animals, as demonstrated by a lowered seizure threshold in aged (24 months) Wistar rats (Hirvonen *et al.*, 1993). In contrast to the effects of strychnine and pilocarpine, nicotine-induced convulsions were delayed and abbreviated in 24-month-old Wistar rats when compared to nicotine-induced convulsions in younger animals, despite the half-life of nicotine being longer in the older animals. Okamoto *et al.* (1992) hypothesized that the decreased seizure response may be related to a muted corticosterone release in response to nicotine. Aminooxyacetic acid (AOAA) also appears to lose some potency with age. Turski *et al.* (1992) found increasing CD_{50} (convulsive dose in 50% of the animals) values of AOAA over the life span of Wistar rats, although most of the increase appeared to occur within the first 10 months.

Using electrical stimulation to induce acute seizures, Stijnen *et al.* (1992) did not observe an effect of age on the seizure threshold with direct cortical stimulation in BN/BiRij rats. Very old animals (35 months) were more sensitive to the anticonvulsant effect of oxazepam and did not display the proconvulsant effects at higher doses that were seen in younger animals. In another study of age-related changes using Wistar rats, Oztas *et al.* (1990) noted increased blood–brain barrier permeability to macromolecules in 24-month-old animals after 10 electroconvulsive shocks (but not after one), suggesting that the older animals are more susceptible to injury caused by repeated seizure insults.

B. Lesion Models

Work with lesion models in aged rats is limited and falls primarily into two groups: KA lesions and photothrombotic lesions. The majority of the KA lesion work involves using cell grafts to attempt to repair the injury caused by the lesion. As might be expected, repopulation of the lesioned hippocampus with grafted fetal CA3 cells is more successful in young animals than in older animals. The survival of the transplanted cells can be improved by presupplementing them with neurotrophic factors and caspase inhibitors (Zaman and Shetty, 2002) or fibroblast growth factor-2 (FGF-2) (Zaman and Shetty, 2003). Combining two growth factors (brain-derived neurotrophic factor and FGF-2) improved the outcome even more and reduced aberrant mossy fiber sprouting (Rao *et al.*, 2006).

Kelly *et al.* (2001) generated a poststroke epilepsy model by inducing cortical thrombosis. Using a photosensitive dye, F344 rats of various ages received neocortical lesions that spared the hippocampus. A subset of middle-aged and aged animals developed class 3 seizures that appeared to originate in the peri-infarct region. Follow-up studies in this model have been limited to assessments in young adult animals; however, additional work in aged animals is ongoing.

C. Chronic Epilepsy Models

Kindling is commonly used as a model of epilepsy. Some of the earliest work in the geriatric epilepsy field was done in F344 rats using kindling by electrical stimulation. de Toledo-Morrell *et al.* (1984) showed that aged animals (26 months) had a markedly delayed hippocampal kindling speed that correlated with decreased performance on a memory task. Subsequent investigations citing this study have focused mostly on the learning and memory implications of this observation. One exception is an experiment by Grecksch *et al.* (1997) which showed that kindling with PTZ was also delayed in aged Wistar rats. Administering a subconvulsive dose (the ED_{16} for each age group) every 48 h resulted in fully kindled animals when they were 6 months of age or younger; 18- and 24-month-old animals never developed generalized clonic or tonic-clonic convulsions.

Genetically epileptic animals are a valuable resource for modeling epilepsy and are susceptible to seizures induced by environmental stimuli (Jobe and Laird, 1981). Genetically epilepsy-prone rats (GEPRs) are susceptible to AGS and have a lower threshold for electroshock and PTZ seizures than do other strains (Reigel *et al.*, 1986). There are two distinct GEPR colonies: GEPR-9 (severe seizures) and GEPR-3 (moderate seizures). In a study by Thompson *et al.* (1991), seizure severity increased and latency decreased in GEPR-9s with increasing age when they were juveniles (<65 days), but no significant differences were seen in these parameters between young adult and senescent animals (480–540 days).

Older Wistar rats have spontaneous spike-wave discharges (SWDs) that have been used to model absence epilepsy in the elderly. Coenen and Van Luijtelaar (1987) showed that the number of SWDs and the number of animals with absences increased with age. In another study, the L-type calcium channel agonist BAY K 8644 significantly reduced the number of these discharges ($p < 0.05$) (van Luijtelaar *et al.*, 1995). Puigcerver *et al.* (1996) showed that CGP 35348, a $GABA_B$ antagonist, strongly suppressed the number and duration of SWDs. Ritchie *et al.* (2003) showed that aged F344 rats also had spontaneous SWDs, which could be useful for studies of therapy efficacy. Using the same $GABA_B$ antagonist, they demonstrated that this compound, alone and in combination with other antiepileptic drugs, could reduce spontaneous SWDs as well as SWDs enhanced by the addition of the convulsant trimethylolpropane phosphate.

In vitro epilepsy models, while not substitutes for whole animal testing, have the advantage of requiring less investigational drug when determining efficacy. Two groups have used slice preparations of hippocampal-entorhinal cortex to study geriatric epilepsy. Using Wistar rats, Holtkamp *et al.* (2003) induced epileptiform activity by withdrawing Mg^{2+} from the media; compared to slices from adults (12–14 weeks), slices from aged animals (24–26 months) had reduced seizure spread and conduction velocity as assessed by optical imaging. This observation was confirmed in a similar study using 4-aminopyridine as the convulsive agent. Weissinger *et al.* (2005) showed that propagation velocities were slower in aged animals than in adults, and the location of seizure onset became more focused in the neocortex in aged animals.

V. Conclusions

The work described in this chapter provides a foundation for future exploration and development of animal models of geriatric epilepsy. As the population ages, researchers will need to advance the understanding of geriatric epilepsy in order to expand and improve current treatment.

References

Ben-Ari, Y. (1985). Limbic seizure and brain damage produced by kainic acid: Mechanisms and relevance to human temporal lobe epilepsy. *Neuroscience* **14,** 375–403.

Benkovic, S. A., O'Callaghan, J. P., and Miller, D. B. (2006). Regional neuropathology following kainic acid intoxication in adult and aged C57BL/6J mice. *Brain Res.* **1070,** 215–231.

Bomze, H. M., Bulsara, K. R., Iskandar, B. J., Caroni, P., and Skene, J. H. (2001). Spinal axon regeneration evoked by replacing two growth cone proteins in adult neurons. *Nat. Neurosci.* **4,** 38–43.

Cheal, M. L. (1986). The gerbil: A unique model for research on aging. *Exp. Aging Res.* **12,** 3–21.

Coenen, A. M., and Van Luijtelaar, E. L. (1987). The WAG/Rij rat model for absence epilepsy: Age and sex factors. *Epilepsy Res.* **1,** 297–301.

D'Costa, A., Breese, C. R., Boyd, R. L., Booze, R. M., and Sonntag, W. E. (1991). Attenuation of Fos-like immunoreactivity induced by a single electroconvulsive shock in brains of aging mice. *Brain Res.* **567,** 204–211.

Darbin, O., Naritoku, D., and Patrylo, P. R. (2004). Aging alters electroencephalographic and clinical manifestations of kainate-induced status epilepticus. *Epilepsia* **45,** 1219–1227.

Dawson, R., Jr., and Wallace, D. R. (1992). Kainic acid-induced seizures in aged rats: Neurochemical correlates. *Brain Res. Bull.* **29,** 459–468.

de Toledo-Morrell, L., Morrell, F., and Fleming, S. (1984). Age-dependent deficits in spatial memory are related to impaired hippocampal kindling. *Behav. Neurosci.* **98,** 902–907.

Dickie, N. M. (1964). Lethargic (1h). *Mouse News Lett.* **30,** 31.

Engstrom, F. L., White, H. S., Kemp, J. W., and Woodbury, D. M. (1986). Acute and chronic acetazolamide administration in DBA and C57 mice: Effects of age. *Epilepsia* **27,** 19–26.

Erickson, J. C., Hollopeter, G., Thomas, S. A., Froelick, G. J., and Palmiter, R. D. (1997). Disruption of the metallothionein-III gene in mice: Analysis of brain zinc, behavior, and neuron vulnerability to metals, aging, and seizures. *J. Neurosci.* **17,** 1271–1281.

Fuller, J. L., and Sjursen, F. H., Jr. (1967). Audiogenic seizures in eleven mouse strains. *J. Hered.* **58,** 135–140.

Goddard, G. V., McIntyre, D. C., and Leech, C. K. (1969). A permanent change in brain function resulting from daily electrical stimulation. *Exp. Neurol.* **25,** 295–330.

Grecksch, G., Becker, A., and Rauca, C. (1997). Effect of age on pentylenetetrazol-kindling and kindling-induced impairments of learning performance. *Pharmacol. Biochem. Behav.* **56,** 595–601.

Hall, C. S. (1947). Genetic differences in fatal audiogenic seizures between two inbred strains of house mice. *J. Hered.* **38,** 2–6.

Hirvonen, M. R., Paljarvi, L., and Savolainen, K. M. (1993). Sustained effects of pilocarpine-induced convulsions on brain inositol and inositol monophosphate levels and brain morphology in young and old male rats. *Toxicol. Appl. Pharmacol.* **122,** 290–299.

Holtkamp, M., Buchheim, K., Siegmund, H., and Meierkord, H. (2003). Optical imaging reveals reduced seizure spread and propagation velocities in aged rat brain *in vitro*. *Neurobiol. Aging* **24,** 345–353.

Hosford, D. A., Clark, S., Cao, Z., Wilson, W. A., Jr., Lin, F. H., Morrisett, R. A., and Huin, A. (1992). The role of GABA$_B$ receptor activation in absence seizures of lethargic (lh/lh) mice. *Science* **257,** 398–401.

Hunter, C., Chung, E., and Van Woert, M. H. (1989). Age-dependent changes in brain glycine concentration and strychnine-induced seizures in the rat. *Brain Res.* **482,** 247–251.

Imaizumi, K., and Nakano, T. (1964). Mutant stocks, strain: El. *Mouse News Lett.* **31,** 57.

Jobe, P. C., and Laird, H. E. (1981). Neurotransmitter abnormalities as determinants of seizure susceptibility and intensity in the genetic models of epilepsy. *Biochem. Pharmacol.* **30,** 3137–3144.

Kelly, K. M., Kharlamov, A., Hentosz, T. M., Kharlamova, E. A., Williamson, J. M., Bertram, E. H., III, Kapur, J., and Armstrong, D. M. (2001). Photothrombotic brain infarction results in seizure activity in aging Fischer 344 and Sprague Dawley rats. *Epilepsy Res.* **47,** 189–203.

Kelly, K. M., Ikonomovic, M. D., Abrahamson, E. E., Kharlamov, E. A., Hentosz, T. M., and Armstrong, D. M. (2003). Alterations in hippocampal voltage-gated calcium channel alpha 1 subunit expression patterns after kainate-induced status epilepticus in aging rats. *Epilepsy Res.* **57,** 15–32.

Kim, H. C., Bing, G., Jhoo, W. K., Kim, W. K., Shin, E. J., Park, E. S., Choi, Y. S., Lee, D. W., Shin, C. Y., Ryu, J. R., and Ko, K. H. (2002). Oxidative damage causes formation of lipofuscin-like substances in the hippocampus of the senescence-accelerated mouse after kainate treatment. *Behav. Brain Res.* **131,** 211–220.

Kitani, K., Sato, Y., Kanai, S., Ohta, M., Nokubo, M., and Masuda, Y. (1988). The neurotoxicity of phenobarbital and its effect in preventing pentylenetetrazole-induced maximal seizure in aging mice. *Arch. Gerontol. Geriatr.* **7,** 261–271.

Kitani, K., Klotz, U., Kanai, S., Sato, Y., Ohta, M., and Nokubo, M. (1989). Age-related differences in the coordination disturbance and anticonvulsant effect of oxazepam in mice. *Arch. Gerontol. Geriatr.* **9,** 31–43.

Kondziella, D., Bidar, A., Urfjell, B., Sletvold, O., and Sonnewald, U. (2002). The pentylenetetrazole-kindling model of epilepsy in SAMP8 mice: Behavior and metabolism. *Neurochem. Int.* **40,** 413–418.

Kondziella, D., Hammer, J., Sletvold, O., and Sonnewald, U. (2003). The pentylenetetrazole-kindling model of epilepsy in SAMP8 mice: Glial-neuronal metabolic interactions. *Neurochem. Int.* **43,** 629–637.

Leppik, I. E., Kelly, K. M., deToledo-Morrell, L., Patrylo, P. R., DeLorenzo, R. J., Mathern, G. W., and White, H. S. (2006). Basic research in epilepsy and aging. *Epilepsy Res.* **68**(Suppl. 1), S21–S37.

Lin, F., Wang, Y., and Hosford, D. A. (1999). Age-related relationship between mRNA expression of GABA(B) receptors and calcium channel beta4 subunits in Cacnb4lh mice. *Brain Res. Mol. Brain Res.* **71,** 131–135.

Manev, H., Pericic, D., and Anic-Stojiljkovic, S. (1987). Sex differences in the sensitivity of CBA mice to convulsions induced by GABA antagonists are age-dependent. *Psychopharmacology (Berl.)* **91,** 226–229.

Nagatomo, I., Akasaki, Y., Uchida, M., Tominaga, M., Hashiguchi, W., Kuchiiwa, S., Nakagawa, S., and Takigawa, M. (2000). Age-related alterations of nitric oxide production in the brains of seizure-susceptible EL mice. *Brain Res. Bull.* **53,** 301–306.

Nokubo, M., and Kitani, K. (1988). Age-dependent decrease in the lethal threshold of pentylenetetrazole in mice. *Life Sci.* **43,** 41–47.

Nokubo, M., Kitani, K., Ohta, M., Kanai, S., Sato, Y., and Masuda, Y. (1986). Age-dependent increase in the threshold for pentylenetetrazole induced maximal seizure in mice. *Life Sci.* **38,** 1999–2007.

Okamoto, M., Kita, T., Okuda, H., Tanaka, T., and Nakashima, T. (1992). Effects of acute administration of nicotine on convulsive movements and blood levels of corticosterone in old rats. *Jpn. J. Pharmacol.* **60,** 381–384.

Oztas, B., Kaya, M., and Camurcu, S. (1990). Age related changes in the effect of electroconvulsive shock on the blood brain barrier permeability in rats. *Mech. Ageing Dev.* **51,** 149–155.

Popa-Wagner, A., Fischer, B., Platt, D., Neubig, R., Schmoll, H., and Kessler, C. (1999). Anomalous expression of microtubule-associated protein 1B in the hippocampus and cortex of aged rats treated with pentylenetetrazole. *Neuroscience* **94,** 395–403.

Popa-Wagner, A., Fischer, B., Platt, D., Schmoll, H., and Kessler, C. (2000). Delayed and blunted induction of mRNA for tissue plasminogen activator in the brain of old rats following pentylenetetrazole-induced seizure activity. *J. Gerontol. A Biol. Sci. Med. Sci.* **55,** B242–B248.

Puigcerver, A., van Luijtelaar, E. L., Drinkenburg, W. H., and Coenen, A. L. (1996). Effects of the GABA_B antagonist CGP 35348 on sleep-wake states, behaviour, and spike-wave discharges in old rats. *Brain Res. Bull.* **40,** 157–162.

Racine, R. J. (1972). Modification of seizure activity by electrical stimulation. II. Motor seizure. *Electroencephalogr. Clin. Neurophysiol.* **32,** 281–294.

Rao, M. S., Hattiangady, B., and Shetty, A. K. (2006). Fetal hippocampal CA3 cell grafts enriched with FGF-2 and BDNF exhibit robust long-term survival and integration and suppress aberrant mossy fiber sprouting in the injured middle-aged hippocampus. *Neurobiol. Dis.* **21,** 276–290.

Reigel, C. E., Dailey, J. W., and Jobe, P. C. (1986). The genetically epilepsy-prone rat: An overview of seizure-prone characteristics and responsiveness to anticonvulsant drugs. *Life Sci.* **39,** 763–774.

Retchkiman, I., Fischer, B., Platt, D., and Wagner, A. P. (1996). Seizure induced C-Fos mRNA in the rat brain: Comparison between young and aging animals. *Neurobiol. Aging* **17,** 41–44.

Ritchie, G. D., Hulme, M. E., and Rossi, J., III (2003). Treatment of spontaneous and chemically induced EEG paroxysms in the Fischer-344 rat with traditional antiepileptic drugs or AED + CGP 35348 polytherapy. *Prog. Neuropsychopharmacol. Biol. Psychiatry* **27,** 847–862.

Schmoll, H., Ramboiu, S., Platt, D., Herndon, J. G., Kessler, C., and Popa-Wagner, A. (2005). Age influences the expression of GAP-43 in the rat hippocampus following seizure. *Gerontology* **51,** 215–224.

Scotti, A. L., Bollag, O., and Nitsch, C. (1998). Seizure patterns of Mongolian gerbils subjected to a prolonged weekly test schedule: Evidence for a kindling-like phenomenon in the adult population. *Epilepsia* **39,** 567–576.

Spangler, E. L., Hengemihle, J., Blank, G., Speer, D. L., Brzozowski, S., Patel, N., and Ingram, D. K. (1997). An assessment of behavioral aging in the Mongolian gerbil. *Exp. Gerontol.* **32,** 707–717.

Stijnen, A. M., Postel-Westra, I., Langemeijer, M. W., Hoogerkamp, A., Voskuyl, R. A., van Bezooijen, C. F., and Danhof, M. (1992). Pharmacodynamics of the anticonvulsant effect of oxazepam in aging BN/BiRij rats. *Br. J. Pharmacol.* **107,** 165–170.

Suzuki, J., and Nakamoto, Y. (1977). Seizure patterns and electroencephalograms of El mouse. *Electroencephalogr. Clin. Neurophysiol.* **43,** 299–311.

Takeda, T., Hosokawa, M., Takeshita, S., Irino, M., Higuchi, K., Matsushita, T., Tomita, Y., Yasuhira, K., Hamamoto, H., Shimizu, K., Ishii, M., and Yamamuro, T. (1981). A new murine model of accelerated senescence. *Mech. Ageing Dev.* **17,** 183–194.

Thompson, J. L., Carl, F. G., and Holmes, G. L. (1991). Effects of age on seizure susceptibility in genetically epilepsy-prone rats (GEPR-9s). *Epilepsia* **32,** 161–167.

Tsuda, H., Ito, M., Oguro, K., Mutoh, K., Shiraishi, H., Shirasaka, Y., and Mikawa, H. (1993). Age- and seizure-related changes in noradrenaline and dopamine in several brain regions of epileptic El mice. *Neurochem. Res.* **18,** 111–117.

Turski, W., Dziki, M., Parada, J., Kleinrok, Z., and Cavalheiro, E. A. (1992). Age dependency of the susceptibility of rats to aminooxyacetic acid seizures. *Brain Res. Dev. Brain Res.* **67,** 137–144.

van Luijtelaar, E. L., Ates, N., and Coenen, A. M. (1995). Role of L-type calcium channel modulation in nonconvulsive epilepsy in rats. *Epilepsia* **36,** 86–92.

Weissinger, F., Buchheim, K., Siegmund, H., and Meierkord, H. (2005). Seizure spread through the life cycle: Optical imaging in combined brain slices from immature, adult, and senile rats *in vitro*. *Neurobiol. Dis.* **19,** 84–95.

Wozniak, D. F., Stewart, G. R., Miller, J. P., and Olney, J. W. (1991). Age-related sensitivity to kainate neurotoxicity. *Exp. Neurol.* **114,** 250–253.

Yamazaki, K., Kumazawa, A., Ito, K., Kurihara, K., Nakayama, M., and Wakabayashi, T. (1992). Convulsions in senescence-accelerated mice (SAM-R/1/Eis). *Lab. Anim. Sci.* **42,** 378–381.

Zaman, V., and Shetty, A. K. (2002). Combined neurotrophic supplementation and caspase inhibition enhances survival of fetal hippocampal CA3 cell grafts in lesioned CA3 region of the aging hippocampus. *Neuroscience* **109,** 537–553.

Zaman, V., and Shetty, A. K. (2003). Fetal hippocampal CA3 cell grafts enriched with fibroblast growth factor-2 exhibit enhanced neuronal integration into the lesioned aging rat hippocampus in a kainate model of temporal lobe epilepsy. *Hippocampus* **13,** 618–632.

LIFE AND DEATH OF NEURONS IN THE AGING CEREBRAL CORTEX

John H. Morrison and Patrick R. Hof

Department of Neuroscience, Mount Sinai School of Medicine
New York, New York 10029, USA

The transition from age-associated memory impairment (AAMI) to the dramatic loss of cognitive abilities accompanying Alzheimer's disease (AD) requires progressive development of neocortical pathology that results in neuron death. The selective vulnerability of this neuron death is reflected in the characteristics of cortical pyramidal neurons that are prone to form neurofibrillary tangles. Loss of the neurons that form long corticocortical projections in the association neocortex emerges as the pathological outcome most directly related to the dementia observed in AD. AAMI likely involves alterations of neuronal spines and synapses without neuron death. Interestingly, the same circuits that are vulnerable to degeneration in AD are vulnerable to synaptic alterations short of neuron death. These synaptic alterations likely impact cognitive function in normal aging in a manner consistent with the more modest cognitive decline typically seen in aging. Estrogen levels affect spine density on pyramidal neurons in the prefrontal cortex; these neurons may provide many of the same circuits implicated in AAMI. This association demonstrates an important interface between reproductive and neural senescence and suggests that the synaptic alterations prevalent in normal aging may be responsive to therapy.

INTERNATIONAL REVIEW OF
NEUROBIOLOGY, VOL. 81
DOI: 10.1016/S0074-7742(06)81004-4

41

I. Introduction

In order to understand the neurobiological foundations of epilepsy in the geriatric population, it is important to review the neuropathological events associated with the functional decline observed in age-associated memory impairment (AAMI) and the events that lead to neurodegenerative disorders such as Alzheimer's disease (AD) and distinguish these events from the neurobiological changes associated with geriatric epilepsy. Three key principles of aging with respect to cortical circuitry will be discussed: (1) the loss of synaptic connections in AD due to death of selectively vulnerable neurons and circuit deterioration; (2) the occurrence of AAMI without significant neuron loss and the possible involvement of synaptic alteration in the same circuits; and (3) the interaction of endocrine senescence (i.e., decrease in circulating estrogen) with neural aging and its potential impact on both cognition and related neural circuits.

II. Cortical Circuitry and Alzheimer's Disease

A. OVERVIEW

The pathological hallmarks of AD are neurofibrillary tangles (NFTs) and senile plaques (SPs). Significant neuron death and synapse loss occur in certain cortical regions and layers, and NFTs in particular are related to the loss of neurons. The circuits most vulnerable to degeneration are the perforant path that interconnects the hippocampus with the entorhinal cortex, and the long corticocortical projections linking association cortices such as the prefrontal cortex and the inferior temporal cortex. Initially, a memory defect occurs as a result of perforant path degeneration, and that is followed by a substantial loss of cognitive ability as the corticocortical circuits degenerate. The degeneration of these circuits follows from the selective vulnerability of certain pyramidal cells that furnish corticocortical connections between the association cortices required for cognition.

B. HIPPOCAMPAL PATHOLOGY IN AD

The entorhinal cortex, the subiculum, and CA1 within the hippocampus represent particularly vulnerable regions that consistently display very high NFT densities and resultant neuron death in AD. In fact, the most consistent observation in AD patients is the presence of large numbers of NFTs in layers II and IV of the entorhinal cortex (Hyman *et al.*, 1984; Morrison and Hof, 1997), with the increase in the number of NFTs corresponding to the number of lost neurons

(Hof *et al.*, 2003). The distribution of SPs in the hippocampal formation is variable but often parallels the termination of vulnerable projections, such as a high density of SPs in the molecular layer of the dentate gyrus, which is the terminal zone of the projection from the entorhinal cortex (Hyman *et al.*, 1990). Thus, the perforant pathway, which projects from layers II and III of the entorhinal cortex, is severely compromised in AD (Hyman *et al.*, 1984; Morrison and Hof, 1997). Extensive amyloid accumulation in the terminal zone of the perforant path has also been described in a transgenic mouse model of AD (Reilly *et al.*, 2003).

In the absence of AD, there does not appear to be a significant neuron loss in the entorhinal cortex in aged humans (Gomez-Isla *et al.*, 1996; Hof *et al.*, 2003; West, 1993), nor is there loss of neurons in the entorhinal cortex in aged monkeys (Gazzaley *et al.*, 1997). However, a lack of quantifiable neuron loss does not necessarily mean that degenerative changes are not occurring in the entorhinal cortex as humans age. In fact, virtually all humans over the age of 55 have some NFTs in layer II of the entorhinal cortex (Bouras *et al.*, 1994), making it difficult to distinguish qualitatively between age-related degenerative events in the entorhinal cortex that represent early progressive AD and those that are more stable. Many of the NFTs in this region in normal adults or patients with mild cognitive impairment (MCI) are "transitional neurons," that is, neurons that are still intact and are included in an analysis of total neuron counts, yet have transitional intraneuronal pathology (Hof *et al.*, 2003). Such transitional neurons may represent only 10% of the total neuron count in layer II of patients with MCI, but they represent the majority of neurons in AD. The degree to which transitional neurons are functional or able to be rescued is unclear, yet this will be increasingly important to determine as potential treatments are introduced. Paths of degeneration are not the only risk with aging: this same connection between the entorhinal cortex and the dentate gyrus displays age-related changes short of degeneration, which could also impact function. In addition, as described below, the fully developed dementia of AD is unlikely to result from the mass destruction of this circuit if the destruction occurs in the absence of neocortical degeneration.

C. Neocortical Pathology in AD

Both NFTs and SPs are prevalent in the cerebral cortex of patients with AD; NFTs are located in the perikarya of large pyramidal neurons and SPs are distributed throughout the cortical regions but are particularly numerous in association areas (Arnold *et al.*, 1991; Hof *et al.*, 1999; Rogers and Morrison, 1985). These visible reflections of pathology are particularly important in that each leads to neuron or synapse loss amounting to the circuit disruption underlying dementia. Considerable neuronal loss occurs in the association regions of the neocortex in AD, leaving primary sensory and motor areas relatively spared (Morrison and Hof, 1997). Synapse loss is also extensive in association cortices (Masliah *et al.*, 1991), and the extent of synapse loss is highly correlated with cognitive impairment (Terry *et al.*, 1991). Many studies have demonstrated strong correlations between the regional and laminar distribution of SPs and NFTs and the presumed neurons of origin in certain long corticocortical and hippocampal projections (Hof and Morrison, 1990; Hof *et al.*, 1990; Hyman *et al.*, 1986; Lewis *et al.*, 1987; Pearson *et al.*, 1985; Rogers and Morrison, 1985). Overall, the distribution and severity of neuron loss follow closely the distribution and severity of NFTs (Gomez-Isla *et al.*, 1996, 1997; Morrison and Hof, 1997), and it has been proposed that SPs reflect the degeneration of the terminations of projections from NFT-containing neurons (Lewis *et al.*, 1987; Pearson *et al.*, 1985; Vickers *et al.*, 2000). In the neocortex, the pyramidal cells that furnish the long corticocortical projections are thought to be particularly vulnerable to degeneration in AD (Hof and Morrison, 2004; Morrison and Hof, 1997). It is possible for elderly individuals to maintain a high level of cognitive performance, with accompanying memory deficits, while sustaining significant compromise of hippocampal circuits. Thus, they may rely more on neocortical than on hippocampal circuits for memories essential for daily activities (Albert, 1996).

D. Neurofilament Protein Is a Marker of Neuronal Vulnerability in AD

Certain pyramidal neurons in the neocortex and in the visual cortex of both humans and monkeys have been shown to be enriched in nonphosphorylated neurofilament protein (NPNFP) (Campbell and Morrison, 1989; Hof and Morrison, 1990, 1995, 2004; Hof *et al.*, 1990). Interestingly, neurofilaments as well as other cytoskeletal proteins have been implicated in NFT formation (Hof and Morrison, 2004; Hof *et al.*, 1999; Morrison and Hof, 1997; Morrison *et al.*, 1987; Trojanowski *et al.*, 1993), and pyramidal cells with a high content of NPNFP emerge as a neuron type highly susceptible to NFT formation (Bussière *et al.*, 2003; Hof and Morrison, 2004). Of direct relevance to AD, many long association corticocortical projections in the macaque monkey originate from neurofilament protein–enriched neurons. In fact, circuits interconnecting key prefrontal and temporal regions consistently showed a high percentage (up to 90%) of the neurons of origin being neurofilament protein enriched (Hof *et al.*, 1995). The association cortices are known to be involved in many aspects of cognition (Goldman-Rakic, 1988). In AD, neurofilament protein–enriched neurons in certain neocortical and hippocampal areas are dramatically affected and die during NFT formation (Hof and Morrison, 1990, 2004; Hof *et al.*, 1990; Morrison and Hof, 1997; Morrison *et al.*, 1987; Vickers, 1997; Vickers *et al.*, 1992, 1994, 2000). If the monkey data are considered within the context of the distribution of neurofilament protein–enriched neurons and NFTs in humans, it is likely that the human homologues of the neurofilament protein–enriched, corticocortically projecting neurons in the macaque monkey are those that are highly vulnerable in human AD. To test this correlation more directly in human aging and AD, layer III of area 9 in the prefrontal cortex was targeted for quantitative analysis; counts of total neurons, NPNFP-immunoreactive neurons with no trace of tau accumulation, and NPNFP-immunoreactive neurons with tau accumulation (i.e., NFT formation) were obtained. All profile counts were correlated with clinical assessment as reflected in the Clinical Dementia Rating Scale (CDR) (Morris, 1993). A CDR score of 0.5 amounts to MCI, CDR 1 reflects probable AD, CDR 2 is mild AD, CDR 3 is moderate AD, and CDR 5 is severe AD. The data obtained indicate that in area 9, neurofilament protein–enriched pyramidal cells are little affected in patients with CDR 0.5 scores, yet these cells undergo notable tau accumulation in patients with CDR 2 scores, and virtually all neurofilament protein–enriched neurons are in the process of degeneration during NFT formation in patients with CDR 3 scores (Fig. 1) (Bussière *et al.*, 2003). However, the remaining non–neurofilament protein–labeled neurons remain relatively un-affected, even with well-established AD (Bussière *et al.*, 2003; Hof and Morrison, 1990, 2004; Hof *et al.*, 1990). This result suggests that NFT formation affects a well-defined subset of neurons and that a CDR 2 score, corresponding to mild AD, represents the clinical transition stage at which this subset of highly vulnerable neurons is profoundly affected in the prefrontal cortex (Hof *et al.*, 2003).

E. Summary: AD and Cortical Circuitry

The relevance and impact of pathological changes in AD have to be under-stood within the context of organized systems that underlie neocortical function. Cognition presumably depends strongly on the complex communication among neocortical regions provided by the corticocortical circuits. These circuits degen-erate in AD, leading to a global neocortical disconnection syndrome that presents clinically as dementia (Hof and Morrison, 2004; Morrison and Hof, 1997). Clearly, other degenerative processes occur in AD that may also contribute to the clinical characteristics of the disease. However, the generalized loss of long corticocortical projections emerges functionally as the most devastating component of AD and the pathological outcome most directly related to dementia.

This interpretation of the pathological features of AD suggests that the debilitating dementia results from changes restricted to the association neocortex. Extensive hippocampal alterations existing in the absence of neocortical involvement result in only minor disruptions in activities of daily living. These memory deficits could be revealed by formal testing, but are not compatible with the diagnosis of dementing illness. Because the elements of the biochemical and anatomical phenotypes linked to differential cellular vulnerability in AD are increasingly recognized, it may be possible to develop therapeutic interventions to protect or rescue the high-risk neurons in AD. The protection of these neurons appears to be an attractive strategy for the management of AD and may be more achievable than the development of a cure.

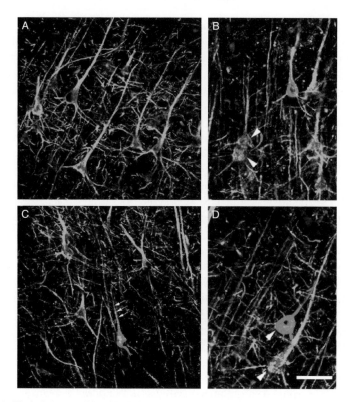

Fig. 1. Typical immunohistochemical patterns observed with antibodies to nonphosphorylated neurofilament protein (SMI-32, red fluorescence) and phosphorylated tau antibody (antibody 988, green fluorescence). Large SMI-32–immunoreactive neurons are seen in layer IIIc from a patient with a CDR 0.5 score, with many neuropil threads, but no NFTs (A). Panels B–D display neurons with different degrees of NFT formation. Two intracellular NFTs are seen in a patient with a CDR 2 score (layer IIIc; B). The tangled neuron in panel C shows faint immunoreactivity in its apical dendrite (arrows; patient with a CDR 2 score, layer Va). Panel D shows a large end-stage NFT next to intact SMI-32–immunoreactive neurons (patient with a CDR 2 score, layer IIIc). Scale bar = 30 μm in D (applies to A–D). CDR, Clinical Dementia Rating Scale; NFT, neurofibrillary tangle. Adapted from Bussière *et al.* (2003), with permission from John Wiley & Sons, Inc., 2006. (See Color Insert.)

III. AAMI: Functional Decline Without Neuron Loss

A. INTRODUCTION

The neocortex as well as CA1 and other hippocampal fields do not display significant neuron loss in normal aging (Peters *et al.*, 1998a). If neuron loss cannot account for memory loss with aging, then the neurobiological complement of such functional decline has yet to be found. Insight on this issue has been gained from animal studies of aging, where AAMI can be analyzed without early AD being a potential confounding factor. Studies from both nonhuman primates and rodents have illuminated both neocortical and hippocampal changes associated with aging. These studies, which are reviewed below, suggest that age-related changes in key excitatory synaptic connections in the absence of significant circuit degeneration may be the primary neurobiological correlate of AAMI.

B. The Aging Cerebral Cortex: Nonhuman Primate Studies

1. Age-Related Alterations of Corticocortical Circuits

As described above, a distinct subpopulation of neurons forming long cortico-cortical projections in the association neocortex is highly vulnerable to the degen-erative process of AD (Lewis et al., 1987; Morrison and Hof, 1997). However, the degree to which sublethal alterations of these same circuits might lead to functional decline has been investigated only recently. Neuronal loss in the neocortex is not observed in the normal course of nonhuman primate aging (Peters et al., 1998a); however, significant cognitive changes can be observed in animals older than 19 years of age (Gallagher and Rapp, 1997). These observations have given rise to questions about whether the same corticocortical circuits that degenerate in human AD are vulnerable to sublethal alterations in normal aging. Electron microscopic investigations have demonstrated consistent changes in oligodendrocyte and axonal myelin sheath morphology in aged nonhuman primates (Peters et al., 1996, 1998b, 2000), suggesting the possible involvement of certain cortical projection systems in aging. Results from several studies indicate that the cognitive processes mediated by the prefrontal cortex are impaired during normal aging processes (Bachevalier et al., 1991; Bartus et al., 1978; O'Donnell et al., 1999; Peters et al., 1996; Rapp and Gallagher, 1997). In particular, aged macaque monkeys consistently show lower performance in delayed response tasks, which are sensitive to prefrontal cortical damage, when compared to younger animals (Rapp and Gallagher, 1997). These and other data indicate that age-related cognitive deficits in nonhuman primates may be a consequence of abnormalities in cortical circuits that affect connectivity, but not by neuron death.

In order to identify potential sublethal circuit-specific alterations, age-related morphological changes at the single-cell level in corticocortically projecting neurons in monkeys were quantified (Duan et al., 2003; Page et al., 2002). Neurons furnishing long corticocortical projections from parietal and temporal regions to area 46 in the prefrontal cortex were targeted for these studies. These neurons were targeted by retrograde transport of fluorescent dyes after injections in area 46, followed by intracellular filling and three-dimensional reconstruction of the neurons (Fig. 2). Such analyses generated detailed quantitative data on the dendritic arbor and spine

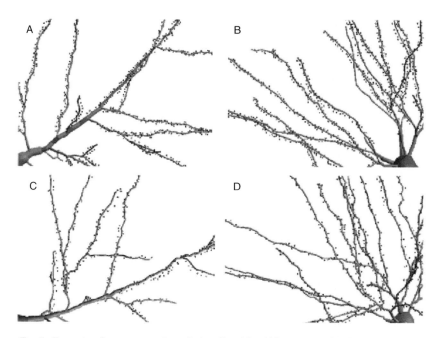

FIG. 2. Examples of macaque monkey apical and basal dendritic segments with all spines plotted, from pyramidal neurons in the temporal association cortex (superior temporal sulcus) that project to the dorsolateral prefrontal cortex, showing that these corticocortical circuits are vulnerable to age-related loss of spines. (A) and (B) are from the apical and basal dendritic trees, respectively, of a neuron from a young animal. (C) and (D) are from apical and basal dendritic trees, respectively, of a neuron from an aged animal. These corticocortically projecting neurons in aged animals consistently had 30–40% fewer spines in both the apical and basal dendritic trees, as shown by quantitative analyses. From Duan *et al.* (2003), by permission of Oxford University Press.

density of the same corticocortically projecting neurons hypothesized to be vulnerable to AD in the human cortex. The total dendritic length and dendritic segment number of these neurons were similar across ages, demonstrating that general dendritic morphology was not vulnerable to aging. However, total spine number on these corticocortically projecting neurons decreased by approximately 35% in the basal and apical dendrites of aged animals compared to young ones (Fig. 2) (Duan *et al.*, 2003; Page *et al.*, 2002). The observed change in spine numbers may lead to a potential deficit in the excitatory drive and sensitivity of these neurons, leading to cognitive deficits such as those described in aged monkeys in the absence of neuron loss. These neurons also show decreases in glutamate receptors with aging (Hof *et al.*, 2002), a further reflection of alterations in excitatory inputs to these neurons that may be related to the spine loss.

2. Circuit-Specific Shifts in N-methyl-D-aspartate (NMDA) Receptors in Hippocampus of Aged Monkeys

Spatial memory is vulnerable to aging (Gallagher and Rapp, 1997) and is also disrupted by pharmacological blockade of NMDA receptor function (Kentros et al., 1998) or hippocampal knockout of the NR1 subunit of the NMDA receptor (Tsien et al., 1996). NMDA receptor–mediated functions, such as maintenance and induction of long-term potentiation and maintenance of stability of spatial information coding by "place cells," are compromised in aging (Barnes et al., 1997). Given the selective vulnerability displayed by both neocortical and hippocampal circuits in aging and AD, high-resolution analyses of receptors can be particularly informative, because the changes are very likely to be cell, circuit, and synapse specific and therefore difficult to resolve at the regional level.

Age-related shifts in NR1, the obligatory subunit of the NMDA receptor, in the molecular layer of the dentate gyrus have been reported in monkeys (Gazzaley et al., 1996b). The projection from the entorhinal cortex to the dentate gyrus is strictly confined to the outer molecular layer (OML), that is, to the distal dendrites of granule cells, whereas other excitatory inputs terminate in a nonoverlapping fashion in the inner molecular layer (IML) on the proximal dendrites (Witter and Amaral, 1991). Aged monkeys, compared to young adult monkeys, exhibit a decrease in the fluorescence intensity for NR1 in the OML of the dentate gyrus as compared to the IML (Gazzaley et al., 1996a). The tight laminar organization of these circuits suggests that the decreased NR1 levels primarily affect the input from the entorhinal cortex, a further reflection of this circuit's vulnerability in aging. These findings suggest that the intradendritic distribution of a neurotransmitter receptor can be modified in an age-related and circuit-specific manner.

3. Age-Related Changes in Rat Hippocampus

A comprehensive analysis of several synaptic indices was performed in the hippocampus of behaviorally characterized young and aged rats (Adams et al., 2001; Smith et al., 2000). Young and aged rats were tested on a hippocampal-dependent version of the Morris water maze; the tests revealed substantial variability in spatial learning ability among aged rats. Immunofluorescence intensity for synaptophysin, a presynaptic vesicle glycoprotein and an established marker for presynaptic terminals, was quantified in the same animals. Interestingly, individual differences in spatial learning capacity correlated with levels of synaptophysin staining in three of the regions examined: the OML and the middle molecular level of the dentate gyrus and the CA3 stratum lacunosum-moleculare. These changes in relative synaptophysin immunofluorescence intensity occurred in the absence of structural degeneration of the innervated dendrites, and thus affected synaptic transmission by compromised glutamate release rather than by degeneration of presynaptic or postsynaptic elements. Most importantly, all three of the regions displaying the decrease received a major projection from layer II of the entorhinal cortex, offering further evidence that this circuit is exquisitely sensitive to aging. These findings suggest that circuit-specific alterations in glutamate release in the rat hippocampus contribute to the effects of aging on learning and memory, in the absence of degeneration.

IV. Interactions Between Neural and Endocrine Senescence

One of the most difficult challenges in research on brain aging is to determine the critical points of interaction and influence between neural senescence and the aging of other systems. For example, it is particularly critical to understand the interaction of reproductive senescence with the aging of the nervous system. Currently, there is a 30-year difference between the life expectancy of an American woman and the average onset of menopause, making the issue of endocrine senescence particularly relevant to human aging (Lamberts *et al.*, 1997).

The role of estrogen in controlling the reproductive axis at the level of the hypothalamus has been studied for many years and characterized in great detail (Fink, 1986). However, estrogens also impact synaptic communication in brain regions involved in cognitive processing, such as the hippocampus (Woolley, 1998), and these effects may be of particular importance in the context of aging, when both circulating estrogen levels change and hippocampal-dependent functions decline (Sherwin, 2000). Until recently, our understanding of estrogen effects on synaptic plasticity in the hippocampus was based primarily on data from young animals. For example, dendritic spine density in CA1 pyramidal cells is sensitive to naturally occurring estrogen fluctuations in young animals (Woolley *et al.*, 1990) and to experimentally induced estrogen depletion and replacement (Gould *et al.*, 1990; Woolley and McEwen, 1992, 1993).

These effects of estrogen on hippocampal circuitry are NMDA receptor dependent (McEwen, 2002), and estrogen replacement directly increases NMDA receptor levels in CA1 dendrites and soma (Gazzaley *et al.*, 1996b). Using quantitative postembedding immunogold electron microscopy, a study investigated estrogen's effects on axospinous synapse density and the synaptic distribution of the NMDA receptor subunit, NR1, within the context of aging (Adams *et al.*, 2001). As shown before (Woolley and McEwen, 1992), estrogen induced an increase in axospinous synapse density in young animals. However, it did not alter the synaptic representation of NR1 in young animals in that the number of NR1 molecules per synapse was equivalent across groups (Adams *et al.*, 2001). Estrogen replacement in aged female rats failed to increase axospinous synapse density; however, estrogen upregulated synaptic NR1 expression when these animals were compared to aged animals not receiving estrogen (Adams *et al.*, 2001). Therefore, the young and aged hippocampus

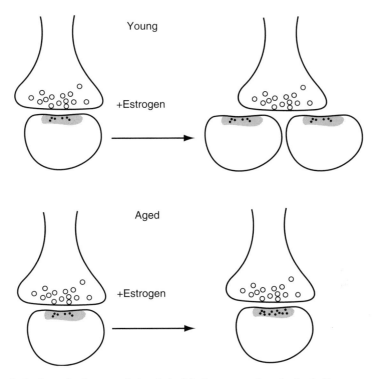

FIG. 3. A schematic of estrogen-induced plasticity in young and aged animals. Estrogen treatment increases *N*-methyl-D-aspartate receptor NR1 subunit expression per synapse in aged hippocampus, whereas it increases spine number but not synaptic NR1 in young female rat hippocampus. Small black dots represent immunogold particles labeling NR1 associated with the postsynaptic density, and open circles are synaptic vesicles. The gray zone indicates the postsynaptic density. From Adams *et al.* (2001), with permission; copyright (2006) National Academy of Sciences, USA.

react differently to estrogen replacement, with the aged animals unable to generate additional synapses, yet able to respond to estrogen by increases in the number of NMDA receptors per synapse (Fig. 3) (Adams *et al.*, 2001). This raises questions concerning the changes in CA1 spines during aging that make the aged rat CA1 less responsive to estrogen with respect to synapse number. It was demonstrated that the key estrogen receptor, ERα, is present in CA1 spines and synapses (Adams *et al.*, 2002; Milner *et al.*, 2001) and presumably has local effects on synaptic transmission in this case rather than its more traditional role in gene regulation in the nucleus (McEwen, 2002). Interestingly, it was also demonstrated that the number of spines in CA1 that contain ERα decreases dramatically with age (Adams *et al.*, 2002), and this decrease may be the cause of the decreased sensitivity to estrogen with respect to formation of new dendritic spines in aged rats.

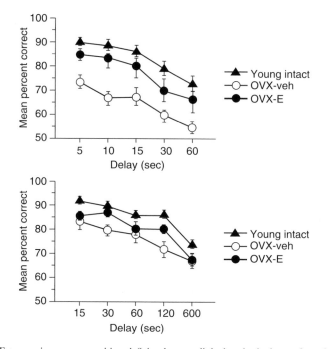

Fig. 4. Estrogen improves cognitive deficits that are linked to both the prefrontal cortex and the hippocampus in aged monkeys. The neuropsychological results shown compare aged ovariecto-mized female monkeys receiving estrogen replacement and aged ovariectomized monkeys receiving vehicle alone to young intact animals. Results from the prefrontal cortex–dependent delayed response task show a substantive increase in performance with estrogen treatment (top panel). Data from the hippocampus-dependent delayed nonmatching to sample task show that the estrogen-treated animals performed significantly better than the nontreated animals at delays of 30 and 120 seconds (bottom panel). OVX-veh, aged ovariectomized monkeys receiving vehicle; OVX-E, aged ovariectomized monkeys receiving estrogen replacement. Note: scales of the x-axes are not proportional. Modified from Rapp *et al.* (2003), with permission; copyright 2006 by the Society for Neuroscience.

Importantly, early results in nonhuman primates show a similar increase in spines in CA1 in response to estrogen; however, aged monkeys were just as responsive as young monkeys (Hao *et al.*, 2003). In addition, the prefrontal cortex in monkeys also displays an estrogen-induced increase in spine number (Hao *et al.*, 2006; Tang *et al.*, 2004). A behavioral analysis demonstrated a significant estrogen-induced enhance-ment of cognitive function in aged rhesus monkeys (Rapp *et al.*, 2003). Aged female rhesus monkeys were ovariectomized and treated with either 100 µg of estradiol cypionate or vehicle alone every 21 days during an extended course of behavioral testing. The estrogen-treated group showed enhanced performance both on a hippocampal-dependent task, delayed nonmatching to sample, and on delayed re-sponse, the prefrontal task demonstrated to be sensitive to age (Fig. 4). These data not only demonstrate estrogen's impact on cognition in a nonhuman primate model but also suggest that age-related decline in cognitive performance is potentially reversible by pharmaceutical intervention.

V. Conclusions

The neurobiological changes associated with epilepsy can be superimposed on other neuropathological events that occur commonly in geriatric patients, and it is useful to understand the processes that can result in these comorbid conditions. The transition from fairly circumscribed memory impairment, sometimes referred to as MCI in humans, to the dramatic loss of cognitive abilities that accompanies AD likely requires progressive development of neocortical pathology. The key vulnerable neocortical cell class has been characterized in some detail, and the demise of this cell class begins early in the transition to dementia. Age-related memory impairment likely involves alternative mechanisms that do not involve neuron death, but rather involve alterations of neuronal spines and synapses. A number of potential molecular and synaptic targets for such age-related decline have been identified, and the key circuits in which such changes occur have been delineated. An important interface exists between reproductive and neural senescence, with estrogen levels influencing many of the same circuits implicated in age-related memory impairment, and some of the synaptic alterations at this interface have been identified.

Acknowledgments

Collective thanks to members of the Morrison and Hof laboratories. Supported by National Institutes of Health grants AG02219, AG05138, AG06647, and AG16765.

References

Adams, M. M., Smith, T. D., Moga, D., Gallagher, M., Wang, Y., Wolfe, B. B., Rapp, P. R., and Morrison, J. H. (2001). Hippocampal dependent learning ability correlates with N-methyl-D-aspartate (NMDA) receptor levels in CA3 neurons of young and aged rats. *J. Comp. Neurol.* **432,** 230–243.

Adams, M. M., Fink, S. E., Shah, R. A., Janssen, W. G., Hayashi, S., Milner, T. A., McEwen, B. S., and Morrison, J. H. (2002). Estrogen and aging affect the subcellular distribution of estrogen receptor-alpha in the hippocampus of female rats. *J. Neurosci.* **22,** 3608–3614.

Albert, M. S. (1996). Cognitive and neurobiologic markers of early Alzheimer disease. *Proc. Natl. Acad. Sci. USA* **93,** 13547–13551.

Arnold, S. E., Hyman, B. T., Flory, J., Damasio, A. R., and Van Hoesen, G. W. (1991). The topographical and neuroanatomical distribution of neurofibrillary tangles and neuritic plaques in the cerebral cortex of patients with Alzheimer's disease. *Cereb. Cortex* **1,** 103–116.

Bachevalier, J., Landis, L. S., Walker, L. C., Brickson, M., Mishkin, M., Price, D. L., and Cork, L. C. (1991). Aged monkeys exhibit behavioral deficits indicative of widespread cerebral dysfunction. *Neurobiol. Aging* **12,** 99–111.

Barnes, C. A., Suster, M. S., Shen, J., and McNaughton, B. L. (1997). Multistability of cognitive maps in the hippocampus of old rats. *Nature* **388,** 272–275.

Bartus, R. T., Fleming, D., and Johnson, H. R. (1978). Aging in the rhesus monkey: Debilitating effects on short-term memory. *J. Gerontol.* **33,** 858–871.

Bouras, C., Hof, P. R., Giannakopoulos, P., Michel, J. P., and Morrison, J. H. (1994). Regional distribution of neurofibrillary tangles and senile plaques in the cerebral cortex of elderly patients: A quantitative evaluation of a one-year autopsy population from a geriatric hospital. *Cereb. Cortex* **4,** 138–150.

Bussière, T., Giannakopoulos, P., Bouras, C., Perl, D. P., Morrison, J. H., and Hof, P. R. (2003). Progressive degeneration of nonphosphorylated neurofilament protein-enriched pyramidal neurons predicts cognitive impairment in Alzheimer's disease: Stereologic analysis of prefrontal cortex area 9. *J. Comp. Neurol.* **463,** 281–302.

Campbell, M. J., and Morrison, J. H. (1989). Monoclonal antibody to neurofilament protein (SMI-32) labels a subpopulation of pyramidal neurons in the human and monkey neocortex. *J. Comp. Neurol.* **282,** 191–205.

Duan, H., Wearne, S. L., Rocher, A. B., Macedo, A., Morrison, J. H., and Hof, P. R. (2003). Age-related dendritic and spine changes in corticocortically projecting neurons in macaque monkeys. *Cereb. Cortex* **13,** 950–961.

Fink, G. (1986). The endocrine control of ovulation. *Sci. Prog.* **70,** 403–423.

Gallagher, M., and Rapp, P. R. (1997). The use of animal models to study the effects of aging on cognition. *Annu. Rev. Psychol.* **48,** 339–370.

Gazzaley, A. H., Siegel, S. J., Kordower, J. H., Mufson, E. J., and Morrison, J. H. (1996a). Circuit-specific alterations of N-methyl-D-aspartate receptor subunit 1 in the dentate gyrus of aged monkeys. *Proc. Natl. Acad. Sci. USA* **93,** 3121–3125.

Gazzaley, A. H., Weiland, N. G., McEwen, B. S., and Morrison, J. H. (1996b). Differential regulation of NMDAR1 mRNA and protein by estradiol in the rat hippocampus. *J. Neurosci.* **16,** 6830–6838.

Gazzaley, A. H., Thakker, M. M., Hof, P. R., and Morrison, J. H. (1997). Preserved number of entorhinal cortex layer II neurons in aged macaque monkeys. *Neurobiol. Aging* **18,** 549–553.

Goldman-Rakic, P. S. (1988). Topography of cognition: Parallel distributed networks in primate association cortex. *Annu. Rev. Neurosci.* **11,** 137–156.

Gomez-Isla, T., Price, J. L., McKeel, D. W., Jr., Morris, J. C., Growdon, J. H., and Hyman, B. T. (1996). Profound loss of layer II entorhinal cortex neurons occurs in very mild Alzheimer's disease. *J. Neurosci.* **16,** 4491–4500.

Gomez-Isla, T., Hollister, R., West, H., Mui, S., Growdon, J. H., Petersen, R. C., Parisi, J. E., and Hyman, B. T. (1997). Neuronal loss correlates with but exceeds neurofibrillary tangles in Alzheimer's disease. *Ann. Neurol.* **41,** 17–24.

Gould, E., Woolley, C. S., Frankfurt, M., and McEwen, B. S. (1990). Gonadal steroids regulate dendritic spine density in hippocampal pyramidal cells in adulthood. *J. Neurosci.* **10,** 1286–1291.

Hao, J., Janssen, W. G., Tang, Y., Roberts, J. A., McKay, H., Lasley, B., Allen, P. B., Greengard, P., Rapp, P. R., Kordower, J. H., Hof, P. R., and Morrison, J. H. (2003). Estrogen increases the number of spinophilin-immunoreactive spines in the hippocampus of young and aged female rhesus monkeys. *J. Comp. Neurol.* **465,** 540–550.

Hao, J., Rapp, P. R., Leffler, A. E., Leffler, S. R., Janssen, W. G., Lou, W., McKay, H., Roberts, J. A., Wearne, S. L., Hof, P. R., and Morrison, J. H. (2006). Estrogen alters spine number and morphology in prefrontal cortex of aged female rhesus monkeys. *J. Neurosci.* **26,** 2571–2578.

Hof, P. R., and Morrison, J. H. (1990). Quantitative analysis of a vulnerable subset of pyramidal neurons in Alzheimer's disease: II. Primary and secondary visual cortex. *J. Comp. Neurol.* **301,** 55–64.

Hof, P. R., and Morrison, J. H. (1995). Neurofilament protein defines regional patterns of cortical organization in the macaque monkey visual system: A quantitative immunohistochemical analysis. *J. Comp. Neurol.* **352,** 161–186.

Hof, P. R., and Morrison, J. H. (2004). The aging brain: Morphomolecular senescence of cortical circuits. *Trends Neurosci.* **27,** 607–613.

Hof, P. R., Cox, K., and Morrison, J. H. (1990). Quantitative analysis of a vulnerable subset of pyramidal neurons in Alzheimer's disease: I. Superior frontal and inferior temporal cortex. *J. Comp. Neurol.* **301,** 44–54.

Hof, P. R., Nimchinsky, E. A., and Morrison, J. H. (1995). Neurochemical phenotype of corticocortical connections in the macaque monkey: Quantitative analysis of a subset of neurofilament protein-immunoreactive projection neurons in frontal, parietal, temporal, and cingulate cortices. *J. Comp. Neurol.* **362,** 109–133.

Hof, P. R., Bouras, C., and Morrison, J. H. (1999). Cortical neuropathology in aging and dementing disorders: Neuronal typology, connectivity, and selective vulnerability. *In* "Cerebral Cortex: Vol. 14: Neurodegenerative and Age-Related Changes in Structure and Function of Cerebral Cortex" (A. Peters and J. H. Morrison, Eds.), pp. 175–312. Kluwer Academic/Plenum, New York.

Hof, P. R., Duan, H., Page, T. L., Einstein, M., Wicinski, B., He, Y., Erwin, J. M., and Morrison, J. H. (2002). Age-related changes in GluR2 and NMDAR1 glutamate receptor subunit protein immunoreactivity in corticocortically projecting neurons in macaque and patas monkeys. *Brain Res.* **928,** 175–186.

Hof, P. R., Bussière, T., Gold, G., Kovari, E., Giannakopoulos, P., Bouras, C., Perl, D. P., and Morrison, J. H. (2003). Stereologic evidence for persistence of viable neurons in layer II of the entorhinal cortex and the CA1 field in Alzheimer disease. *J. Neuropathol. Exp. Neurol.* **62,** 55–67.

Hyman, B. T., Van Hoesen, G. W., Damasio, A. R., and Barnes, C. L. (1984). Alzheimer's disease: Cell-specific pathology isolates the hippocampal formation. *Science* **225,** 1168–1170.

Hyman, B. T., Van Hoesen, G. W., Kromer, L. J., and Damasio, A. R. (1986). Perforant pathway changes and the memory impairment of Alzheimer's disease. *Ann. Neurol.* **20,** 472–481.

Hyman, B. T., Van Hoesen, G. W., and Damasio, A. R. (1990). Memory-related neural systems in Alzheimer's disease: An anatomic study. *Neurology* **40,** 1721–1730.

Kentros, C., Hargreaves, E., Hawkins, R. D., Kandel, E. R., Shapiro, M., and Muller, R. V. (1998). Abolition of long-term stability of new hippocampal place cell maps by NMDA receptor blockade. *Science* **280,** 2121–2126.

Lamberts, S. W., van den Beld, A. W., and van der Lely, A. J. (1997). The endocrinology of aging. *Science* **278,** 419–424.

Lewis, D. A., Campbell, M. J., Terry, R. D., and Morrison, J. H. (1987). Laminar and regional distributions of neurofibrillary tangles and neuritic plaques in Alzheimer's disease: A quantitative study of visual and auditory cortices. *J. Neurosci.* **7,** 1799–1808.

Masliah, E., Hansen, L., Albright, T., Mallory, M., and Terry, R. D. (1991). Immunoelectron microscopic study of synaptic pathology in Alzheimer's disease. *Acta Neuropathol. (Berl.)* **81,** 428–433.

McEwen, B. (2002). Estrogen actions throughout the brain. *Recent Prog. Horm. Res.* **57,** 357–384.

Milner, T. A., McEwen, B. S., Hayashi, S., Li, C. J., Reagan, L. P., and Alves, S. E. (2001). Ultrastructural evidence that hippocampal alpha estrogen receptors are located at extranuclear sites. *J. Comp. Neurol.* **429,** 355–371.

Morris, J. C. (1993). The clinical dementia rating (CDR): Current version and scoring rules. *Neurology* **43,** 2412–2414.

Morrison, J. H., and Hof, P. R. (1997). Life and death of neurons in the aging brain. *Science* **278,** 412–419.

Morrison, J. H., Lewis, D. A., Campbell, M. J., Huntley, G. W., Benson, D. L., and Bouras, C. (1987). A monoclonal antibody to non-phosphorylated neurofilament protein marks the vulnerable cortical neurons in Alzheimer's disease. *Brain Res.* **416,** 331–336.

O'Donnell, K. A., Rapp, P. R., and Hof, P. R. (1999). Preservation of prefrontal cortical volume in behaviorally characterized aged macaque monkeys. *Exp. Neurol.* **160,** 300–310.

Page, T. L., Einstein, M., Duan, H., He, Y., Flores, T., Rolshud, D., Erwin, J. M., Wearne, S. L., Morrison, J. H., and Hof, P. R. (2002). Morphological alterations in neurons forming corticocortical projections in the neocortex of aged Patas monkeys. *Neurosci. Lett.* **317,** 37–41.

Pearson, R. C., Esiri, M. M., Hiorns, R. W., Wilcock, G. K., and Powell, T. P. (1985). Anatomical correlates of the distribution of the pathological changes in the neocortex in Alzheimer disease. *Proc. Natl. Acad. Sci. USA* **82,** 4531–4534.

Peters, A., Rosene, D. L., Moss, M. B., Kemper, T. L., Abraham, C. R., Tigges, J., and Albert, M. S. (1996). Neurobiological bases of age-related cognitive decline in the rhesus monkey. *J. Neuropathol. Exp. Neurol.* **55,** 861–874.

Peters, A., Morrison, J. H., Rosene, D. L., and Hyman, B. T. (1998a). Feature article: Are neurons lost from the primate cerebral cortex during normal aging? *Cereb. Cortex* **8,** 295–300.

Peters, A., Sethares, C., and Moss, M. B. (1998b). The effects of aging on layer 1 in area 46 of prefrontal cortex in the rhesus monkey. *Cereb. Cortex* **8,** 671–684.

Peters, A., Moss, M. B., and Sethares, C. (2000). Effects of aging on myelinated nerve fibers in monkey primary visual cortex. *J. Comp. Neurol.* **419,** 364–376.

Rapp, P. R., and Gallagher, M. (1997). Toward a cognitive neuroscience of normal aging. *In* "Advances in Cell Aging and Gerontology" (P. S. Timiras and E. E. Bittar, Eds.), Vol. 2, pp. 1–21. JAI Press, Greenwich, CT.

Rapp, P. R., Morrison, J. H., and Roberts, J. A. (2003). Cyclic estrogen replacement improves cognitive function in aged ovariectomized rhesus monkeys. *J. Neurosci.* **23,** 5708–5714.

Reilly, J. F., Games, D., Rydel, R. E., Freedman, S., Schenk, D., Young, W. G., Morrison, J. H., and Bloom, F. E. (2003). Amyloid deposition in the hippocampus and entorhinal cortex: Quantitative analysis of a transgenic mouse model. *Proc. Natl. Acad. Sci. USA* **100,** 4837–4842.

Rogers, J., and Morrison, J. H. (1985). Quantitative morphology and regional and laminar distributions of senile plaques in Alzheimer's disease. *J. Neurosci.* **5,** 2801–2808.

Sherwin, B. B. (2000). Oestrogen and cognitive function throughout the female lifespan. *Novartis Found. Symp.* **230,** 188–196; discussion 196–201.

Smith, T. D., Adams, M. M., Gallagher, M., Morrison, J. H., and Rapp, P. R. (2000). Circuit-specific alterations in hippocampal synaptophysin immunoreactivity predict spatial learning impairment in aged rats. *J. Neurosci.* **20,** 6587–6593.

Tang, Y., Janssen, W. G., Hao, J., Roberts, J. A., McKay, H., Lasley, B., Allen, P. B., Greengard, P., Rapp, P. R., Kordower, J. H., Hof, P. R., and Morrison, J. H. (2004). Estrogen replacement increases spinophilin-immunoreactive spine number in the prefrontal cortex of female rhesus monkeys. *Cereb. Cortex* **14,** 215–223.

Terry, R. D., Masliah, E., Salmon, D. P., Butters, N., DeTeresa, R., Hill, R., Hansen, L. A., and Katzman, R. (1991). Physical basis of cognitive alterations in Alzheimer's disease: Synapse loss is the major correlate of cognitive impairment. *Ann. Neurol.* **30,** 572–580.

Trojanowski, J. Q., Schmidt, M. L., Shin, R. W., Bramblett, G. T., Rao, D., and Lee, V. M. (1993). Altered tau and neurofilament proteins in neuro-degenerative diseases: Diagnostic implications for Alzheimer's disease and Lewy body dementias. *Brain Pathol.* **3,** 45–54.

Tsien, J. Z., Huerta, P. T., and Tonegawa, S. (1996). The essential role of hippocampal CA1 NMDA receptor-dependent synaptic plasticity in spatial memory. *Cell* **87,** 1327–1338.

Vickers, J. C. (1997). A cellular mechanism for the neuronal changes underlying Alzheimer's disease. *Neuroscience* **78,** 629–639.

Vickers, J. C., Delacourte, A., and Morrison, J. H. (1992). Progressive transformation of the cytoskeleton associated with normal aging and Alzheimer's disease. *Brain Res.* **594,** 273–278.

Vickers, J. C., Riederer, B. M., Marugg, R. A., Buee-Scherrer, V., Buee, L., Delacourte, A., and Morrison, J. H. (1994). Alterations in neurofilament protein immunoreactivity in human hippocampal neurons related to normal aging and Alzheimer's disease. *Neuroscience* **62,** 1–13.

Vickers, J. C., Dickson, T. C., Adlard, P. A., Saunders, H. L., King, C. E., and McCormack, G. (2000). The cause of neuronal degeneration in Alzheimer's disease. *Prog. Neurobiol.* **60,** 139–165.

West, M. J. (1993). Regionally specific loss of neurons in the aging human hippocampus. *Neurobiol. Aging* **14,** 287–293.

Witter, M. P., and Amaral, D. G. (1991). Entorhinal cortex of the monkey: V. Projections to the dentate gyrus, hippocampus, and subicular complex. *J. Comp. Neurol.* **307,** 437–459.

Woolley, C. S. (1998). Estrogen-mediated structural and functional synaptic plasticity in the female rat hippocampus. *Horm. Behav.* **34,** 140–148.

Woolley, C. S., and McEwen, B. S. (1992). Estradiol mediates fluctuation in hippocampal synapse density during the estrous cycle in the adult rat. *J. Neurosci.* **12,** 2549–2554.

Woolley, C. S., and McEwen, B. S. (1993). Roles of estradiol and progesterone in regulation of hippocampal dendritic spine density during the estrous cycle in the rat. *J. Comp. Neurol.* **336,** 293–306.

Woolley, C. S., Gould, E., Frankfurt, M., and McEwen, B. S. (1990). Naturally occurring fluctuation in dendritic spine density on adult hippocampal pyramidal neurons. *J. Neurosci.* **10,** 4035–4039.

AN *IN VITRO* MODEL OF STROKE-INDUCED EPILEPSY: ELUCIDATION OF THE ROLES OF GLUTAMATE AND CALCIUM IN THE INDUCTION AND MAINTENANCE OF STROKE-INDUCED EPILEPTOGENESIS

Robert J. DeLorenzo,*[,†,‡] David A. Sun,[§]
Robert E. Blair,[†] and Sompong Sombati[†]

*Department of Pharmacology and Toxicology, Virginia Commonwealth University
Richmond, Virginia 23298, USA
[†]Department of Neurology, Virginia Commonwealth University
Richmond, Virginia 23298, USA
[‡]Department of Biochemistry and Molecular Biophysics, Virginia Commonwealth
University, Richmond, Virginia 23298, USA
[§]Department of Neurological Surgery, Vanderbilt University
Medical Center, Nashville, Tennessee 37232, USA

INTERNATIONAL REVIEW OF
NEUROBIOLOGY, VOL. 81
DOI: 10.1016/S0074-7742(06)81005-6

59

Stroke is a major risk factor for developing acquired epilepsy (AE). Although the underlying mechanisms of ischemia-induced epileptogenesis are not well understood, glutamate has been found to be associated with both epileptogenesis and ischemia-induced injury in several research models. This chapter discusses the development of an *in vitro* model of epileptogenesis induced by glutamate injury in hippocampal neurons, as found in a clinical stroke, and the implementation of this model of stroke-induced AE to evaluate calcium's role in the induction and maintenance of epileptogenesis.

To monitor the acute effects of glutamate on neurons and chronic alterations in neuronal excitability up to 8 days after glutamate exposure, whole-cell current-clamp electrophysiology was employed. Various durations and concentrations of glutamate were applied to primary hippocampal cultures. A single 30-min, 5-μM glutamate exposure produced a subset of neurons that died or had a stroke-like injury, and a larger population of injured neurons that survived. Neurons that survived the injury manifested spontaneous, recurrent, epileptiform discharges (SREDs) in neural networks characterized by paroxysmal depolarizing shifts (PDSs) and high-frequency spike firing that persisted for the life of the culture. The neuronal injury produced in this model was evaluated by determining the magnitude of the prolonged, reversible membrane depolarization, loss of synaptic activity, and neuronal swelling.

The permanent epileptiform phenotype expressed as SREDs that resulted from glutamate injury was found to be dependent on the presence of extracellular calcium. The "epileptic" neurons manifested elevated intracellular calcium levels when compared to control neurons, independent of neuronal activity and seizure discharge, demonstrating that alterations in calcium homeostatic mechanisms occur in association with stroke-induced epilepsy. Findings from this investigation present the first *in vitro* model of glutamate injury–induced epileptogenesis that may help elucidate some of the mechanisms that underlie stroke-induced epilepsy.

I. Introduction

Epilepsy is a common neurological disorder and has been estimated to affect 40–50 million people worldwide (Delgado-Escueta and Porter, 1999). Acquired epilepsy (AE) accounts for a significant percentage of all epilepsy cases and is characterized by a known cause of the epilepsy, usually an injury or disease process (Hauser and Hesdorffer, 1990). The most common cause of AE is stroke or cerebral ischemia, representing approximately 40% of these cases (Hauser *et al.*, 1991). Very little is known concerning the molecular mechanisms by which an ischemic insult produces epilepsy, despite the important role of stroke in the development of AE. Thus, to study the molecular events involved in epileptogenesis and to investigate the underlying causes of AE, it is essential to develop models of injury-induced epileptogenesis.

Recent studies characterized and developed an *in vitro* model of injury-induced epileptogenesis to replicate the development of epilepsy as a consequence of stroke (Sun *et al.*, 2001). Injury produced by exposure of cultured hippocampal neurons to glutamate in this model was essentially identical to the glutamate injury observed in stroke. Glutamate exposure produced a mixed population of injured neurons; some developing excitotoxic death and a larger population surviving the injury, analogous to the neurons in the ischemic penumbra (Witte *et al.*, 2000). Neurons that survived the injury phase of glutamate exposure were found to manifest spontaneous, recurrent, epileptiform discharges (SREDs) in synchronized neural networks for the remaining life of the cultures. This *in vitro* model of glutamate injury–induced AE provides a powerful tool to evaluate the molecular mechanisms mediating stroke-induced epileptogenesis (Sun *et al.*, 2001). Using this model of stroke-induced AE, injury-induced epileptogenesis was demonstrated to be calcium dependent (Sun *et al.*, 2002). Furthermore, the maintenance phase of stroke-induced AE was associated with permanent alterations in intracellular calcium levels and altered calcium homeostatic mechanisms (Sun *et al.*, 2004). This chapter is a review of the data demonstrating the development of this *in vitro* model of stroke-induced AE as well as the use of this model to elucidate the role of Ca^{2+} in the induction and maintenance of the epileptic condition in stroke-induced AE.

II. Role of Glutamate in the Pathophysiology of Stroke

Studies on the basic mechanisms of stroke have elucidated some of the pathological events underlying ischemia-induced AE. Developing models of stroke-induced epileptogenesis is important to our understanding of the pathophysiology of this form of AE. During a stroke, ischemia and anoxia cause massive release of the excitatory amino acid neurotransmitter glutamate (Bullock *et al.*, 1995; Davalos *et al.*, 1997), resulting in the excessive activation of postsynaptic glutamate receptors. Overproduction of glutamate is widely recognized to be the major cause of neuronal injury in stroke (Choi, 2000; Dirnagl *et al.*, 1999). Most neurons in the core of the injury of the stroke die after undergoing an irreversible membrane depolarization. This process of cell death produces anoxic depolarization (Hansen and Nedergaard, 1988). The core of dead neurons form the basis of the characteristic irreversible infarct associated with stroke (Lipton, 1999). The peri-infarct penumbra is a term used to describe the area around the infarct of dead neurons. In the peri-infarct penumbra, the injury is less severe and produces a mixed population of neurons that either undergo irreversible anoxic depolarization and die or undergo reversible depolarization and survive

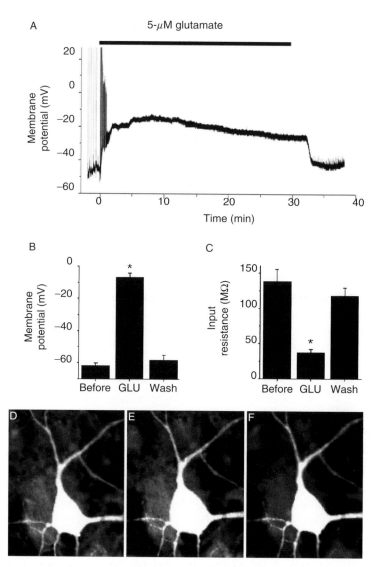

Fig. 1. Application of 5-μM glutamate (GLU) caused hippocampal neuronal injury characterized by depolarization of membrane potential, reduction in membrane input resistance, and somatic swelling. (A) Current-clamp recording of a neuron before, during, and after glutamate application of 30 min (representative of 12 experiments). In the presence of glutamate (black bar), this neuron depolarized from −52 to −17 mV and synaptic activity ceased. On washout, the neuron repolarized to −47 mV and excitatory postsynaptic potentials (EPSPs) returned. (B) Neuronal membrane potential before [−61.7±1.4 (mean ± SE) mV, $n = 12$ neurons], during (GLU, −6.9 ± 2.9 mV, $n = 12$ neurons), and after (−58.4 ± 3.4 mV, $n = 19$ neurons) glutamate application [*$p < 0.05$, repeated measures (RM) analysis of variance (ANOVA) and Tukey test]. (C) Neuronal input resistance before (137.6 ± 7.4 MΩ,

(Hossmann, 1994). The neurons that survive in the penumbra region are the underlying substrates for ischemia-induced epileptogenesis because dead neurons do not seize.

The process of epileptogenesis has also been associated with glutamate receptor activation (Anderson *et al.*, 1986; Croucher and Bradford, 1990; Croucher *et al.*, 1988; Dingledine *et al.*, 1990; Rice and DeLorenzo, 1998; Sombati and DeLorenzo, 1995; Stasheff *et al.*, 1989). On the basis of these observations, it was hypothesized that a less severe glutamate insult that produces prolonged, reversible neuronal depolarization like the glutamate injury in the penumbra of an injury could induce epileptogenesis in surviving neurons. This hypothesis was tested by developing an *in vitro* model of stroke-induced epilepsy to evaluate whether an excitotoxic glutamate injury that produces prolonged, reversible depolarization can cause SREDs that last for the life of hippocampal neurons in culture. This model was developed to gain insights into the molecular mechanisms underlying stroke-induced epileptogenesis.

III. Developing an *In Vitro* Model of Glutamate Injury That Causes a Mixed Population of Injured and Dead Neurons in Preparations of Hippocampal Neurons in Culture

To model the core and penumbra of an ischemic injury, the effect of different concentrations and durations of glutamate exposure on cultured hippocampal neurons was evaluated (Sun *et al.*, 2001). Exposure of cultured neurons to different glutamate concentrations and exposure times produced both cell death (40% of neurons) and reversible injury (60% of neurons) (Sun *et al.*, 2001). The effects of glutamate injury were evaluated by quantitating membrane depolarization, neuronal swelling, and development of neuronal excitotoxicity.

Using whole-cell current-clamp recording, membrane depolarization was evaluated on neurons before, during, and after exposure to 5-μM glutamate for 30 min (Sun *et al.*, 2001). With glutamate exposure, neurons began to depolarize and spike briefly before succumbing to a larger prolonged depolarization (Fig. 1A). Glutamate treatment did not produce continuous spiking. However, glutamate treatment caused an initial membrane depolarization analogous to the depolarization produced during ischemic or anoxic brain injury (Hossmann, 1994; Sombati *et al.*, 1991). The glutamate depolarization was also associated with a loss of synaptic

---

$n = 10$ neurons), during (GLU, 37.2 $\pm$ 5.0 MΩ, $n = 10$ neurons), and after (117.8 $\pm$ 11.2 MΩ, $n = 10$ neurons) glutamate application (*$p < 0.05$, RM ANOVA and Tukey test). (D) A representative fluorescein diacetate (FDA)-stained neuron before exposure to 5-μM glutamate for 30 min. (E) The same FDA-stained neuron in the presence of glutamate, increasing in somatic area by 31%. (F) The same FDA-stained neuron within an hour of washout with restored pre-exposure morphology, only 4% greater in area compared to pretreatment (panel D). From Sun *et al.* (2001) with permission.

potentials (Fig. 1A), and reversible depolarization was quantitated (Fig. 1B). Some neurons (16%, 3 of 19 neurons) did not return to baseline potentials and manifested extended neuronal depolarization (Sombati *et al.*, 1991). These neurons did not survive and represented the neurons in the culture that died following glutamate exposure. Glutamate induced a significant and reversible decrease in membrane input resistance in the neurons undergoing reversible depolarization (Fig. 1C).

By direct examination of fluorescein diacetate (FDA)-stained pyramidal-shaped neurons, neuronal swelling was evaluated (Fig. 1D–F) in the culture (Sun *et al.*, 2001). Neurons were found to swell significantly based on increased somatic area [$37 \pm 8\%$ (mean $\pm$ SE) compared to baseline area, $n = 6$ neurons, $p < 0.05$]. Washout of glutamate was associated with a marked decrease in neuronal swelling, decreasing to $14 \pm 8\%$ (mean $\pm$ SE) of baseline ($n = 6$ neurons) within hours, which was not significantly different from pre-exposure.

The FDA-propidium iodide technique (Didier *et al.*, 1990) was used to evaluate excitotoxicity (Fig. 2A). Glutamate exposure (5 μM) of 5 min resulted in $27 \pm 4\%$ (mean $\pm$ SE) neuronal death ($n = 8$ cultures), not significantly different from sham controls ($18 \pm 2\%$, $n = 13$ cultures), that is, cultures that were treated with the same media and conditions, but were not exposed to glutamate. Neuronal death increased significantly compared to control neurons by increasing exposure from a duration of 30 min ($47 \pm 4\%$, $n = 10$ cultures, $p < 0.05$) to 90 min ($64 \pm 2\%$, $n = 8$ cultures, $p < 0.05$). Over the course of 8 days following glutamate exposure, further neuronal death in experimental groups was similar to controls. By manipulating glutamate exposure times and concentrations, it was possible to develop excitotoxic conditions that produced a small population of neurons that died and a larger population of neurons that recovered after glutamate exposure. This level of injury is very similar to the pathological conditions observed in the ischemic penumbra.

IV. Development of SREDs in Neurons Surviving Injury

Because of the close association between stroke and epilepsy (Hauser *et al.*, 1991), it was hypothesized that neurons surviving glutamate injury would display long-term changes in excitability. Current-clamp recordings from sham control neurons and neurons 1–8 days after exposure to glutamate were evaluated. SREDs were observed in $86 \pm 4\%$ (mean $\pm$ SE) of neurons per culture in the 30-min treatment group ($n = 17$ cultures). These "epileptic" neurons were not observed in the sham control group and were observed significantly more often in the 30-min treatment group than in the 5-min or 90-min glutamate treatment groups (Fig. 2B). SREDs were fully developed following a latency period of 24 h after the injury, but were not observed immediately after glutamate exposure (Fig. 1A). The percentage of epileptic neurons in the 30-min treatment group ranged from

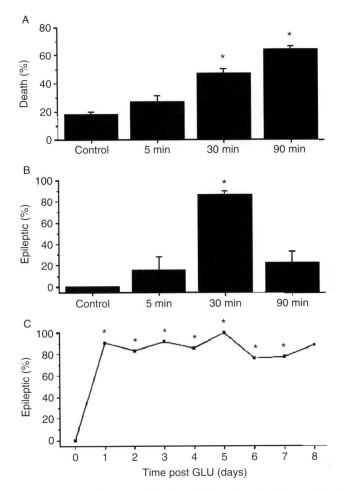

FIG. 2. Glutamate-induced neuronal death and chronic epileptiform activity in cultured hippocampal neurons. (A) Exposure to 5 μM for 30 and 90 min induced significantly more neuronal death 24 h after injury than sham controls and 5-min exposure (mean $\pm$ SE, *$p < 0.05$, ANOVA and Tukey test). (B) The 30-min treatment group [86 $\pm$ 4% (mean $\pm$ SE), $n = 17$ cultures] manifested significantly more "epileptic" neurons 24 h after glutamate injury than sham control (0 $\pm$ 0%, $n = 8$ cultures), 5-min treatment (17 $\pm$ 11%, $n = 5$ cultures), and 90-min treatment groups (22 $\pm$ 11%, $n = 3$ cultures, *$p < 0.05$, ANOVA and Tukey test). (C) Epileptiform activity induced by a 30-min, 5-μM glutamate exposure persisted for the life of the culture. One through 8 days after glutamate treatment, 77–100% of neurons demonstrated epileptiform activity. The percentage of neurons manifesting spontaneous, recurrent, epileptiform discharges (SREDs) was significantly different from the Day 0 control (before exposure, *$p < 0.01$ chi-square test, $n = 15$ neurons). No statistical differences were observed in the percentage of epileptic neurons on Days 1 through 8 ($n = 32, 48, 12, 7, 2, 17, 18, 9$ neurons, respectively) after glutamate treatment, demonstrating the persistence of SREDs throughout the life of the culture ($p = 0.841$, chi-square test). From Sun *et al.* (2001) with permission.

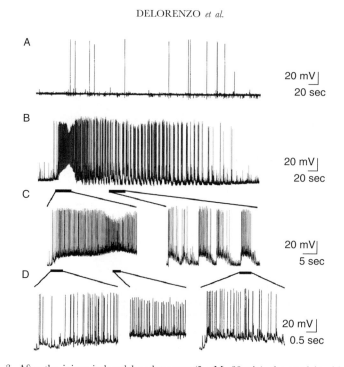

FIG. 3. After the injury induced by glutamate (5 μM, 30 min), the surviving hippocampal neurons manifested spontaneous, recurrent, epileptiform discharges (SREDs). (A) Representative current-clamp recording from a control neuron with a resting potential of −53 mV. Spontaneous action potentials, excitatory postsynaptic potentials, and inhibitory postsynaptic potentials were observed, indicative of normal neurophysiology. (B) One of the seven SREDs from a neuron (resting potential of −63 mV) 24 h after glutamate exposure. This SRED started and terminated spontaneously, lasting approximately 6 min. (C) Regions at the initiation and near the termination of the SRED in panel above (bars) displayed at a faster timescale. The SRED started abruptly as a sustained membrane depolarization of approximately 20 mV. As the episode began to terminate, discrete paroxysmal depolarizing shifts (PDSs) became apparent. (D) Regions of high-frequency spike firing throughout the SRED (bars) displayed at a faster timescale. The initial depolarization of the SRED triggered high-frequency spike firing of approximately 7 Hz. As the prolonged depolarization maximized, spike firing reached a frequency of 12 Hz, sufficient to reduce spike amplitude, presumably due to sodium channel inactivation. As the SRED began to terminate, PDSs still maintained high-frequency spike firing of 7 Hz. From Sun *et al.* (2001) with permission.

77% to 100% over the 8 days after exposure and was not significantly different from day to day (Fig. 2C).

Sham control neurons demonstrated spontaneous action potentials (spikes), excitatory postsynaptic potentials (EPSPs), and inhibitory postsynaptic potentials (IPSPs), typical of normal synaptic activity (Fig. 3A). Evaluation of spike discharges in control neurons ($n = 12$ neurons, over 4 h of recording) demonstrated that the great majority of spike firings (88%) occurred in a frequency range less than 3 Hz (Sun *et al.*, 2001). SREDs were never observed in control neurons.

Only the hippocampal neurons that survived glutamate injury manifested SREDs (Fig. 3B). This supports the hypothesis that dead neurons do not seize. The SREDs that developed in the surviving neurons manifested depolarizing events (Fig. 3C) that started abruptly and were typical of the PDSs characteristic of epileptiform discharges (Matsumoto and Marsan, 1964). With the development of the epileptiform activity, the discharges became shorter with apparent discrete PDSs (Fig. 3C). The PDSs triggered high-frequency spike discharges throughout the SREDs (Fig. 3D). The spike frequency was evaluated during the SREDs, and analysis of 10 representative epileptic neurons (totaling over 7 h of recording) revealed that more than 61% of the spikes in epileptic neurons had an instantaneous frequency greater than 3 Hz (Sun *et al.*, 2001). The majority of the high-frequency epileptiform discharges (greater than 60%) lasted longer than 20 sec and were, therefore, considered SREDs or electrographic seizures. No significant differences were observed in membrane potential, input resistance, or spike characteristics (amplitude, rise time, 50% rise, 90% rise) between control and epileptic neurons. These findings demonstrated that SREDs were present in neurons surviving glutamate injury, although intrinsic membrane properties were not significantly different from control neurons.

SREDs or electrographic seizures occurred spontaneously, randomly, and recurrently in neurons surviving glutamate injury. The mean ($\pm$SE) duration and the average time ($\pm$SE) interval between SREDs were 2.1 $\pm$ 0.3 min and 7.2 $\pm$ 1.0 min, respectively ($n = 10$ neurons, over 7 h of recording). In a 2.5-h recording, one neuron demonstrated nine independent SREDs, ranging in duration from 1.08 to 4.83 min (Fig. 4).

V. Neuronal Networks Display Synchronized SREDs and Respond to Anticonvulsant Treatment

To show that neurons surviving glutamate injury and displaying SREDs were bursting in synchronized neural networks, the possible synchrony of the discharges was evaluated employing whole-cell current-clamp recordings on pairs of neurons 1–8 days after glutamate injury. Distances ranging from immediately adjacent to as far as 800-μm separated neuronal pairs. The SREDs or epileptiform discharges occurred simultaneously in 90% of neuron pairs ($n = 29$ pairs), demonstrating that the onset and termination of the epileptiform discharges were highly synchronized. In addition, individual PDSs were synchronized between neurons (Sun *et al.*, 2001). High-frequency spikes occurred simultaneously and were associated with both the prolonged initial depolarization and discrete PDSs near the end of SREDs.

To determine the response of the SREDs in this new model to anticonvulsant treatment, both phenobarbital and ethosuximide were evaluated in the cultured

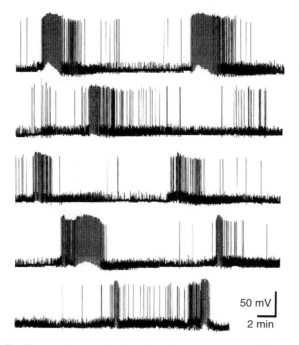

Fig. 4. Epileptiform activity persisted for the duration of the recording session. A current-clamp recording from a glutamate-treated neuron showed that spontaneous, recurrent, epileptiform discharges (SREDs) occurred throughout the recording period. The five tracings shown were part of one continuous, 2.5-h current-clamp experiment. This neuron (−66-mV resting potential) exhibited nine independent SREDs during the recording period. From Sun *et al.* (2001) with permission.

neuronal populations. SREDs were terminated by phenobarbital (Sun *et al.*, 2001). Conversely, ethosuximide, a T-type voltage-gated Ca^{2+} channel inhibitor effective in treating generalized absence seizures (Rogawski and Porter, 1990), had no effect on SREDs in the model (Sun *et al.*, 2001). The seizure-like SREDs induced by glutamate injury responded to therapeutically relevant concentrations of anticonvulsants, analogous to the setting of generalized tonic-clonic and partial complex seizures (Macdonald and Kelly, 1993).

VI. Use of the *In Vitro* Model of Stroke-Induced AE to Evaluate the Calcium Hypothesis of Epileptogenesis

The calcium hypothesis of epileptogenesis was developed to explain the role of Ca^{2+} in causing and maintaining epileptogenesis. This hypothesis states that alterations in neuronal intracellular calcium concentrations ($[Ca^{2+}]_i$) and neuronal Ca^{2+}

homeostatic mechanisms play an essential signaling role in the induction and maintenance of the epileptic phenotype (DeLorenzo *et al.*, 1998; Perlin and DeLorenzo, 1992). Overactivation of the *N*-methyl-D-aspartate (NMDA) receptor subtype of glutamate receptors and irreversible elevations in free $[Ca^{2+}]_i$ have been implicated in excitotoxic neuronal death (Choi, 1994). Furthermore, NMDA receptor activation and prolonged, but reversible, elevations in $[Ca^{2+}]_i$ have been implicated in the induction of epilepsy (Croucher and Bradford, 1990; Croucher *et al.*, 1988; DeLorenzo *et al.*, 1998; Dingledine *et al.*, 1990; Rice and DeLorenzo, 1998; Stasheff *et al.*, 1989). The *in vitro* stroke-induced AE model was used to test the hypothesis that prolonged elevations in $[Ca^{2+}]_i$ and NMDA receptor activation are required for glutamate injury–induced epileptogenesis in cultured hippocampal neurons. Fluorescent Ca^{2+} imaging and electrophysiological techniques were employed to characterize the ability of injury induced by glutamate exposure to cause acute changes in neuronal $[Ca^{2+}]_i$ and chronic changes in neuronal excitability in the presence of different Ca^{2+} conditions and various glutamate receptor antagonists (Sun *et al.*, 2002).

VII. Role of Ca^{2+} and NMDA Receptor Activation in Epileptogenesis

To test the role of Ca^{2+} in mediating epileptogenesis, neurons were reversibly injured, with the same epileptogenic glutamate exposure that induced SREDs, in a reduced extracellular calcium concentration ($[Ca^{2+}]_e$) solution (2.0 mM reduced to 0.2 mM). Neurons survived the glutamate injury in this low $[Ca^{2+}]_e$ environment, exhibited normal levels of activity similar to controls (Fig. 5), and never manifested SREDs ($p = 1.0$, $n = 11$). The reduced $[Ca^{2+}]_e$ solution was shown to also significantly lower the $[Ca^{2+}]_i$ load of the neuron following glutamate exposure in comparison to the glutamate-only treatment (Fig. 6A–C). To evaluate if glutamate injury–induced epileptogenesis was associated with changes in neuronal $[Ca^{2+}]_i$, neuronal $[Ca^{2+}]_i$ before, during, and after epileptogenic glutamate exposure was monitored using the ratiometric, high-affinity calcium indicator fura-2 (Grynkiewicz *et al.*, 1985). The control neurons that were exposed to solution changes without glutamate did not manifest elevations in $[Ca^{2+}]_i$ (Fig. 6A). Conversely, $[Ca^{2+}]_i$ increased rapidly following glutamate exposures that were epileptogenic (Fig. 6B). The majority of neurons slowly restored basal Ca^{2+} levels after removal of glutamate (Fig. 6B). Some neurons (6 of 21), however, still maintained $[Ca^{2+}]_i$ greater than 400 nM even after 120 min of recording. Neurons that manifested sustained elevations in $[Ca^{2+}]_i$ greater than 400 nM for durations of 120 min or more were categorized as having an inability to restore resting $[Ca^{2+}]_i$ (IRRC). Neurons with IRRC have been shown to undergo delayed excitotoxic neuronal death (Limbrick *et al.*, 1995). Neurons that died were excluded from statistical analysis. In surviving

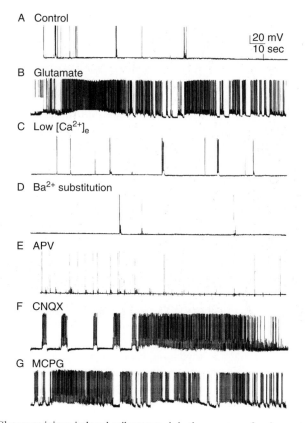

FIG. 5. Glutamate injury–induced epileptogenesis in the presence of various extracellular Ca^{2+} conditions and glutamate receptor antagonists. Ninety percent of neurons surviving an epileptogenic glutamate injury (B, $n = 30$) manifested spontaneous, recurrent, epileptiform discharges (SREDs), which were not observed in controls (A, $n = 37$). Glutamate injury performed in low (0.2 mM) extracellular Ca^{2+} (C, $n = 11$), 2.0-mM Ba^{2+} substitution for Ca^{2+} (D, $n = 11$), and in the presence of the competitive N-methyl-D-aspartate (NMDA) receptor antagonist 2-amino-5-phosphonovalerate (APV) (E, $n = 12$) inhibited the induction of epileptogenesis. Glutamate injury–induced epileptiform activity was observed in 80% of neurons injured in the presence of the non-NMDA receptor antagonists 6-cyano-7-nitroquinoxaline-2,3-dione (CNQX) (F, $n = 10$) and α-methyl-4-carboxyphenylglycine (MCPG) (G, $n = 10$). From Sun *et al.* (2002), with permission from Blackwell Publishing.

neurons, $[Ca^{2+}]_i$ was statistically elevated over basal levels from the time of exposure through 76 min of recording ($p < 0.05$, $n = 15$, Fig. 6B). The neurons associated with glutamate injury–induced epileptogenesis also manifested prolonged, reversible increases in neuronal free calcium.

Calcium is a major second messenger in neurons (Berridge, 1998; Rogawski and Porter, 1990). Thus, the role of Ca^{2+} as a second messenger in glutamate

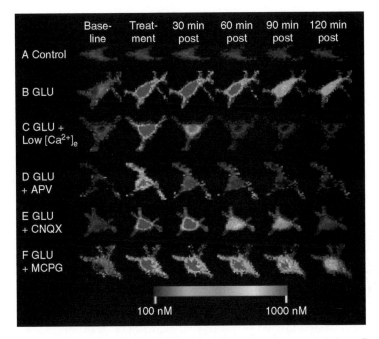

Fig. 6. Representative pseudocolor digital images of fura-2-loaded neurons during epileptogenic glutamate (GLU) injury in the presence of various extracellular Ca^{2+} conditions and glutamate receptor antagonists. The representative control neuron (A) did not undergo changes in intracellular Ca^{2+} concentrations ($[Ca^{2+}]_i$) during the recording period. In contrast, neuronal $[Ca^{2+}]_i$ increased on treatment with glutamate (B) and in all other experimental conditions, as indicated by the change in color from blue to red. The representative neurons treated in nonepileptogenic conditions of low extracellular Ca^{2+} concentrations ($[Ca^{2+}]_e$) (C) and NMDA receptor antagonism by APV (D) restored basal neuronal $[Ca^{2+}]_i$ within 60 and 30 min, respectively. In contrast, neuronal $[Ca^{2+}]_i$ measurements in epileptogenic conditions of glutamate alone (B) and in the presence of CNQX, an antagonist of AMPA and KA receptors (E), and MCPG, an antagonist of mGluRs (F), were elevated for 90 min or longer. AMPA, α-amino-3-hydroxy-5-methyl-4-isoxazolepropionic acid; APV, 2-amino-5-phosphonovalerate; CNQX, 6-cyano-7-nitroquinoxaline-2,3-dione; KA, kainate; mGluRs, metabotropic G-protein–coupled glutamate receptors; MCPG, α-methyl-4-carboxyphenylglycine; NMDA, *N*-methyl-D-aspartate. From Sun *et al.* (2002), with permission from Blackwell Publishing. (See Color Insert.)

injury–induced epileptogenesis was evaluated. Glutamate exposure that produced epileptogenesis was performed in Ca^{2+}-free, barium (Ba^{2+})-substituted solutions. Barium carries a divalent positive charge across the neuronal membrane through conventional routes of Ca^{2+} entry (Mayer and Westbrook, 1987; Nelson *et al.*, 1984), but it does not activate Ca^{2+}-sensitive second messenger pathways (DeLorenzo *et al.*, 1998). Glutamate injury–induced epileptogenesis was blocked by substituting extracellular solutions with 2.0-mM Ba^{2+} for Ca^{2+} ($p = 1.0$, $n = 11$, Fig. 5D). Glutamate injury–induced epileptogenesis required extracellular Ca^{2+}

and activation of intracellular Ca^{2+} pathways leading to the development of persistent neuronal hyperexcitability.

To further evaluate the calcium hypothesis of epileptogenesis, cultured hippocampal neurons were exposed to epileptogenic glutamate concentrations in the presence of the competitive NMDA receptor antagonist 2-amino-5-phosphono-valerate (APV) (Davies and Watkins, 1982). In the presence of 50-μM APV, injured neurons exhibited activity similar to that of control neurons and never manifested SREDs ($p = 1.0$, $n = 10$, Fig. 5E). However, APV treatment reduced the duration of increased $[Ca^{2+}]_i$ levels (Fig. 6D). The noncompetitive NMDA receptor antagonist MK-801 (Wong *et al.*, 1986) at 10 μM also significantly blocked glutamate injury–induced epileptogenesis. These results indicate that activation of the NMDA receptor subtype of glutamate receptors was required for glutamate injury–induced epileptogenesis.

VIII. Antagonism of Non-NMDA Receptor Subtypes of Glutamate Receptors Does Not Inhibit Glutamate Injury–Induced AE

The ion channels formed by certain configurations of the α-amino-3-hydroxy-5-methyl-4-isoxazole propionic acid (AMPA) and kainate (KA) receptor subtypes of glutamate receptors are also permeable to Ca^{2+} (Dingledine *et al.*, 1999). On the basis of these results, the role of AMPA and KA receptors in glutamate injury–induced epileptogenesis was tested by exposing cultured hippocampal neurons to epileptogenic concentrations of glutamate in the presence of the competitive AMPA and KA receptor antagonist 6-cyano-7-nitroquinoxaline-2,3-dione (CNQX) (Honore *et al.*, 1988). The neurons surviving epileptogenic glutamate injury in 2-μM CNQX manifested SREDs (Fig. 5F) at statistically greater levels than in controls ($p < 0.001$, $n = 10$). Thus, CNQX did not prevent the development of epileptogenesis ($p = 0.584$, $n = 10$) and did not have a major effect on $[Ca^{2+}]_i$ levels when compared to glutamate treatment alone (Fig. 6E).

The metabotropic G-protein–coupled glutamate receptors (mGluRs) alter neuronal $[Ca^{2+}]_i$ through the modulation of intracellular Ca^{2+} stores, ligand-gated ion channels, and voltage-gated ion channels (Conn and Pin, 1997). To investigate the potential role of mGluR-mediated Ca^{2+} signaling during epileptogenic glutamate injury, mGluRs were blocked with the competitive antagonist α-methyl-4-carboxyphenylglycine (MCPG) (Watkins and Collingridge, 1994). At 250 μM, MCPG had no effect on glutamate injury–induced epileptogenesis ($p < 0.001$, $n = 10$, Fig. 5G), and it did not significantly decrease the rise in $[Ca^{2+}]_i$ levels during injury (Fig. 6F).

IX. Stroke-Induced AE Is Associated With Prolonged Elevations in Neuronal $[Ca^{2+}]_i$ Levels and Alterations in Ca^{2+} Homeostatic Mechanisms

It was further examined whether persistent increases in $[Ca^{2+}]_i$ levels and alterations in Ca^{2+} homeostatic mechanisms in the *in vitro* model of stroke-induced AE were associated with glutamate injury–induced epileptogenesis (Sun *et al.*, 2002, 2004), as observed in status epilepticus (SE)-induced epilepsy models (Pal *et al.*, 2000, 2001; Raza *et al.*, 2001). On the basis of the results of studies in SE-induced epilepsy, it was hypothesized that increases in $[Ca^{2+}]_i$ and alterations in Ca^{2+} homeostatic mechanisms represented an underlying mechanism mediating the induction and maintenance of epileptogenesis in AE. This hypothesis was tested in the stroke model of AE. Finding a common pathophysiological mechanism for injury-induced epilepsy would not only expand our understanding of the pathophysiology of epileptogenesis, but might offer a broader therapeutic window to prevent or reverse injury-induced epilepsy.

X. Glutamate Injury–Induced Epileptogenesis Causes Long-Lasting Elevations in Basal Neuronal $[Ca^{2+}]_i$ Levels

Using the indicator fura-2 (Grynkiewicz *et al.*, 1985), basal $[Ca^{2+}]_i$ was measured in control and epileptic neurons, including after all synaptic and epileptiform activity was prevented with tetrodotoxin (Fig. 7). The control neurons exhibited an average ($\pm$SE) basal $[Ca^{2+}]_i$ of 131.04 ± 5.87 nM ($n = 147$) 1 day after control treatment. Conversely, epileptic neurons manifested a statistically higher average basal $[Ca^{2+}]_i$ of 278.13 ± 13.36 nM ($n = 130$, $p < 0.001$ Student t test) 1 day after epileptogenic glutamate injury (Sun *et al.*, 2004). Basal $[Ca^{2+}]_i$ in control neurons was 120.04 ± 4.81 nM ($n = 97$) 5 days after control treatment. Five days after glutamate treatment, epileptic neurons manifested a statistically higher average basal $[Ca^{2+}]_i$ of 299.12 ± 24.31 nM [$n = 94$, $p < 0.001$ analysis of variance (ANOVA)] (Sun *et al.*, 2004). Low $[Ca^{2+}]_e$ media during glutamate injury that is known to inhibit epileptogenesis (Sun *et al.*, 2002) was found to block the rise in $[Ca^{2+}]_i$ during glutamate injury, and the basal $[Ca^{2+}]_i$ levels in these neurons were the same as in the controls (106.63 ± 1.85 nM, $n = 94$) (Sun *et al.*, 2004). This study demonstrated the Ca^{2+} dependence of both the induction of epilepsy and the prolonged elevations in basal neuronal $[Ca^{2+}]_i$ in this model (Sun *et al.*, 2004). Studies (Sun *et al.*, 2002, 2004) have also shown that the induction of epilepsy in this model caused a permanent (for the life of the neurons in culture) elevation of $[Ca^{2+}]_i$ in epileptic neurons that was independent of the Ca^{2+} influx associated with epileptiform activity and was due to alterations in Ca^{2+} homeostatic mechanisms.

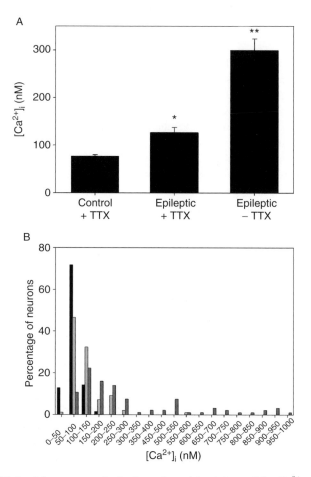

Fig. 7. "Epileptic" neurons maintained an elevated basal intracellular Ca^{2+} concentrations ($[Ca^{2+}]_i$) in the absence of spontaneous, recurrent, epileptiform discharges (SREDs). (A) Blockade of epileptiform activity by 600-nM tetrodotoxin (TTX) significantly reduced the long-lasting elevations in basal $[Ca^{2+}]_i$ in epileptic neurons ($n = 99$) in comparison to epileptic neurons ($n = 94$) manifesting SREDs (epileptic − TTX). However, basal $[Ca^{2+}]_i$ in epileptic neurons ($n = 99$) was still significantly elevated compared to control neurons ($n = 71$) in the presence of 600-nM TTX. (B) Histograms of basal $[Ca^{2+}]_i$ with bins of 50 nM in control ($n = 71$, black) and epileptic ($n = 99$, light gray) neurons in TTX and epileptic neurons without TTX ($n = 94$, dark gray). In TTX, 99% of control and 80% of epileptic neurons had basal $[Ca^{2+}]_i \leq 150$ nM. In contrast, only 33% of epileptic neurons without TTX had basal $[Ca^{2+}]_i \leq 150$ nM. *$p < 0.05$, analysis of variance (ANOVA) and Tukey post hoc test versus Control + TTX. **$p < 0.05$, ANOVA and Tukey post hoc test versus epileptic + TTX. Data are represented by mean ± SEM. From Sun *et al.* (2004), with permission.

XI. Epileptic Neurons Demonstrate Impaired Recovery of Resting $[Ca^{2+}]_i$ After Brief Glutamate-Induced Ca^{2+} Loading

Epileptic and control cultures were exposed to glutamate to challenge neurons with elevated $[Ca^{2+}]_i$ levels, allowing the examination of the epileptogenesis-induced alterations in Ca^{2+} homeostasis in glutamate injury–induced epilepsy. Levels of $[Ca^{2+}]_i$ in the neurons were monitored for the ability of neurons to restore resting $[Ca^{2+}]_i$. Tetrodotoxin was used to inhibit epileptiform activity in the monitoring studies. Responses to $[Ca^{2+}]_i$ were measured using the low-affinity, ratiometric Ca^{2+} indicator fura-2FF (Hyrc *et al.*, 2000). Using this technique, it was possible to evaluate the Ca^{2+} homeostatic mechanisms independent of endogenous neuronal activity. The $[Ca^{2+}]_i$ levels were determined as ratiometric values (340/380) (Sun *et al.*, 2004). The evaluation of images from epileptic and control neurons following a brief $[Ca^{2+}]_i$ load (Sun *et al.*, 2004) demonstrated that epileptic neurons could not restore $[Ca^{2+}]_i$ levels to basal levels as rapidly as control neurons (Fig. 8). The induction of epileptogenesis in the stroke-induced AE model caused a significant alteration in neuronal Ca^{2+} homeostatic mechanisms. In addition, alterations in the Ca^{2+} homeostatic mechanisms of the epileptic neurons observed in the glutamate injury–induced AE model were seen throughout the cytoplasm of the neurons and into the dendritic processes (Sun *et al.*, 2004). Previous studies demonstrated that these injury-induced abnormalities in Ca^{2+} homeostatic mechanisms were also apparent in the nucleus (Sun *et al.*, 2004).

The $[Ca^{2+}]_i$ response of control neurons (Fig. 8A) and epileptic neurons (Fig. 8B) increased significantly during glutamate exposure. Epileptic neurons demonstrated an impaired ability to restore resting $[Ca^{2+}]_i$ after brief glutamate exposure. The ratio value was still elevated after 5 min of glutamate washout and remained elevated at 15 min and 30 min post-glutamate washout. The average $[Ca^{2+}]_i$ is depicted in Fig. 8B. Following the glutamate challenge, the epileptic neurons responded to glutamate with significant increases in $[Ca^{2+}]_i$ that were sustained for more than 30 min of recording [$n = 44$, $p < 0.05$ repeated measures (RM) ANOVA and Tukey post hoc test].

XII. The Importance of *In Vitro* Models of Stroke-Induced AE

Excitotoxic glutamate injury in the *in vitro* model of stroke-induced AE produced a mixed population of neurons characterized by both cell survival and cell death (Sun *et al.*, 2001), analogous to the excitotoxic glutamate injuries associated with both ischemic and anoxic events (Choi, 2000). Prolonged, reversible membrane depolarization, decreased membrane input resistance, loss of synaptic potentials, and neuronal swelling were characteristics of the glutamate

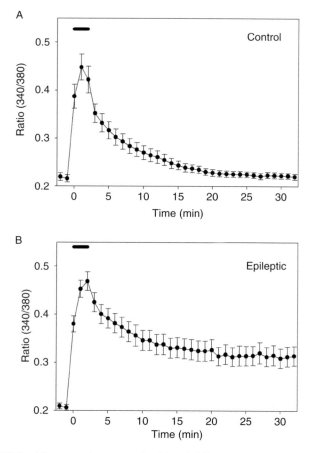

FIG. 8. "Epileptic" neurons demonstrated a delayed ability to restore resting intracellular Ca^{2+} concentrations ($[Ca^{2+}]_i$) after brief glutamate exposure. (A) Ratiometric response of control neurons ($n = 43$) to brief glutamate exposure (50 μM, 2 min; black bar, $t = 0$ min). Glutamate-induced $[Ca^{2+}]_i$ elevations were significantly higher than basal levels for 20 min, but returned to basal levels within 30 min. (B) Ratiometric $[Ca^{2+}]_i$ response of epileptic neurons ($n = 44$) to brief glutamate exposure (50 μM, 2 min; black bar, $t = 0$ min). Glutamate-induced $[Ca^{2+}]_i$ elevations were significantly higher than basal levels for more than 30 min [$p < 0.05$, repeated measures (RM) analysis of variance (ANOVA) and Tukey post hoc test]. Data are represented by mean $\pm$ SEM. From Sun *et al.* (2004), with permission.

injury in this model. The SREDs that were observed in approximately 86% of the neurons that survived the glutamate exposure persisted for the life of the cultures. Thus, neurons that survive the glutamate injury are the substrates for the development of SREDs. This novel model of glutamate injury–induced epileptogenesis may represent a useful model to study stroke-induced AE.

The SREDs produced in the stroke-induced model of AE express many characteristics of overt epileptic seizures: SREDs start and terminate spontaneously (Figs. 3 and 4) and are synchronized in nature. PDSs, hallmarks of typical epileptiform activity (Matsumoto and Marsan, 1964), were found in the SREDs produced by excitotoxic glutamate injury. Like seizures, these SREDs were characterized by bursts of high-frequency spike firing (Fig. 3) greater than 3 Hz for 20 sec or longer. SREDs were never observed in sham control neurons. Finally, the SREDs produced by glutamate injury responded to anticonvulsants as partial complex seizures would. Phenobarbital, but not ethosuximide, inhibited the SREDs. These findings indicate that hippocampal cultures subjected to nonlethal injury by glutamate exposure are transformed into neuronal networks manifesting SREDs for the life of the culture, producing an *in vitro* model of epilepsy.

Glutamate has been implicated in the pathophysiology of epileptogenesis in whole animal (Croucher and Bradford, 1990; Croucher *et al.*, 1988; Rice and DeLorenzo, 1998), slice (Anderson *et al.*, 1986; Stasheff *et al.*, 1989), and tissue culture models (Sombati and DeLorenzo, 1995) of SE-induced AE. All of these models utilized continuous neuronal spiking produced by seizures (Rice and DeLorenzo, 1998), repeated high-frequency excitation (Croucher and Bradford, 1990; Croucher *et al.*, 1988; Stasheff *et al.*, 1989), or low extracellular magnesium environments (Anderson *et al.*, 1986; Sombati and DeLorenzo, 1995) to induce epileptogenesis rather than the glutamate-induced, prolonged, reversible depolarization described in this chapter. Activation of the NMDA receptor has been implicated in these models for the induction of epileptogenesis (Croucher and Bradford, 1990; Croucher *et al.*, 1988; DeLorenzo *et al.*, 1998; Rice and DeLorenzo, 1998; Stasheff *et al.*, 1989). In addition, epileptiform discharges have been produced by growing neurons in culture in the presence of agents that block glutamate receptors and synaptic transmission. Removal of these agents resulted in the expression of seizure-like activity (Furshpan and Potter, 1989; Hoffmann *et al.*, 2000; Koroshetz and Furshpan, 1990). Thus, altering glutamate receptor function can induce epileptiform discharges.

Glutamate injury–induced epileptogenesis is clearly distinct from the low magnesium model of epileptogenesis and represents a separate cause of AE. These models utilize distinct mechanisms to induce SREDs. The glutamate injury model utilizes a prolonged, reversible glutamate-induced depolarization to cause epileptogenesis, analogous to the peri-infarct ischemic depolarization in ischemic stroke. The low magnesium model uses continuous spike firing, analogous to the seizure activity in SE (Sombati and DeLorenzo, 1995) to produce SREDs. Another difference in these models is the duration of the injury phase. Thirty min of glutamate exposure produces epileptic neurons in the glutamate injury model. The low-magnesium model requires 3 h of spike activity to induce SREDs. Both models manifest neuronal cell death, but the stroke model has a higher percentage of injury-induced cell death (Sombati and DeLorenzo, 1995; Sun *et al.*, 2002). Although the injuries used to induce epileptogenesis are different in these models, these models may share common underlying mechanisms.

Stroke is the most common cause of AE, and the association between stroke and epilepsy has been demonstrated clinically (Hauser *et al.*, 1991). The glutamate injury produced in this model of epileptogenesis resembles some of the phenomena associated with stroke. Several features of the ischemic penumbra (Dirnagl *et al.*, 1999) are present in this stroke-induced AE model, including increases in extracellular glutamate (Bullock *et al.*, 1995; Davalos *et al.*, 1997), reversible depolarization with loss of synaptic activity (Hossmann, 1994), acute neuronal swelling, and excitotoxic delayed neuronal death (Choi, 2000). This glutamate-induced AE model demonstrates for the first time a reproducible model of glutamate injury that produces spontaneous, recurrent, epileptiform activity in hippocampal neurons. This model of glutamate injury–induced AE may offer new insights into the development and maintenance of the epileptic condition after a neurological trauma like stroke. This model may also provide therapeutic strategies to develop both novel antiepileptogenic and anticonvulsant agents to prevent stroke-induced epilepsy.

XIII. Calcium Plays a Role in the Induction of Stroke-Induced Epileptogenesis

Calcium has been implicated in the development of AE, leading to the calcium hypothesis of epileptogenesis, which suggests that prolonged elevations in $[Ca^{2+}]_i$ play a role in mediating some of the long-term plasticity changes associated with epileptogenesis and the persistent manifestation of seizure activity (DeLorenzo *et al.*, 1998; Perlin and DeLorenzo, 1992). Epileptogenesis has also been associated with activation of the Ca^{2+}-permeable NMDA receptor (Croucher and Bradford, 1990; Croucher *et al.*, 1988; Dingledine *et al.*, 1990; Rice and DeLorenzo, 1998; Stasheff *et al.*, 1989) and this activation has been implicated as a major source of the epileptogenic Ca^{2+} signal (DeLorenzo *et al.*, 1998). The studies reviewed in this chapter strongly support this calcium hypothesis of epileptogenesis as an initiating mechanism in glutamate injury–induced epileptogenesis. The findings provide direct evidence that persistent elevations in $[Ca^{2+}]_i$ following glutamate exposure contribute to the development of persistent epileptiform discharges in the *in vitro*, glutamate injury–induced epileptogenesis model of stroke-induced epilepsy (Pal *et al.*, 1999; Sun *et al.*, 2002, 2004). The results from these studies also indicate a positive correlation between the induction of epileptogenesis and the duration of the $[Ca^{2+}]_i$ elevation. Also, the positive correlation between the induction of epileptogenesis and the total $[Ca^{2+}]_i$ load indicates that Ca^{2+} influx through the NMDA receptor and prolonged elevations in $[Ca^{2+}]_i$ were required for the induction of epileptogenesis by glutamate injury.

Glutamate injury–induced epileptogenesis required an injury sufficient to produce plasticity changes, but not so severe that it caused cell death (Sun *et al.*, 2001). Studies have shown that glutamate injury–induced epileptogenesis required parameters of total calcium load and duration of significantly elevated calcium sufficient to produce long-lasting hyperexcitability, but not sufficient to produce neuronal death.

There is a continuum of NMDA receptor–Ca^{2+} transduction during neuronal injury ranging from complete recovery of function to neuronal survival with persistently altered function, and finally, to neuronal death. Inhibiting the NMDA receptor–Ca^{2+} transduction pathway, as indicated by short duration elevations in $[Ca^{2+}]_i$ (low $[Ca^{2+}]_e$, APV, and MK-801), blocked the altered neuronal function in the form of SREDs and neuronal death in the form of IRRC, an acute hallmark of excitotoxicity (Limbrick *et al.*, 1995). Conversely, experimental conditions that activated the NMDA receptor–Ca^{2+} transduction pathway (glutamate alone, CNQX, and MCPG), as evidenced by prolonged but reversible alterations in Ca^{2+} homeostasis, induced epileptogenesis. Some neurons in these epileptogenic experimental conditions with glutamate, CNQX, and MCPG (29%, 20%, and 28%, respectively) manifested NMDA receptor–Ca^{2+}-dependent IRRC and underwent delayed excitotoxic neuronal death (Limbrick *et al.*, 1995).

The prolonged elevations in neuronal $[Ca^{2+}]_i$ and activation of the NMDA receptor–Ca^{2+} transduction pathway may underlie potential alterations in enzyme regulation and gene expression during glutamate injury–induced epileptogenesis, since NMDA receptor activation, cytoplasmic Ca^{2+} signals, and nuclear Ca^{2+} signals can regulate gene transcription through multiple Ca^{2+}-activated enzyme and transcription factor pathways (Ghosh and Greenberg, 1995; Hardingham *et al.*, 1997; Platenik *et al.*, 2000). Long-term modulation of gene expression in epilepsy has been implicated as a molecular mechanism mediating long-term plasticity changes associated with epileptogenesis (DeLorenzo and Morris, 1999; Morris *et al.*, 1999, 2000). Persistent increases in DNA binding and expression of serum response factor and ΔFosB have been associated with long-term plasticity changes in the pilocarpine model of epilepsy (Morris *et al.*, 1999, 2000). It is reasonable to suggest that the NMDA receptor–mediated Ca^{2+} changes observed in the glutamate injury model induce epileptogenesis through the long-term modulation of gene expression, since these transcription factors are activated by the NMDA–Ca^{2+} transduction pathway. Future investigations are needed to evaluate the role of Ca^{2+}-activated gene expression in glutamate injury–induced epileptogenesis.

The results with this model have demonstrated that a less severe glutamate exposure that induced epileptogenesis in surviving neurons also caused prolonged elevations in neuronal $[Ca^{2+}]_i$ that were dependent on the presence of extracellular Ca^{2+} and activation of the NMDA receptor. Rather than triggering excitotoxic neuronal death pathways, this calcium load activates Ca^{2+}-regulated pathways that initiate long-term plasticity changes in neurons that ultimately lead to the development of epilepsy.

XIV. Long-Lasting Changes in $[Ca^{2+}]_i$ Levels and Ca^{2+} Homeostatic Mechanisms Play a Role in Maintaining AE

Calcium plays an important role as a messenger in neurons, transducing external signals to the internal cellular environment (Ghosh and Greenberg, 1995). Thus, Ca^{2+} can initiate several pathways in numerous ways, including the regulation of Ca^{2+}-sensitive enzyme systems and the direct and indirect regulation of gene transcription (Bading, 1999; Carafoli *et al.*, 1997). Long-lasting changes in neuronal Ca^{2+} homeostasis have been observed in epileptic brain cells in whole animal and in *in vitro* models of epileptogenesis induced by SE (Pal *et al.*, 1999, 2000, 2001; Parsons *et al.*, 2001; Raza *et al.*, 2001). Alterations in the expression of transcription factors, serum response factor, and ΔFosB (Morris *et al.*, 1999, 2000) and in the activity of the major protein kinase, Ca^{2+}/calmodulin-dependent kinase II (Blair *et al.*, 1999; Bronstein *et al.*, 1993; Churn *et al.*, 1991, 2000), have been observed chronically in these same models. A more comprehensive understanding of the altered Ca^{2+} homeostasis of epileptic neurons may help elucidate the underlying mechanisms leading to the persistent manifestation of epileptiform activity in stroke-induced epileptogenesis.

Similar to the SE-induced epilepsy models (Raza *et al.*, 2001), the stroke-induced model of epilepsy produced long-lasting alterations in Ca^{2+} homeostatic mechanisms. Both SE and stroke produce prolonged, reversible increases in neuronal $[Ca^{2+}]_i$ that have been implicated in the induction of epileptogenesis (Sun *et al.*, 2002). Inhibiting Ca^{2+} influx by glutamate exposure in low $[Ca^{2+}]_e$ environments blocked epileptogenesis (Sun *et al.*, 2002) and prevented chronically elevated basal $[Ca^{2+}]_i$ (Sun *et al.*, 2004). Thus, the long-lasting alterations in Ca^{2+} homeostasis seen in epileptic neurons are also Ca^{2+} dependent.

A complex balance of Ca^{2+} influx, efflux, sequestration, and extrusion mediates Ca^{2+} homeostasis in neurons. In both whole animal and *in vitro* models of seizures, there is decreased activity of the sarco(endo)plasmic reticular Ca^{2+} ATPase (SERCA) (Pal *et al.*, 2001; Parsons *et al.*, 2001). Enhanced Ca^{2+}-induced Ca^{2+} release via the inositol triphosphate receptor has been observed in the *in vitro* SE-induced model of epileptogenesis (Pal *et al.*, 2001). Mitochondria serve to buffer rapid elevations in $[Ca^{2+}]_i$ (Vandecasteele *et al.*, 2001). Alterations in the function of mitochondria may also contribute to observed alterations in epileptic neurons. Results have indicated that the glutamate-induced epileptic neurons demonstrate a decreased activity of the SERCA and enhanced Ca^{2+}-induced Ca^{2+} release via the inositol triphosphate receptor in the stroke-induced model of epileptogenesis. These results suggest that the stroke-induced model of AE may be producing the same underlying alterations in some of the molecular mechanisms that regulate calcium homeostasis that were observed to be altered in the SE-induced epilepsy models (Pal *et al.*, 2001; Parsons *et al.*, 2001). Understanding the mechanisms altering Ca^{2+} homeostasis in this model of glutamate injury–induced epileptogenesis may offer insight into the pathophysiology of injury-induced epileptogenesis.

The chronic manifestation of SREDs in glutamate injury–induced epileptogenesis may be maintained by the alterations in the basal levels of $[Ca^{2+}]_i$. Although the increases in basal $[Ca^{2+}]_i$ in epileptic neurons are small, they may play a significant role in the perpetuation of SREDs owing to the major regulatory and signaling roles of Ca^{2+} in neurons (Berridge, 1998; Ghosh and Greenberg, 1995). The role of calcium in both the induction and maintenance of epileptogenesis has been suggested by evidence from multiple models of AE (DeLorenzo *et al.*, 2005; Loscher and Schmidt, 2004; Raza *et al.*, 2004). By understanding the long-term alterations in neuronal Ca^{2+} homeostasis, it may be possible to gain an insight into the second messenger signaling mechanisms mediating the maintenance phase of epilepsy and direct therapies to the reversal of the epileptic condition. By understanding the role of calcium in epileptogenesis, we may be able to gain significant insights into the molecular mechanisms underlying the initiation and persistence of AE.

Acknowledgments

This study was supported by National Institutes of Health, National Institute of Neurological Disorders and Stroke grants R01-NS052520 and R01-NS051505 to R.J.D.

References

Anderson, W. W., Lewis, D. V., Swartzwelder, H. S., and Wilson, W. A. (1986). Magnesium-free medium activates seizure-like events in the rat hippocampal slice. *Brain Res.* **398,** 215–219.

Bading, H. (1999). Nuclear calcium-activated gene expression: Possible roles in neuronal plasticity and epileptogenesis. *Epilepsy Res.* **36,** 225–231.

Berridge, M. J. (1998). Neuronal calcium signaling. *Neuron* **21,** 13–26.

Blair, R. E., Churn, S. B., Sombati, S., Lou, J. K., and DeLorenzo, R. J. (1999). Long-lasting decrease in neuronal Ca^{2+}/calmodulin-dependent protein kinase II activity in a hippocampal neuronal culture model of spontaneous recurrent seizures. *Brain Res.* **851,** 54–65.

Bronstein, J. M., Farber, D. B., and Wasterlain, C. G. (1993). Regulation of type-II calmodulin kinase: Functional implications. *Brain Res. Brain Res. Rev.* **18,** 135–147.

Bullock, R., Zauner, A., Woodward, J., and Young, H. F. (1995). Massive persistent release of excitatory amino acids following human occlusive stroke. *Stroke* **26,** 2187–2189.

Carafoli, E., Nicotera, P., and Santella, L. (1997). Calcium signalling in the cell nucleus. *Cell Calcium* **22,** 313–319.

Choi, D. (2000). Stroke. *Neurobiol. Dis.* **7,** 552–558.

Choi, D. W. (1994). Glutamate receptors and the induction of excitotoxic neuronal death. *Prog. Brain Res.* **100,** 47–51.

Churn, S. B., Anderson, W. W., and DeLorenzo, R. J. (1991). Exposure of hippocampal slices to magnesium-free medium produces epileptiform activity and simultaneously decreases calcium and calmodulin-dependent protein kinase II activity. *Epilepsy Res.* **9,** 211–217.

Churn, S. B., Kochan, L. D., and DeLorenzo, R. J. (2000). Chronic inhibition of Ca(2+)/calmodulin kinase II activity in the pilocarpine model of epilepsy. *Brain Res.* **875,** 66–77.

Conn, P. J., and Pin, J. P. (1997). Pharmacology and functions of metabotropic glutamate receptors. *Annu. Rev. Pharmacol. Toxicol.* **37,** 205–237.

Croucher, M. J., and Bradford, H. F. (1990). NMDA receptor blockade inhibits glutamate-induced kindling of the rat amygdala. *Brain Res.* **506,** 349–352.

Croucher, M. J., Bradford, H. F., Sunter, D. C., and Watkins, J. C. (1988). Inhibition of the development of electrical kindling of the prepyriform cortex by daily focal injections of excitatory amino acid antagonists. *Eur. J. Pharmacol.* **152,** 29–38.

Davalos, A., Castillo, J., Serena, J., and Noya, M. (1997). Duration of glutamate release after acute ischemic stroke. *Stroke* **28,** 708–710.

Davies, J., and Watkins, J. C. (1982). Actions of D and L forms of 2-amino-5-phosphonovalerate and 2-amino-4-phosphonobutyrate in the cat spinal cord. *Brain Res.* **235,** 378–386.

Delgado-Escueta, A. V., and Porter, R. (1999). Symptomatic lesional epilepsies: Introduction. *Adv. Neurol.* **79,** 433–435.

DeLorenzo, R. J., and Morris, T. A. (1999). Long-term modulation of gene expression in epilepsy. *Neuroscientist* **5,** 86–99.

DeLorenzo, R. J., Pal, S., and Sombati, S. (1998). Prolonged activation of the N-methyl-D-aspartate receptor-Ca^{2+} transduction pathway causes spontaneous recurrent epileptiform discharges in hippocampal neurons in culture. *Proc. Natl. Acad. Sci. USA* **95,** 14482–14487.

DeLorenzo, R. J., Sun, D. A., and Deshpande, L. S. (2005). Cellular mechanisms underlying acquired epilepsy: The calcium hypothesis of the induction and maintainance of epilepsy. *Pharmacol. Ther.* **105,** 229–266.

Didier, M., Heaulme, M., Soubrie, P., Bockaert, J., and Pin, J. P. (1990). Rapid, sensitive, and simple method for quantification of both neurotoxic and neurotrophic effects of NMDA on cultured cerebellar granule cells. *J. Neurosci. Res.* **27,** 25–35.

Dingledine, R., McBain, C. J., and McNamara, J. O. (1990). Excitatory amino acid receptors in epilepsy. *Trends Pharmacol. Sci.* **11,** 334–338.

Dingledine, R., Borges, K., Bowie, D., and Traynelis, S. F. (1999). The glutamate receptor ion channels. *Pharmacol. Rev.* **51,** 7–61.

Dirnagl, U., Iadecola, C., and Moskowitz, M. A. (1999). Pathobiology of ischaemic stroke: An integrated view. *Trends Neurosci.* **22,** 391–397.

Furshpan, E. J., and Potter, D. D. (1989). Seizure-like activity and cellular damage in rat hippocampal neurons in cell culture. *Neuron* **3,** 199–207.

Ghosh, A., and Greenberg, M. E. (1995). Calcium signaling in neurons: Molecular mechanisms and cellular consequences. *Science* **268,** 239–247.

Grynkiewicz, G., Poenie, M., and Tsien, R. Y. (1985). A new generation of Ca^{2+} indicators with greatly improved fluorescence properties. *J. Biol. Chem.* **260,** 3440–3450.

Hansen, A. J., and Nedergaard, M. (1988). Brain ion homeostasis in cerebral ischemia. *Neurochem. Pathol.* **9**, 195–209.

Hardingham, G. E., Chawla, S., Johnson, C. M., and Bading, H. (1997). Distinct functions of nuclear and cytoplasmic calcium in the control of gene expression. *Nature* **385**, 260–265.

Hauser, W. A., and Hesdorffer, D. C. (1990). "Epilepsy: Frequency, Causes, and Consequences." Demos, New York.

Hauser, W. A., Annegers, J. F., and Kurland, L. T. (1991). Prevalence of epilepsy in Rochester, Minnesota: 1940–1980. *Epilepsia* **32**, 429–445.

Hoffmann, H., Gremme, T., Hatt, H., and Gottmann, K. (2000). Synaptic activity-dependent developmental regulation of NMDA receptor subunit expression in cultured neocortical neurons. *J. Neurochem.* **75**, 1590–1599.

Honore, T., Davies, S. N., Drejer, J., Fletcher, E. J., Jacobsen, P., Lodge, D., and Nielsen, F. E. (1988). Quinoxalinediones: Potent competitive non-NMDA glutamate receptor antagonists. *Science* **241**, 701–703.

Hossmann, K. A. (1994). Viability thresholds and the penumbra of focal ischemia. *Ann. Neurol.* **36**, 557–565.

Hyrc, K. L., Bownik, J. M., and Goldberg, M. P. (2000). Ionic selectivity of low-affinity ratiometric calcium indicators: mag-Fura-2, Fura-2FF and BTC. *Cell Calcium* **27**, 75–86.

Koroshetz, W. J., and Furshpan, E. J. (1990). Seizure-like activity and glutamate receptors in hippocampal neurons in culture. *Neurosci. Res. Suppl.* **13**, S65–S74.

Limbrick, D. D., Jr., Churn, S. B., Sombati, S., and DeLorenzo, R. J. (1995). Inability to restore resting intracellular calcium levels as an early indicator of delayed neuronal cell death. *Brain Res.* **690**, 145–156.

Lipton, P. (1999). Ischemic cell death in brain neurons. *Physiol. Rev.* **79**, 1431–1568.

Loscher, W., and Schmidt, D. (2004). New horizons in the development of antiepileptic drugs: The search for new targets. *Epilepsy Res.* **60**, 77–159.

Macdonald, R. L., and Kelly, K. M. (1993). Antiepileptic drug mechanisms of action. *Epilepsia* **34** (Suppl. 5), S1–S8.

Matsumoto, H., and Marsan, C. A. (1964). Cortical cellular phenomena in experimental epilepsy: Interictal manifestations. *Exp. Neurol.* **9**, 286–304.

Mayer, M. L., and Westbrook, G. L. (1987). Permeation and block of N-methyl-D-aspartic acid receptor channels by divalent cations in mouse cultured central neurones. *J. Physiol.* **394**, 501–527.

Morris, T. A., Jafari, N., Rice, A. C., Vasconcelos, O., and DeLorenzo, R. J. (1999). Persistent increased DNA-binding and expression of serum response factor occur with epilepsy-associated long-term plasticity changes. *J. Neurosci.* **19**, 8234–8243.

Morris, T. A., Jafari, N., and DeLorenzo, R. J. (2000). Chronic DeltaFosB expression and increased AP-1 transcription factor binding are associated with the long term plasticity changes in epilepsy. *Brain Res. Mol. Brain Res.* **79**, 138–149.

Nelson, M. T., French, R. J., and Krueger, B. K. (1984). Voltage-dependent calcium channels from brain incorporated into planar lipid bilayers. *Nature* **308**, 77–80.

Pal, S., Sombati, S., Limbrick, D. D., Jr., and DeLorenzo, R. J. (1999). *In vitro* status epilepticus causes sustained elevation of intracellular calcium levels in hippocampal neurons. *Brain Res.* **851**, 20–31.

Pal, S., Limbrick, D. D., Jr., Rafiq, A., and DeLorenzo, R. J. (2000). Induction of spontaneous recurrent epileptiform discharges causes long-term changes in intracellular calcium homeostatic mechanisms. *Cell Calcium* **28**, 181–193.

Pal, S., Sun, D., Limbrick, D., Rafiq, A., and DeLorenzo, R. J. (2001). Epileptogenesis induces long-term alterations in intracellular calcium release and sequestration mechanisms in the hippocampal neuronal culture model of epilepsy. *Cell Calcium* **30**, 285–296.

Parsons, J. T., Churn, S. B., and DeLorenzo, R. J. (2001). Chronic inhibition of cortex microsomal Mg(2+)/Ca(2+) ATPase-mediated Ca(2+) uptake in the rat pilocarpine model following epileptogenesis. *J. Neurochem.* **79,** 319–327.

Perlin, J. B., and DeLorenzo, R. J. (1992). Calcium and epilepsy. *In* "Recent Advances in Epilepsy, 5" (T. A. Pedley and B. S. Meldrum, Eds.), pp. 15–36. Churchill Livingstone, Edinburgh.

Platenik, J., Kuramoto, N., and Yoneda, Y. (2000). Molecular mechanisms associated with long-term consolidation of the NMDA signals. *Life Sci.* **67,** 335–364.

Raza, M., Pal, S., Rafiq, A., and DeLorenzo, R. J. (2001). Long-term alteration of calcium homeostatic mechanisms in the pilocarpine model of temporal lobe epilepsy. *Brain Res.* **903,** 1–12.

Raza, M., Blair, R. E., Sombati, S., Carter, D. S., Deshpande, L. S., and DeLorenzo, R. J. (2004). Evidence that injury-induced changes in hippocampal neuronal calcium dynamics during epileptogenesis cause acquired epilepsy. *Proc. Natl. Acad. Sci. USA* **101,** 17522–17527.

Rice, A. C., and DeLorenzo, R. J. (1998). NMDA receptor activation during status epilepticus is required for the development of epilepsy. *Brain Res.* **782,** 240–247.

Rogawski, M. A., and Porter, R. J. (1990). Antiepileptic drugs: Pharmacological mechanisms and clinical efficacy with consideration of promising developmental stage compounds. *Pharmacol. Rev.* **42,** 223–286.

Sombati, S., and DeLorenzo, R. J. (1995). Recurrent spontaneous seizure activity in hippocampal neuronal networks in culture. *J. Neurophysiol.* **73,** 1706–1711.

Sombati, S., Coulter, D. A., and DeLorenzo, R. J. (1991). Neurotoxic activation of glutamate receptors induces an extended neuronal depolarization in cultured hippocampal neurons. *Brain Res.* **566,** 316–319.

Stasheff, S. F., Anderson, W. W., Clark, S., and Wilson, W. A. (1989). NMDA antagonists differentiate epileptogenesis from seizure expression in an *in vitro* model. *Science* **245,** 648–651.

Sun, D. A., Sombati, S., and DeLorenzo, R. J. (2001). Glutamate injury-induced epileptogenesis in hippocampal neurons: An *in vitro* model of stroke-induced "epilepsy." *Stroke* **32,** 2344–2350.

Sun, D. A., Sombati, S., Blair, R. E., and DeLorenzo, R. J. (2002). Calcium-dependent epileptogenesis in an *in vitro* model of stroke-induced "epilepsy." *Epilepsia* **43,** 1296–1305.

Sun, D. A., Sombati, S., Blair, R. E., and DeLorenzo, R. J. (2004). Long-lasting alterations in neuronal calcium homeostasis in an *in vitro* model of stroke-induced epilepsy. *Cell Calcium* **35,** 155–163.

Vandecasteele, G., Szabadkai, G., and Rizzuto, R. (2001). Mitochondrial calcium homeostasis: Mechanisms and molecules. *IUBMB Life* **52,** 213–219.

Watkins, J., and Collingridge, G. (1994). Phenylglycine derivatives as antagonists of metabotropic glutamate receptors. *Trends Pharmacol. Sci.* **15,** 333–342.

Witte, O. W., Bidmon, H. J., Schiene, K., Redecker, C., and Hagemann, G. (2000). Functional differentiation of multiple perilesional zones after focal cerebral ischemia. *J. Cereb. Blood Flow Metab.* **20,** 1149–1165.

Wong, E. H., Kemp, J. A., Priestley, T., Knight, A. R., Woodruff, G. N., and Iversen, L. L. (1986). The anticonvulsant MK-801 is a potent N-methyl-D-aspartate antagonist. *Proc. Natl. Acad. Sci. USA* **83,** 7104–7108.

MECHANISMS OF ACTION OF ANTIEPILEPTIC DRUGS

H. Steve White, Misty D. Smith, and Karen S. Wilcox

Anticonvulsant Drug Development Program, Department of Pharmacology and Toxicology
University of Utah, Salt Lake City, Utah 84108, USA

The management of seizures in the patient with epilepsy relies heavily on antiepileptic drug (AED) therapy. Fortunately, for a large percentage of patients, AEDs provide excellent seizure control at doses that do not adversely affect normal function. At the molecular level, the majority of AEDs are thought to modify excitatory and inhibitory neurotransmission through effects on voltage-gated ion channels (e.g., sodium and calcium) and γ-aminobutyric acid $(GABA)_A$ receptors, respectively. In addition to these effects, two of the "second-generation" AEDs have been found to limit glutamate-mediated excitatory neurotransmission (i.e., felbamate and topiramate). Not surprisingly, those AEDs with broad spectrum clinical activity are often found to exert an action at more than one molecular target. Emerging evidence suggests that receptor and voltage-gated subunits are modified by chronic seizures. Thus, attempts to understand the relationship between target and effect continue to provide important information about the neuropathology of the epileptic network and to facilitate the development of novel therapies for the treatment of refractory epilepsy.

I. Introduction

For the vast majority of people with epilepsy, initial therapy consists of pharmacological treatment with one or more of the established antiepileptic drugs (AEDs). These medications include phenytoin (PHT), carbamazepine (CBZ), and valproate (VPA, valproic acid). Barbiturates such as phenobarbital (PB), certain benzodiazepines (BZDs), and ethosuximide (ESM) are now used infrequently. The mechanisms of action of currently marketed AEDs are not fully understood. Although numerous molecular targets exist wherein AEDs may exert their actions, the common link among the various proposed mechanisms involves the ability of a drug to modulate excitatory and inhibitory neurotransmission through an effect on ion channels, neurotransmitter receptors, and neurotransmitter metabolism (Table I). This chapter focuses primarily on the proposed mechanisms of action of the first-generation AEDs (PHT, CBZ, VPA, ESM, BZDs, and barbiturates) and the second-generation AEDs introduced into the US market since 1993 (i.e., felbamate, FBM; gabapentin, GBP; lamotrigine, LTG; topiramate, TPM; pregabalin, PGB; tiagabine, TGB; oxcarbazepine, OCBZ; levetiracetam, LEV; and zonisamide, ZNS). In the majority of patients, AEDs provide seizure control at doses that do not have adverse effects. Depending on the concentration employed, multiple effects have been reported for most of these drugs; however, the discussion that follows will be limited to those drug effects that are thought to result at therapeutic concentrations and thereby are presumed to represent the primary mechanisms of action. A number of adequate reviews have been published on this topic, and the reader is referred to these for more in-depth discussion and additional references (Macdonald and Kelly, 1995; Meldrum, 1996; Rho and Sankar, 1999; Rogawski and Porter, 1990; White, 1997). It should be noted that many of the older established and newer AEDs have been observed to exert a number of different pharmacological actions that could account for their anticonvulsant properties. In cases where multiple actions have been defined, it is highly likely that the separate mechanisms offer some degree of synergy.

II. Modulation of Voltage-Gated Ion Channels

Actions at both the voltage-gated sodium and calcium channels stabilize neuronal membranes, block action potential firing and propagation, reduce neurotransmitter release, and prevent seizure spread (Table I). Interestingly, the

TABLE I

Functional Implications of Proposed Mechanisms of Action of Antiepileptic Drugs

Molecular mechanism of action	Antiepileptic drug	Consequences of action
Sodium channel blockers	PHT, CBZ, LTG, FBM, OCBZ, TPM, VPA, ZNS	Block action potential propagation
		Stabilize neuronal membranes
		Decrease neurotransmitter release
		Decrease focal firing
		Decrease seizure spread
Calcium channel blockers	ESM, VPA, GBP, LTG, TPM, ZNS, LEV, PGB, FBM	Decrease neurotransmitter release (N- and P-types)
		Decrease slow-depolarization (T-type)
		Decrease spike-wave discharges
GABA_A receptor allosteric modulators	BZDs, PB, FBM, TPM, ZNS	Increase membrane hyperpolarization
		Elevate seizure threshold
		Attenuate (BZDs) spike-wave discharges
		Aggravate (barbiturates) spike-wave discharges
		Decrease focal firing
Inhibition of GABA metabolism and reuptake	VPA, TGB	Increase synaptic GABA levels
		Increase membrane hyperpolarization
		Decrease focal firing
		Aggravate spike-wave discharges
NMDA receptor antagonist	FBM	Decrease slow excitatory neurotransmission
		Decrease excitatory amino acid neurotoxicity
		Delay epileptogenesis
AMPA/kainate receptor antagonists	PB, TPM	Decrease fast excitatory neurotransmission
		Attenuate focal firing
Binds to synaptic vesicle protein 2A (SV2A)	LEV	Currently unknown

AMPA, α-amino-3-hydroxy-5-methyl-4-isoxazolepropionic acid; BZDs, benzodiazepines; CBZ, carbamazepine; ESM, ethosuximide; FBM, felbamate; GABA, γ-aminobutyric acid; GBP, gabapentin; LEV, levetiracetam; LTG, lamotrigine; NMDA, N-methyl-D-aspartate; OCBZ, oxcarbazepine; PB, phenobarbital; PGB, pregabalin; PHT, phenytoin; TGB, tiagabine; TPM, topiramate; VPA, valproic acid; ZNS, zonisamide. From Leppik et al. (2006), with permission.

vast majority of clinically available AEDs are thought to exert their anticonvulsant effects at voltage-gated ion channels including sodium (PHT, CBZ, VPA, FBM, LTG, TPM, OCBZ, and ZNS), calcium (VPA, ESM, FBM, LTG, TPM, and GBP), and potassium (TPM) channels (Guerrini and Parmeggiani, 2006; Rho and Sankar, 1999; White, 1999). However, VPA, ESM, and perhaps ZNS are unique in that they appear to attenuate voltage-dependent, low-threshold T-type calcium currents in thalamocortical neurons, thereby interrupting the thalamic oscillatory firing patterns associated with spike-wave seizures. In contrast to VPA, ESM, and ZNS, several other AEDs or their active metabolites modulate N-type (LTG and OCBZ), L-type (FBM and TPM), and P-type (LTG) calcium currents (Stefani *et al.*, 1997; Zhang *et al.*, 2000). In addition, LEV has been found to modulate a high-voltage activated calcium current (HVACC) (Niespodziany *et al.*, 2001); however, the specific type has yet to be defined. The ability of TPM to activate potassium currents is somewhat unique among the AEDs and this effect may contribute to membrane hyperpolarization (Herrero *et al.*, 2002).

III. Enhanced Inhibition

Once released from presynaptic nerve terminals, the inhibitory neurotransmitter γ-aminobutyric acid (GABA) binds to both $GABA_A$ and $GABA_B$ receptors (Macdonald and Olsen, 1994). The $GABA_A$ receptor complex is a multimeric, macromolecular protein that forms a chloride-selective ion pore (Macdonald and Olsen, 1994). Several of the available AEDs have been demonstrated to potentiate GABAergic neurotransmission mediated by the $GABA_A$ receptor complex (e.g., BZDs, barbiturates, FBM, and TPM) (Kume *et al.*, 1996; Macdonald and Kelly, 1995; White *et al.*, 1997). Furthermore, several AEDs have been found to increase GABAergic tone by decreasing GABA metabolism (i.e., VPA) (Loscher, 2002), preventing GABA reuptake (i.e., TGB) (During *et al.*, 1992), or increasing GABA synthesis (i.e., VPA and GBP) (Loscher, 1981; Loscher *et al.*, 1991).

IV. Excitation Reduction

Excessive excitatory neurotransmission mediated by glutamate and other excitatory amino acids can contribute to seizure activity. Four glutamate receptor types have been identified within the central nervous system (CNS) (Michaelis, 1998). Three of these receptors, α-amino-3-hydroxy-5-methyl-4-isoxazolepropionic acid (AMPA), N-methyl-D-aspartate (NMDA), and kainate, are coupled to ion channels

(Michaelis, 1998). The fourth type of glutamate receptor is the metabotropic glutamate receptor. Thus far, there have been eight metabotropic receptors identified (Michaelis, 1998); however, none of the AEDs have been found to modulate neurotransmission via action at a metabotropic glutamate receptor.

Among the available AEDs, FBM and TPM are novel in that they possess an ability to reduce glutamate-mediated excitatory neurotransmission by modulation of NMDA (FBM), AMPA (TPM), and kainate (TPM) receptors. Actions of FBM at the NMDA receptor would be expected to reduce slow excitatory neurotransmission and perhaps decrease the excitatory amino acid neurotoxicity associated with excessive glutamate release. FBM modulates the glutamate receptor by preferentially binding to the NR2B subunit of the NMDA receptor (Harty and Rogawski, 2000). These mechanisms may contribute to the ability of FBM to reduce sustained repetitive firing in mouse spinal cord neurons (White et al., 1992) and to its neuroprotective properties (Wallis and Panizzon, 1995). In contrast, the action of TPM at AMPA and kainate receptors would be predicted to modify fast excitatory neurotransmission and perhaps attenuate focal firing (Table I).

V. First-Generation AEDs

A. BENZODIAZEPINES AND BARBITURATES

Multiple binding sites for GABA, anticonvulsant BZDs, barbiturates, neurosteroids, convulsant β-carbolines, and the chemoconvulsant picrotoxin have been identified (Macdonald and Olsen, 1994). The GABA$_B$ receptor is coupled via a guanosine 5'-triphosphate–binding protein to calcium or potassium channels; this receptor does not form an ion pore and does not appear to contribute to the anticonvulsant action of either the BZDs or barbiturates. Thus, the principal anticonvulsant action of the BZDs and barbiturates is thought to be related to their ability to enhance inhibitory neurotransmission by allosterically modulating the GABA$_A$ receptor complex (Macdonald, 2002a; Olsen, 2002).

GABA receptor current can be enhanced by increasing open and burst frequency, channel conductance, and open and burst duration (Macdonald and Olsen, 1994). Results from several studies have demonstrated that the principal effect of the barbiturates is to increase the mean channel open duration (Macdonald et al., 1989; Twyman et al., 1989), whereas the BZDs enhance GABA receptor current by increasing opening frequency without affecting open or burst duration (Rogers et al., 1994; Study and Barker, 1981; Twyman et al., 1989; Vicini et al., 1987). However, barbiturates have also been shown to reduce currents following activation of excitatory amino acid receptors (Dildy-Mayfield et al., 1996; Ticku et al., 1992; Yamakura et al., 1995).

TABLE II
PROPOSED MECHANISMS OF ACTION OF ANTIEPILEPTIC DRUGS

Antiepileptic drug	Sodium channel blockade	Calcium channel blockade	Glutamate receptor antagonism	GABA potentiation	Carbonic anhydrase inhibition	Synaptic vesicle protein 2A modulation	Potassium channel activity
First generation							
Carbamazepine[a]	X						
Diazepam[b]				X			
Ethosuximide[c]	Noninactivating	T-type					
Phenobarbital[d]			AMPA [high]	X			
Phenytoin[e]	X						
Valproate[f]	X	T-type		X			
Second generation							
Felbamate[g]	X	L-type	NMDA	X			
Gabapentin[h]		$\alpha_2\delta$ subunit		X			
Lamotrigine[i]	X	N- and P-types HVA					
Levetiracetam[j]		N-type				X	
Oxcarbazepine[k]	X	X					Blocks
Pregabalin[l]							
Tiagabine[m]				X			
Topiramate[n]	X	L-type	AMPA/KA	X	X		Activates
Zonisamide[o]	X	T-type		X	X		

[a]Lang et al. (1993); Macdonald (2002b); McLean and Macdonald (1986a); Willow and Catterall (1982); Willow et al. (1984, 1985).

[b]Macdonald (2002a).

[c]Coulter et al. (1989); Holland and Ferrendelli (2002); Leresche et al. (1998).

[d]Dildy-Mayfield et al. (1996); Ticku et al. (1992); Yamakura et al. (1995).

[e]DeLorenzo and Sun (2002); Lang et al. (1993); Schwarz and Grigat (1989).

[f]Kelly et al. (1990); Loscher (1981); Zona and Avoli (1990).

[g]Kume et al. (1996); Pisani et al. (1995); Rho et al. (1994); Stefani et al. (1997); Subramaniam et al. (1995); Taglialatela et al. (1996); Ticku et al. (1991); Wamsley et al. (1994); White et al. (1995b).

[h]Gee et al. (1996); Hommou et al. (1995); Loscher et al. (1991); Petroff et al. (1995); Stefani et al. (1998); Taylor (2002); Taylor et al. (1998).

[i]Cheung et al. (1992); Coulter (1997); Lang et al. (1993); Leach et al. (1986, 2002); Lees and Leach (1993); Riddall et al. (1993); Stefani et al. (1996).

[j]Lynch et al. (2004); Margineanu and Klitgaard (2002); Margineanu and Wulfert (1997); Niespodziany et al. (2001); Rigo et al. (2002).

[k]Calabresi et al. (1995); McLean (2002); Stefani et al. (1995); Wamil (1991).

[l]Micheva et al. (2006); Vartanian et al. (2006).

[m]During et al. (1992); Fink-Jensen et al. (1992); Giardina (2002); Roepstorff and Lambert (1992); Suzdak and Jansen (1995); Thompson and Gahwiler (1992).

[n]Coulter et al. (1993); DeLorenzo et al. (2000); Dodgson et al. (2000); Gibbs et al. (2000); Herrero et al. (2002); McLean et al. (2000); Severt et al. (1995); Shank et al. (1994, 2000); Simeone et al. (2000); Skradski and White (2000); Taverna et al. (1999); White (2002); White et al. (1995a, 1997); Zhang et al. (1998); Zona et al. (1997).

[o]Kito et al. (1996); Macdonald (2002c); Mimaki et al. (1988); Rock et al. (1989); Schauf (1987); Suzuki et al. (1992).

X, activity at target at therapeutic concentrations; AMPA, α-amino-3-hydroxy-5-methyl-4-isoxazolepropionic acid; GABA, γ-aminobutyric acid; HVA, high-voltage activated; KA, kainate; NMDA, N-methyl-D-aspartate.

B. Carbamazepine and Phenytoin

Many mechanisms of action have been ascribed to PHT since its identification by Putnam and Merritt (1937). Both PHT and CBZ have been shown to block posttetanic potentiation (PTP), which is the process wherein high-frequency stimulation results in transiently enhanced responsiveness to a subsequent single stimulation (DeLorenzo *et al.*, 2000; Macdonald, 2002b; Rogawski and Porter, 1990). PTP has been suggested to contribute to the facilitation of local excitatory discharges and to enhance spread from a seizure focus (Dichter and Ayala, 1987). Inhibition of PTP by PHT and CBZ may account for their ability to limit seizure spread.

PHT and CBZ both possess the ability to limit sustained repetitive firing of action potentials of neurons in culture (Macdonald, 1989; McLean and Macdonald, 1986a). This effect occurs at therapeutically relevant free concentrations and correlates with the ability of PHT and CBZ to block tonic extension seizures in animals (Piredda *et al.*, 1985; White *et al.*, 1995c) and tonic-clonic seizures in humans (Mattson *et al.*, 1985). Whether this effect is what underlies the ability of PHT and CBZ to attenuate PTP and ultimately block seizure spread is not known.

PHT and CBZ have both been found to exert an inhibitory effect on voltage-gated sodium channels (Lang *et al.*, 1993; Willow and Catterall, 1982; Willow *et al.*, 1984, 1985) (Tables I and II) that is both use and voltage dependent. These properties account for the unique ability of PHT, CBZ, and other voltage-dependent sodium channel blockers to limit the high-frequency firing characteristic of epileptic discharges without significantly altering normal patterns of neuronal firing. Frequency-, voltage-, and time-dependent inactivation of sodium channels by PHT has also been observed in isolated rat hippocampal neurons and *Xenopus* oocytes injected with messenger ribonucleic acid from human brain tissue (Tomaselli *et al.*, 1989; Wakamori *et al.*, 1989). In mammalian myelinated nerve fibers, PHT and CBZ produce a shift in the steady-state inactivation curve to more negative voltages and delay the rate of sodium channel recovery from inactivation (Schwarz and Grigat, 1989).

C. Ethosuximide

The mechanism of action of ESM remained somewhat elusive until 1989, when it was shown to reduce low-threshold T-type calcium currents (Tables I and II) in thalamic neurons (Coulter *et al.*, 1989). Activation of T-type calcium channels in thalamic relay neurons generates low-threshold calcium spikes that are thought to underlie the abnormal thalamocortical rhythmicity associated with the 3-Hz spike-and-wave discharge visible in an electroencephalogram of absence epilepsy (Coulter *et al.*, 1989). ESM and dimethadione, the active metabolite of the antiabsence drug trimethadione, block positive currents in a voltage-dependent manner (Holland and Ferrendelli, 2002). This effect of ESM is thought to represent the primary mechanism by which it controls absence epilepsy (Holland and Ferrendelli, 2002). However, it is important to note that several other investigators have failed to confirm this observation (Holland and Ferrendelli, 2002), and additional studies are clearly needed. ESM has also been found to inhibit a noninactivating sodium current in thalamocortical neurons (Leresche *et al.*, 1998). This effect may account for the ability of ESM to block spike-and-wave discharges in both animals and humans (Holland and Ferrendelli, 2002).

D. VALPROIC ACID

VPA has perhaps one of the broadest preclinical and clinical profiles of all of the currently available AEDs. Multiple mechanisms of action of VPA likely account for its broad spectrum of activity (Tables I and II) (Loscher, 2002; White *et al.*, 1995c). VPA has been found to block sustained repetitive firing of mouse central neurons in culture (McLean and Macdonald, 1986b) and rat hippocampal slices (Capek and Esplin, 1986). Results from considerable *in vitro* investigation support an effect of VPA on voltage-sensitive sodium channels. For example, VPA has been found to inhibit sodium currents in isolated *Xenopus laevis* myelinated nerves (VanDongen *et al.*, 1986), neocortical neurons *in vitro* (Zona and Avoli, 1990), and rat hippocampal neurons (Van den Berg *et al.*, 1993). These results support an action for VPA at the voltage-sensitive sodium channel.

In addition to its effects on sodium channels, VPA has been observed to produce a modest reduction of T-type calcium currents in primary afferent neurons (Kelly *et al.*, 1990), to elevate whole brain GABA levels (Loscher, 1981), and to potentiate GABA responses at high concentrations (Loscher, 2002; Rho and Sankar, 1999; Rogawski and Porter, 1990). These effects, coupled with its effects on sodium channels, may contribute to the broad anticonvulsant efficacy of VPA.

VI. Second-Generation AEDs

A. FELBAMATE

FBM (2-phenyl-1,3-propanediol dicarbamate) received Food and Drug Administration (FDA) approval in mid-1993 and ushered in the new era of AED development. Indeed, FBM was the first new AED approved in the United States since 1978. Results from numerous preclinical studies have demonstrated that FBM has a broad anticonvulsant profile in animal seizure models (Sofia, 1995; Swinyard *et al.*, 1986; White *et al.*, 1992). Consistent with this preclinical profile,

FBM has demonstrated a broad clinical spectrum. On entry in the US market, FBM received approval for the clinical management of partial seizures, primary and secondary generalized tonic-clonic seizures, and Lennox-Gastaut syndrome (Pellock and Brodie, 1997).

The precise mechanism of action of FBM has not been clearly elucidated; however, it has been found to produce several effects that support its anticonvulsant as well as neuroprotective actions (Tables I and II) (Pellock *et al.*, 2002). FBM reduces sustained repetitive firing in mouse spinal cord neurons in a concentration-dependent manner (White *et al.*, 1992) and current-evoked firing of rat striatal neurons (Pisani *et al.*, 1995). These effects are consistent with its ability to block tonic extension seizures in rodents, and the effects occur at concentrations that are achieved in the brains of animals receiving anticonvulsant doses of FBM (Adusumalli *et al.*, 1993; McCabe *et al.*, 1993) and in the brains of epileptic patients treated with FBM (Adusumalli *et al.*, 1994). Subsequent studies have confirmed a direct interaction of FBM with voltage-dependent sodium channels (Taglialatela *et al.*, 1996).

FBM has also been found to enhance GABA-evoked chloride currents and inhibit NMDA-evoked currents in hippocampal neurons (Rho *et al.*, 1994), and to enhance GABA-mediated chloride currents in cortical neurons (Kume *et al.*, 1996). At concentrations up to 1 mM, FBM does not appear to affect ligand binding to GABA-, BZD-, or picrotoxin-binding sites on the $GABA_A$ receptor ionophore, nor does it enhance GABA-stimulated radiolabeled chloride flux into cultured mouse spinal cord neurons (Ticku *et al.*, 1991). Thus, at this time the mechanism by which FBM enhances GABA-evoked currents is unknown.

FBM also blocks NMDA-evoked currents (Subramaniam *et al.*, 1995), an action that is unique among both the standard and newer AEDs. Reports from several studies suggest an interaction with the glycine binding site of the NMDA receptor (Coffin *et al.*, 1994; De Sarro *et al.*, 1994; McCabe *et al.*, 1993; Wamsley *et al.*, 1994; White *et al.*, 1995b); however, strong experimental evidence suggests that FBM acts as a low-potency open-channel blocker (Rho *et al.*, 1994; Subramaniam *et al.*, 1995). FBM exerts a neuroprotective effect in rat hippocampal slice preparations that is reversed by glycine (Wallis and Panizzon, 1993). Despite its interaction with this receptor complex, FBM does not appear to produce either the behavioral or pathological CNS impairment associated with NMDA antagonists (Sofia, 1995). Finally, FBM at low concentrations has demonstrated an ability to inhibit dihydropyridine-sensitive, high-threshold, voltage-sensitive calcium currents, thus, further attenuating neuronal excitability (Stefani *et al.*, 1997). It is clear that FBM remains a mechanistically diverse and interesting AED therapy with broad clinical utility. However, its clinical usefulness has been somewhat limited by the observation of serious hematologic and hepatic toxicity since its entrance into the US market (Pellock and Brodie, 1997). Better understanding of its underlying mechanisms may help identify patients at risk for these adverse effects and greatly increase clinical use of this highly effective AED in therapy-resistant patient populations.

B. GABAPENTIN

GBP, 1-(aminomethyl)cyclohexaneacetic acid, was originally designed and synthe-sized as a drug to enhance GABA-mediated inhibition by mimicking the steric conformation of the endogenous neurotransmitter GABA (Schmidt, 1989; Taylor, 2002). However, results from a number of studies have essentially excluded this as a possible mechanism of action (Reimann, 1983; Stringer and Lorenzo, 1999). Unlike AEDs that directly modulate voltage- and receptor-gated ion channels, there is a substantial time lag between the appearance of peak plasma and brain GBP concen-trations and the time to peak anticonvulsant effect following intravenous administra-tion (Welty *et al.*, 1993). This delay in anticonvulsant effect suggests that prolonged synaptic and/or cytosolic exposure to GBP is important and supports an indirect mechanism of action for GBP. This hypothesis is supported by both *in vivo* and *in vitro* studies. For example, only after prolonged application of GBP was a reduction in sustained repetitive action potential firing observed (Wamil and McLean, 1994). The ability of GBP to limit sodium-dependent sustained action potential firing in cultured mouse spinal cord neurons was observed to be voltage- and frequency-dependent, but developed slowly with prolonged exposure. The precise mechanism of this effect is not known; however, it is unlikely that GBP inhibits sodium currents in a manner similar to that of the established sodium channel blockers PHT and CBZ.

GBP has also been reported to increase GABA concentrations in discrete brain regions (Loscher *et al.*, 1991); this effect parallels its anticonvulsant time course. Similarly, GBP has been reported to increase the cytosolic concentration of GABA in isolated optic nerves from neonatal rats (Kocsis and Honmou, 1994). Since this preparation contains mostly axons from retinal ganglion cells and glial cells, and lacks neuronal cell bodies and synapses, the majority of GABA is presumed to be localized in the glial compartment. GBP has also been demonstrated to enhance nipecotic acid–induced inward currents in isolated CA1 hippocampal pyramidal neurons in brain slices (Honmou *et al.*, 1995). Thus, it would appear that GBP possesses a unique ability to increase the concentration of releasable GABA in both the glial and neuronal compartments. Furthermore, GBP has been reported to increase *in vivo* occipital lobe GABA levels in epilepsy patients (Petroff *et al.*, 1995).

GBP may increase brain GABA turnover by interacting with a number of different metabolic processes. It has been demonstrated to enhance glutamate dehydrogenase and glutamic acid decarboxylase and inhibit branched-chain amino acid aminotransferase and GABA aminotransferase (Taylor, 2002). Although any one of these effects could contribute to the anticonvulsant action of GBP, it is not clear at this point which effects are important (Taylor, 2002).

Finally, GBP has been reported to bind to a novel site in rat brain (Suman-Chauhan *et al.*, 1993). Binding of [^{3}H]GBP is displaced by unlabeled GBP and several structural GBP analogues, including 3-isobutyl GABA (Suman-Chauhan *et al.*, 1993). It is also displaced stereospecifically by L-amino acids, including L-leucine, L-isoleu-cine, L-methionine, and L-phenylalanine, thereby suggesting an association between the [^{3}H]GBP-binding site and the system L transporter of neuronal cell membranes (Thurlow *et al.*, 1993). Subsequent studies have demonstrated that the GBP-binding site is homologous with the $\alpha_2\delta$ subunit of the L-type voltage-dependent calcium channel (Gee *et al.*, 1996). Although a firm relationship between interaction with the $\alpha_2\delta$ subunit and calcium channel function has not been established, it has been suggested that GBP modulates neurotransmitter release (Table I) (Bertrand *et al.*, 2001). Although modulation of the $\alpha_2\delta$ subunit and calcium channels may be considered the most likely mechanism of GBP action, GBP pharmacology may also involve an unexpected interaction with GABA$_B$ receptors. GBP has been shown to cause the selective reduction of excitatory neurotransmitter release and relative sparing of inhibitory neurotransmission (Parker *et al.*, 2004). This may be related to GBP's activation of a specific subset of GABA$_B$ receptors. If so, GBP may prove to be a useful tool for the pharmacological differentiation of GABA$_B$ receptors at excitatory versus inhibitory nerve terminals. These conclusions are consistent with previous reports that proposed a preferential effect of GBP on GABA$_B$ receptors containing the GABA$_{B1a}$ subunit (Bertrand *et al.*, 2001; Ng *et al.*, 2001).

C. LAMOTRIGINE

LTG (3,5-diamino-6-[2,3-dichlorophenyl]-*as*-triazine) emerged from an anti-folate drug development program that was based on the observations that chronic use of several AEDs (i.e., PB, primidone, and PHT) is associated with reduced folate levels (Reynolds *et al.*, 1966) and that high doses of folic acid induce seizures in laboratory animals (Hommes and Obbens, 1972). However, LTG displays very weak antifolate activity (Sander and Patsalos, 1992). Finally, structure-activity studies have demonstrated that there is little correlation between anticonvulsant activity and antifolate activity (Rogawski and Porter, 1990).

LTG selectively blocks veratrine-evoked glutamate release and is without effect against potassium chloride (KCl)-evoked release (Leach *et al.*, 1986). Thus, it appeared from this study that LTG, much like PHT and CBZ, acted at a voltage-dependent sodium channel, thereby decreasing presynaptic glutamate release (Tables I and II). Subsequent studies demonstrated an ability of LTG to inhibit sustained repetitive firing in a voltage-dependent manner in cultured spinal cord neurons (Cheung *et al.*, 1992). LTG was also found to inhibit sustained burst firing induced by ionophoresis of L-glutamate or potassium, but it did not affect the first action potential of the burst (Lees and Leach, 1993). These results, like those of the release experiments, were highly supportive of an interaction between LTG and voltage-sensitive sodium channels. That LTG inhibits [^{3}H]batrachotoxin binding and veratrine-stimulated [^{14}C]guanidine transport into synaptosomes is consistent with this hypothesis (Cheung *et al.*, 1992; Riddall *et al.*, 1993). Confirmation was provided when LTG, in side-by-side studies with PHT and CBZ, was demonstrated to inhibit sodium currents of

N4TG1 mouse neuroblastoma cells in a concentration-dependent manner (Lang *et al.*, 1993). This effect of LTG, as with PHT and CBZ, was use and voltage dependent. In this study, all three AEDs slowed recovery from inactivation.

On the basis of these studies, it is likely that the anticonvulsant effect of LTG, as with PHT and CBZ, is primarily due to a direct interaction of LTG with the voltage-dependent sodium channel. However, this single mechanism of action is not sufficient to explain the apparent broad clinical profile of LTG. To this end, LTG has been found to modulate neurotransmitter release through an interaction with N- and P-type voltage-gated calcium channels (Stefani *et al.*, 1996, 1997). This effect could certainly contribute to the anticonvulsant activity of LTG and provide a basis for the broad clinical profile afforded by this AED (Tables I and II).

D. LEVETIRACETAM

LEV [(-)-(*S*)-α-ethyl-2-oxo-1-pyrrolidine acetamide] is the *S*-enantiomer of the ethyl analogue of the nootropic drug piracetam (Margineanu and Klitgaard, 2002). LEV displays a unique preclinical profile in animal seizure and epilepsy models. For example, it is inactive in convulsive seizures evoked by the traditional maximal electroshock and subcutaneous pentylenetetrazol seizure models (Margineanu and Klitgaard, 2002). In contrast, LEV is very active against sound-induced and fully kindled partial seizures (Margineanu and Klitgaard, 2002). *In vivo*, LEV is also effective against nonconvulsive, electrographic pentylenetetrazol-induced spike-wave discharges (Gower *et al.*, 1992). Furthermore, LEV is very effective against spontaneous spike-wave discharges in the genetic absence epileptic rats of Strasbourg (GAERS) (Gower *et al.*, 1995). LEV decreases the after-discharge duration in the kindled rat and reduces the amplitude of bicuculline-induced hippocampal population spikes (De Smedt *et al.*, 2005; Margineanu and Wulfert, 1995, 1997). In rat hippocampal slices, LEV reduces the amplitude and number of repetitive population spikes induced by a high potassium–low calcium perfusion medium (Margineanu and Klitgaard, 2000). LEV inhibits bicuculline-induced action potential bursts and decreases the frequency of NMDA-evoked bursting in hippocampal pyramidal neurons (Birnstiel *et al.*, 1997). Despite these observations, a unified molecular mechanism has yet to be clearly defined. This is perhaps not so surprising, given the unique anticonvulsant profile of LEV. LEV does bind to a specific, saturable, and stereoselective binding site in rat brain membranes; the binding site displays a high density in the hippocampus, cortex, and cerebellum (Noyer *et al.*, 1995). The finding that there is an excellent correlation between LEV binding and anticonvulsant activity in the audiogenic mouse suggests a possible functional role for this LEV-binding site. Among the various actions that might be ascribed to LEV, its ability to modulate neuronal high-voltage calcium currents (Tables I and II) appears to be the one most likely to contribute to its ability to dampen neuronal hyperexcitability (Niespodziany *et al.*, 2001). At higher concentrations, LEV has also been found to prevent the negative modulatory effects of zinc and β-carbolines on GABA- and glycine-gated currents (Rigo *et al.*, 2002). These effects would be expected to prolong the hyperpolarization associated with GABA- and glycine-mediated neurotransmission. The LEV-binding site is highly homologous to synaptic vesicle protein 2A (SV2A) (Lynch *et al.*, 2004). The relationship between SV2A binding and anticonvulsant efficacy is not known. SV2A is closely associated with synaptic vesicles and is thought to contribute to the docking and release of neurotransmitter substances. Thus, it is highly possible that LEV modifies neurotransmitter release by binding to SV2A (Table II).

E. OXCARBAZEPINE

OCBZ (10,11-dihydro-10-oxo-carbamazepine) is structurally related to CBZ (Dam and Ostergaard, 1995). The keto substitution at the 10,11 position of the dibenzazepine nucleus contributes to better tolerability in humans without negatively affecting efficacy (Dam *et al.*, 1989; Gram and Philbert, 1986; Houtkooper *et al.*, 1987; Reinikainen *et al.*, 1987). OCBZ is rapidly and completely reduced to its active metabolite 10,11-dihydro-10-hydroxy-carbamazepine (HCBZ) (Jensen *et al.*, 1991). Both enantiomers of the racemate are active and likely contribute to the anticonvulsant activity of HCBZ (Schmutz *et al.*, 1993).

OCBZ, HCBZ, and CBZ share a similar mechanistic profile (McLean, 2002; McLean *et al.*, 1994). Therapeutic concentrations of both OCBZ and HCBZ block sustained repetitive firing in cultured spinal cord neurons (Wamil, 1991) in a voltage- and frequency-dependent fashion. This effect suggests an interaction with voltage-sensitive sodium channels (McLean, 2002).

The metabolite HCBZ and its two enantiomers have been found to inhibit penicillin (1.2 mM)–evoked epileptic-like discharges. This inhibitory effect, which was attenuated by the potassium channel blocker 4-amino-pyridine, suggests an anticonvulsant effect on potassium channels (Schmutz *et al.*, 1993). Finally, HCBZ has been reported to inhibit N-type calcium currents in cortical and striatal neurons (Calabresi *et al.*, 1995; Stefani *et al.*, 1995). Such an action may result in reduced neurotransmitter release and, as such, could contribute to the anticonvulsant activity of OCBZ.

F. PREGABALIN

In 2005, PGB, (S)-3-(aminomethyl)-5-methylhexanoic acid, was approved for use as adjunctive therapy for the treatment of partial seizures. PGB is a structural analogue of GABA, but like GBP, it has no activity at the $GABA_A$ receptor. Although PGB has been demonstrated to bind to the $\alpha_2\delta$ auxillary subunit of the voltage-gated calcium channel (Pfizer, 2006), it is still unclear how it exerts its anticonvulsant activity. Work in cultured hippocampal neurons suggests that PGB may reduce spontaneous and evoked neurotransmitter release, most likely by targeting the readily releasable pool of synaptic vesicles (Micheva *et al.*, 2006).

The preclinical profile of PGB revealed that it is more potent and bioavailable than GBP (Vartanian *et al.*, 2006). In animal models of seizure activity, PGB effectively blocked tonic extension seizures induced by maximal electroshock, clonic seizures induced by subcutaneous pentylenetetrazol, kindled seizures, and audiogenic seizures in the DBA/2 mouse (Vartanian *et al.*, 2006). PGB was only found to block bicuculline- and strychnine-induced seizures at high concentrations, and it was unable to block absence seizures in the GAERS animal model.

G. TIAGABINE

TGB, (-)-(*R*)-1-[4,4-bis(3-methyl-2-thienyl)-3-butenyl]nipecotic acid, is a selective GABA uptake inhibitor that emerged from a mechanistic-based drug discovery program designed to identify lipophilic GABA uptake inhibitors for the treatment of epilepsy (Giardina, 2002; Suzdak and Jansen, 1995). The principal mechanism of action of TGB appears to be related to its ability to inhibit neuronal and glial GABA uptake (Table II) (White, 1999). TGB is a potent inhibitor of GABA uptake in various brain regions (Braestrup *et al.*, 1990). It selectively and reversibly binds to the GAT-1 GABA uptake carrier of both neurons and glia. In both animal (Fink-Jensen *et al.*, 1992) and human brains (During *et al.*, 1992), TGB increases extracellular GABA levels *in vivo*. By inhibiting GABA uptake, TGB increases the synaptic concentration of GABA. This effect is assumed to be the basis of the anticonvulsant activity of TGB against partial seizures and its ability to aggravate spike-wave seizures in rodents and humans. Mechanistically, elevated synaptic concentrations of GABA at the level of the thalamus are likely to potentiate $GABA_B$-mediated slow after-hyperpolarization (White, 1999). This effect enhances deinactivation of T-type calcium currents and increases amplification of thalamocortical rhythms necessary to support spike-wave discharges. Electrophysiologically, TGB has been shown to prolong the pharmacological effects of GABA. For example, in the hippocampal slice preparation, TGB prolongs the inhibitory postsynaptic potentials and current in CA1 and CA3 cells produced by exogenously applied GABA (Rekling *et al.*, 1990; Roepstorff and Lambert, 1992; Thompson and Gahwiler, 1992). Collectively, these studies suggest that inhibition of GABA uptake is the primary mechanism through which TGB exerts its anticonvulsant effect.

H. TOPIRAMATE

TPM (2,3:4,5-di-*O*-isopropylidene-β-D-fructopyranose sulfamate) is a chemically unique AED. TPM is effective against a broad range of seizure subtypes, including partial and generalized tonic-clonic seizures, as well as uncomplicated absence seizures (Tables I and II) (Nieto-Barrera, 2002). In addition, TPM has demonstrated clinical utility for the prophylactic treatment of migraines (Nieto-Barrera, 2002). In studies conducted thus far, TPM has been found to possess multiple potential mechanisms of action (Roepstorff and Lambert, 1992; Suzuki *et al.*, 1992) (Tables I and II), including a unique ability to decrease excitation and enhance inhibition.

At therapeutic concentrations (10-100 μM), TPM inhibits sustained repetitive firing in a use- and concentration-dependent manner and reduces voltage-activated sodium currents in cultured neocortical neurons (DeLorenzo *et al.*, 2000; Shank *et al.*, 2000). TPM has also proven effective at reducing the duration and frequency of action potentials within spontaneous epileptiform bursts (DeLorenzo *et al.*, 2000). Furthermore, TPM has been shown to reduce kainate-evoked inward currents, block kainate-evoked cobalt influx, and block kainic acid receptor-mediated post-synaptic currents in the amygdala, indicating that TPM has antagonistic effects on the kainate/AMPA subtype of glutamate receptor (Dodgson *et al.*, 2000; White, 2002; Zhang *et al.*, 1998). The effect of TPM on kainate-evoked currents appears to be unique to TPM among the newer AEDs and is consistent with a decrease in neuronal excitability. When coupled with its effects against sodium currents, this effect may contribute to its efficacy against partial and generalized convulsive seizures.

Effects at the sodium channel and the AMPA/kainate receptor do not necessarily support the ability of TPM to block absence-like spike-wave discharges in the spontaneously epileptic rat nor the anecdotal reports suggesting efficacy against generalized absence epilepsy. TPM has been reported to enhance GABA-evoked, single-channel chloride currents in cultured neocortical neurons (White *et al.*, 1997). Kinetic analysis of single-channel recordings from excised outside-out patches demonstrated that TPM increased the frequency of channel opening and the burst frequency but was without effect on open-channel duration or burst duration. The effect of TPM on GABA$_A$ channel activity was similar to that observed with BZDs, but it was not reversible by the BZD antagonist, flumazenil (White *et al.*, 1995a). Although consistent with apparent efficacy against absence epilepsy, enhancement of GABA-mediated inhibition by TPM would not be predicted from previous *in vitro* studies, wherein TPM failed to displace radio-labeled ligand binding to known binding sites on GABA$_A$ receptors (Shank *et al.*, 1994). TPM modulation of GABA currents through a novel interaction with specific GABA$_A$ receptor subunits was reported (Simeone *et al.*, 2006; White, 2002).

TPM was shown to inhibit certain carbonic anhydrase isoforms, an activity that may contribute to the TPM mechanism of action through an alteration of bicarbonate homeostasis (McLean *et al.*, 2000). The ability of TPM to reversibly hyperpolarize hippocampal neurons in rat hippocampal slices (Zona *et al.*, 1997) and to inhibit neuronal repetitive firing in rat olfactory cortical neurons by inducing a slow outward membrane current may be a function of its ability to activate an outward current carried by potassium ions (Severt *et al.*, 1995; Zona *et al.*, 1997). Furthermore, TPM has been observed to decrease HVACC in CA1 pyramidal cells (Kuzmiski *et al.*, 2005). This effect at both L- and non-L-type calcium channels required a short preincubation period.

The effects of TPM on sodium channels, GABA$_A$ receptors, HVACC, and AMPA/kainate receptors are unique when compared to prototypical modulators of these processes. For example, the effects of TPM on all four of these protein complexes can be highly variable. TPM has been observed to produce both immediate and delayed effects; sometimes its effect is reversible and sometimes it is not; in some preparations, the action of TPM appears to be dependent on the age of neurons in culture (Skradski and White, 2000). All of these observations, albeit frustrating, are suggestive of a unique interaction with the target protein that may be dependent in part on its molecular structure. Indeed, results suggest that the effects of TPM, at least on GABA$_A$ receptor function, depend on the expression of specific subunits (Simeone *et al.*, 2006; White *et al.*, 1995a). Clearly, additional studies are required to further elucidate its precise mechanism of action.

I. ZONISAMIDE

ZNS (1,2-benzisoxazole-3-methanesulfonamide) was discovered as a result of routine biological screening of 1,2-benzisoxazole derivatives. Clinically, ZNS appears to possess efficacy against a number of seizure types: partial and secondarily generalized seizures, generalized tonic-clonic seizures, generalized tonic seizures, atypical absence seizures, atonic seizures, and myoclonic seizures (Seino *et al.*, 1995).

ZNS has a broad mechanistic profile, which likely accounts for its similarly broad anticonvulsant profile in animal and human studies (Tables I and II) (Macdonald, 2002c). ZNS has been found to block sustained repetitive firing in cultured spinal cord neurons (Rock *et al.*, 1989), inhibit voltage-sensitive sodium channels (Schauf, 1987), and decrease voltage-dependent T-type calcium currents in cultured neurons (Suzuki *et al.*, 1992). Inhibitory effects of ZNS on [^{3}H]flunitrazepam binding suggest actions at the BZD-binding site of GABA$_A$ receptors (Mimaki *et al.*, 1990). However, in electrophysiological studies, ZNS did not positively modulate chloride currents evoked by iontophoretically applied GABA (Rock *et al.*, 1989). Although additional experiments are required, experimental and clinical studies demonstrating efficacy against myoclonic seizures suggest that there may be a GABAergic mechanism of action.

From a mechanistic perspective, effects of ZNS on voltage-sensitive sodium channels are likely to contribute to its ability to block tonic extension seizures in animals and generalized tonic seizures in humans. Moreover, inhibition of low voltage–activated T-type calcium currents and possible interaction at the GABA$_A$

receptor are likely to correlate with its efficacy against generalized absence and myoclonic seizures, respectively.

VII. Summary and Implications for the Management of the Older Patient With Epilepsy

Several AEDs display a number of different pharmacological actions that could account for their anticonvulsant properties (Tables I and II). An action at any one of these may contribute to a drug's efficacy *in vivo*. Interestingly, in addition to their anticonvulsant efficacy, several of the AEDs (e.g., VPA, GBP, LTG, and TPM) have been found useful for the management of other CNS disorders such as bipolar disorder (VPA and LTG), migraine (VPA and TPM), and neuropathic pain (GBP) (Rogawski, 2006). When one considers that these drugs target many of the same pathophysiological mechanisms thought to underlie these other CNS disorders, their efficacy in each of these neurological conditions is not surprising.

At the present time, there is no conclusive experimental evidence to suggest that the neuropathology of a seizure network is markedly dependent on patient age or that the mechanistic basis of a seizure differs with the age of the patient. Unfortunately, the vast majority of mechanistic studies conducted to date have not addressed the influence of age on seizure activity or response to AED therapy. Even though there is no *a priori* reason to suspect that any given AED would be more or less effective for the management of seizures in the older patient, this is an area of experimental medicine that has not yet been thoroughly evaluated. However, emerging evidence suggests that receptor- and voltage-gated ion channel subunits are modified by chronic seizures and their associated pathophysiology. As such, the use of subunit-selective AEDs may be designed with the goal in mind to optimize efficacy in this patient population while minimizing toxicity. In the meantime, other patient factors associated with aging, such as altered renal and hepatic function, and the presence of comorbidities should be considered when selecting an AED for older patients.

References

Adusumalli, V. E., Wichmann, J. K., Kucharczyk, N., and Sofia, R. D. (1993). Distribution of the anticonvulsant felbamate to cerebrospinal fluid and brain tissue of adult and neonatal rats. *Drug Metab. Dispos.* **21,** 1079–1085.

Adusumalli, V. E., Wichmann, J. K., Kucharczyk, N., Kamin, M., Sofia, R. D., French, J., Sperling, M., Bourgeois, B., Devinsky, O., Dreifuss, F. E., Kuzniecky, R. I., Faught, E., *et al.* (1994). Drug concentrations in human brain tissue samples from epileptic patients treated with felbamate. *Drug Metab. Dispos.* **22,** 168–170.

Bertrand, S., Ng, G. Y., Purisai, M. G., Wolfe, S. E., Severidt, M. W., Nouel, D., Robitaille, R., Low, M. J., O'Neill, G. P., Metters, K., Lacaille, J. C., Chronwall, B. M., et al. (2001). The anticonvulsant, antihyperalgesic agent gabapentin is an agonist at brain gamma-aminobutyric acid type B receptors negatively coupled to voltage-dependent calcium channels. J. Pharmacol. Exp. Ther. **298**, 15–24.

Birnstiel, S., Wulfert, E., and Beck, S. G. (1997). Levetiracetam (ucb L059) affects in vitro models of epilepsy in CA3 pyramidal neurons without altering normal synaptic transmission. Naunyn Schmiedebergs Arch. Pharmacol. **356,** 611–618.

Braestrup, C., Nielsen, E. B., Sonnewald, U., Knutsen, L. J., Andersen, K. E., Jansen, J. A., Frederiksen, K., Andersen, P. H., Mortensen, A., and Suzdak, P. D. (1990). (R)-N-[4,4-bis (3-methyl-2-thienyl)but-3-en-1-yl]nipecotic acid binds with high affinity to the brain gamma-aminobutyric acid uptake carrier. J. Neurochem. **54**, 639–647.

Calabresi, P., De Murtas, M., Stefani, A., Pisani, A., Sancesario, G., Mercuri, N. B., and Bernardi, G. (1995). Action of GP 47779, the active metabolite of oxcarbazepine, on the corticostriatal system. I. Modulation of corticostriatal synaptic transmission. Epilepsia **36**, 990–996.

Capek, R., and Esplin, B. (1986). Effects of valproate on action potentials and repetitive firing of CA1 pyramidal cells in the hippocampal slice preparation. Soc. Neurosci. Abstr. **12**, 46. Abstract 17.19.

Cheung, H., Kamp, D., and Harris, E. (1992). An in vitro investigation of the action of lamotrigine on neuronal voltage-activated sodium channels. Epilepsy Res. **13**, 107–112.

Coffin, V., Cohen-Williams, M., and Barnett, A. (1994). Selective antagonism of the anticonvulsant effects of felbamate by glycine. Eur. J. Pharmacol. **256**, R9–R10.

Coulter, D. A. (1997). Antiepileptic drug cellular mechanisms of action: Where does lamotrigine fit in? J. Child Neurol. **12**(Suppl. 1), S2–S9.

Coulter, D. A., Huguenard, J. R., and Prince, D. A. (1989). Characterization of ethosuximide reduction of low-threshold calcium current in thalamic neurons. Ann. Neurol. **25**, 582–593.

Coulter, D. A., Sombati, S., and DeLorenzo, R. J. (1993). Selective effects of topiramate on sustained repetitive firing and spontaneous bursting in cultured hippocampal neurons. Epilepsia **34**(Suppl. 2), 123.

Dam, M., and Ostergaard, L. H. (1995). Other antiepileptic drugs. Oxcarbazepine. In "Antiepileptic Drugs" (R. H. Levy, R. H. Mattson, and B. S. Meldrum, Eds.), 4th ed., pp. 987–995. Raven Press Ltd., New York.

Dam, M., Ekberg, R., Loyning, Y., Waltimo, O., and Jakobsen, K. (1989). A double-blind study comparing oxcarbazepine and carbamazepine in patients with newly diagnosed, previously untreated epilepsy. Epilepsy Res. **3**, 70–76.

DeLorenzo, R. J., and Sun, D. A. (2002). Phenytoin and other hydantoins: Mechanisms of action. In "Antiepileptic Drugs" (R. H. Levy, R. H. Mattson, B. S. Meldrum, and E. Perucca, Eds.), 5th ed., pp. 551–564. Lippincott Williams & Wilkins, Philadelphia.

DeLorenzo, R. J., Sombati, S., and Coulter, D. A. (2000). Effects of topiramate on sustained repetitive firing and spontaneous recurrent seizure discharges in cultured hippocampal neurons. Epilepsia **41**(Suppl. 1), S40–S44.

De Sarro, G., Ongini, E., Bertorelli, R., Aguglia, U., and De Sarro, A. (1994). Excitatory amino acid neurotransmission through both NMDA and non-NMDA receptors is involved in the anticonvulsant activity of felbamate in DBA/2 mice. Eur. J. Pharmacol. **262**, 11–19.

De Smedt, T., Vonck, K., Raedt, R., Dedeurwaerdere, S., Claeys, P., Legros, B., Wyckhuys, T., Wadman, W., and Boon, P. (2005). Rapid kindling in preclinical anti-epileptic drug development: The effect of levetiracetam. Epilepsy Res. **67**, 109–116.

Dichter, M. A., and Ayala, G. F. (1987). Cellular mechanisms of epilepsy: A status report. Science **237**, 157–164.

Dildy-Mayfield, J. E., Eger, E. I., II, and Harris, R. A. (1996). Anesthetics produce subunit-selective actions on glutamate receptors. J. Pharmacol. Exp. Ther. **276**, 1058–1065.

Dodgson, S. J., Shank, R. P., and Maryanoff, B. E. (2000). Topiramate as an inhibitor of carbonic anhydrase isoenzymes. *Epilepsia* **41**(Suppl. 1), S35–S39.

During, M., Mattson, R., Scheyer, R., Rask, C., Pierce, M., McKelvy, J., and Thomas, V. (1992). The effect of tiagabine HCl on extracellular GABA levels in the human hippocampus [abstract]. *Epilepsia* **33**(Suppl. 3), 83.

Fink-Jensen, A., Suzdak, P. D., Swedberg, M. D., Judge, M. E., Hansen, L., and Nielsen, P. G. (1992). The gamma-aminobutyric acid (GABA) uptake inhibitor, tiagabine, increases extracellular brain levels of GABA in awake rats. *Eur. J. Pharmacol.* **220**, 197–201.

Gee, N. S., Brown, J. P., Dissanayake, V. U., Offord, J., Thurlow, R., and Woodruff, G. N. (1996). The novel anticonvulsant drug, gabapentin (Neurontin), binds to the alpha2delta subunit of a calcium channel. *J. Biol. Chem.* **271**, 5768–5776.

Giardina, W. J. (2002). Tiagabine: Mechanisms of action. *In* "Antiepileptic Drugs" (R. H. Levy, R. H. Mattson, B. S. Meldrum, and E. Perucca, Eds.), 5th ed., pp. 675–680. Lippincott Williams & Wilkins, Philadelphia.

Gibbs, J. W., III, Sombati, S., DeLorenzo, R. J., and Coulter, D. A. (2000). Cellular actions of topiramate: Blockade of kainate-evoked inward currents in cultured hippocampal neurons. *Epilepsia* **41**(Suppl. 1), S10–S16.

Gower, A. J., Noyer, M., Verloes, R., Gobert, J., and Wulfert, E. (1992). ucb L059, a novel anticonvulsant drug: Pharmacological profile in animals. *Eur. J. Pharmacol.* **222**, 193–203.

Gower, A. J., Hirsch, E., Boehrer, A., Noyer, M., and Marescaux, C. (1995). Effects of levetiracetam, a novel antiepileptic drug, on convulsant activity in two genetic rat models of epilepsy. *Epilepsy Res.* **22**, 207–213.

Gram, L., and Philbert, A. (1986). Oxcarbazepine. *In* "New Anticonvulsant Drugs" (B. S. Meldrum and R. J. Porter, Eds.), pp. 229–235. John Libbey & Co., London.

Guerrini, R., and Parmeggiani, L. (2006). Topiramate and its clinical applications in epilepsy. *Expert Opin. Pharmacother.* **7**, 811–823.

Harty, T. P., and Rogawski, M. A. (2000). Felbamate block of recombinant N-methyl-D-aspartate receptors: Selectivity for the NR2B subunit. *Epilepsy Res.* **39**, 47–55.

Herrero, A. I., Del Olmo, N., Gonzalez-Escalada, J. R., and Solis, J. M. (2002). Two new actions of topiramate: Inhibition of depolarizing GABA(A)-mediated responses and activation of a potassium conductance. *Neuropharmacology* **42**, 210–220.

Holland, K. D., and Ferrendelli, J. A. (2002). Succinimides: Mechanisms of action. *In* "Antiepileptic Drugs" (R. H. Levy, R. H. Mattson, B. S. Meldrum, and E. Perucca, Eds.), 5th ed., pp. 639–645. Lippincott Williams & Wilkins, Philadelphia.

Hommes, O. R., and Obbens, E. A. (1972). The epileptogenic action of Na-folate in the rat. *J. Neurol. Sci.* **16**, 271–281.

Honmou, O., Kocsis, J. D., and Richerson, G. B. (1995). Gabapentin potentiates the conductance increase induced by nipecotic acid in CA1 pyramidal neurons *in vitro*. *Epilepsy Res.* **20**, 193–202.

Houtkooper, M. A., Lammertsma, A., Meyer, J. W., Goedhart, D. M., Meinardi, H., van Oorschot, C. A., Blom, G. F., Hoppener, R. J., and Hulsman, J. A. (1987). Oxcarbazepine (GP 47.680): A possible alternative to carbamazepine? *Epilepsia* **28**, 693–698.

Jensen, P. K., Gram, L., and Schmutz, M. (1991). Oxcarbazepine. *Epilepsy Res. Suppl.* **3**, 135–140.

Kelly, K. M., Gross, R. A., and Macdonald, R. L. (1990). Valproic acid selectively reduces the low-threshold (T) calcium current in rat nodose neurons. *Neurosci. Lett.* **116**, 233–238.

Kito, M., Maehara, M., and Watanabe, K. (1996). Mechanisms of T-type calcium channel blockade by zonisamide. *Seizure* **5**, 115–119.

Kocsis, J. D., and Honmou, O. (1994). Gabapentin increases GABA-induced depolarization in rat neonatal optic nerve. *Neurosci. Lett.* **169**, 181–184.

Kume, A., Greenfield, L. J. Jr., Macdonald, R. L., and Albin, R. L. (1996). Felbamate inhibits [3H] t-butylbicycloorthobenzoate (TBOB) binding and enhances Cl^- current at the gamma-aminobutyric AcidA (GABAA) receptor. *J. Pharmacol. Exp. Ther.* **277**, 1784–1792.

Kuzmiski, J. B., Barr, W., Zamponi, G. W., and MacVicar, B. A. (2005). Topiramate inhibits the initiation of plateau potentials in CA1 neurons by depressing R-type calcium channels. *Epilepsia* **46,** 481–489.

Lang, D. G., Wang, C. M., and Cooper, B. R. (1993). Lamotrigine, phenytoin and carbamazepine interactions on the sodium current present in N4TG1 mouse neuroblastoma cells. *J. Pharmacol. Exp. Ther.* **266,** 829–835.

Leach, M. J., Marden, C. M., and Miller, A. A. (1986). Pharmacological studies on lamotrigine, a novel potential antiepileptic drug: II. Neurochemical studies on the mechanism of action. *Epilepsia* **27,** 490–497.

Leach, M. J., Randall, A. D., Stefani, A., and Hainsworth, A. H. (2002). Lamotrigine: Mechanisms of action. *In* "Antiepileptic Drugs" (R. H. Levy, R. H. Mattson, B. S. Meldrum, and E. Perucca, Eds.), 5th ed., pp. 363–369. Lippincott Williams & Wilkins, Philadelphia.

Lees, G., and Leach, M. J. (1993). Studies on the mechanism of action of the novel anticonvulsant lamotrigine (Lamictal) using primary neurological cultures from rat cortex. *Brain Res.* **612,** 190–199.

Leppik, I. E., Kelly, K. M., deToledo-Morrell, L., Patrylo, P. R., DeLorenzo, R. J., Mathern, G. W., and White, H. S. (2006). Basic research in epilepsy and aging. *Epilepsy Res.* **68**(Suppl. 1), S21–S37.

Leresche, N., Parri, H. R., Erdemli, G., Guyon, A., Turner, J. P., Williams, S. R., Asprodini, E., and Crunelli, V. (1998). On the action of the anti-absence drug ethosuximide in the rat and cat thalamus. *J. Neurosci.* **18,** 4842–4853.

Loscher, W. (1981). Valproate induced changes in GABA metabolism at the subcellular level. *Biochem. Pharmacol.* **30,** 1364–1366.

Loscher, W. (2002). Valproic acid: Mechanisms of action. *In* "Antiepileptic Drugs" (R. H. Levy, R. H. Mattson, B. S. Meldrum, and E. Perucca, Eds.), 5th ed., pp. 767–779. Lippincott Williams & Wilkins, Philadelphia.

Loscher, W., Honack, D., and Taylor, C. P. (1991). Gabapentin increases aminooxyacetic acid-induced GABA accumulation in several regions of rat brain. *Neurosci. Lett.* **128,** 150–154.

Lynch, B. A., Lambeng, N., Nocka, K., Kensel-Hammes, P., Bajjalieh, S. M., Matagne, A., and Fuks, B. (2004). The synaptic vesicle protein SV2A is the binding site for the antiepileptic drug levetiracetam. *Proc. Natl. Acad. Sci. USA* **101,** 9861–9866.

Macdonald, R. L. (1989). Antiepileptic drug actions. *Epilepsia* **30**(Suppl. 1), S19–S28; discussion S64–S68.

Macdonald, R. L. (2002a). Benzodiazepines: Mechanisms of action. *In* Antiepileptic Drugs, (R. H. Levy, R. H. Mattson, B. S. Meldrum, and E. Perucca, Eds.), 5th ed., pp. 179–186. Lippincott Williams & Wilkins, Philadelphia.

Macdonald, R. L. (2002b). Carbamazepine: Mechanisms of action. *In* "Antiepileptic Drugs" (R. H. Levy, R. H. Mattson, B. S. Meldrum, and E. Perucca, Eds.), 5th ed., pp. 227–235. Lippincott Williams & Wilkins, Philadelphia.

Macdonald, R. L. (2002c). Zonisamide: Mechanisms of action. *In* "Antiepileptic Drugs" (R. H. Levy, R. H. Mattson, B. S. Meldrum, and E. Perucca, Eds.), 5th ed., pp. 867–872. Lippincott Williams & Wilkins, Philadelphia.

Macdonald, R. L., and Kelly, K. M. (1995). Antiepileptic drug mechanisms of action. *Epilepsia* **36** (Suppl. 2), S2–S12.

Macdonald, R. L., and Olsen, R. W. (1994). GABAA receptor channels. *Annu. Rev. Neurosci.* **17,** 569–602.

Macdonald, R. L., Rogers, C. J., and Twyman, R. E. (1989). Barbiturate regulation of kinetic properties of the GABAA receptor channel of mouse spinal neurones in culture. *J. Physiol.* **417,** 483–500.

Margineanu, D. G., and Klitgaard, H. (2000). Inhibition of neuronal hypersynchrony *in vitro* differentiates levetiracetam from classical antiepileptic drugs. *Pharmacol. Res.* **42,** 281–285.

Margineanu, D. G., and Klitgaard, H. (2002). Levetiracetam: Mechanisms of action. *In* "Antiepileptic Drugs" (R. H. Levy, R. H. Mattson, B. S. Meldrum, and E. Perucca, Eds.), 5th ed., pp. 419–427. Lippincott Williams & Wilkins, Philadelphia.

Margineanu, D. G., and Wulfert, E. (1995). ucb L059, a novel anticonvulsant, reduces bicuculline-induced hyperexcitability in rat hippocampal CA3 *in vivo*. *Eur. J. Pharmacol.* **286,** 321–325.

Margineanu, D. G., and Wulfert, E. (1997). Inhibition by levetiracetam of a non-GABAA receptor-associated epileptiform effect of bicuculline in rat hippocampus. *Br. J. Pharmacol.* **122,** 1146–1150.

Mattson, R. H., Cramer, J. A., Collins, J. F., Smith, D. B., Delgado-Escueta, A. V., Browne, T. R., Williamson, P. D., Treiman, D. M., McNamara, J. O., McCutchen, C. B., *et al.* (1985). Comparison of carbamazepine, phenobarbital, phenytoin, and primidone in partial and secondarily generalized tonic-clonic seizures. *N. Engl. J. Med.* **313,** 145–151.

McCabe, R. T., Wasterlain, C. G., Kucharczyk, N., Sofia, R. D., and Vogel, J. R. (1993). Evidence for anticonvulsant and neuroprotectant action of felbamate mediated by strychnine-insensitive glycine receptors. *J. Pharmacol. Exp. Ther.* **264,** 1248–1252.

McLean, M. J. (2002). Oxcarbazepine: Mechanisms of action. *In* "Antiepileptic Drugs" (R. H. Levy, R. H. Mattson, B. S. Meldrum, and E. Perucca, Eds.), 5th ed., pp. 451–458. Lippincott Williams & Wilkins, Philadelphia.

McLean, M. J., and Macdonald, R. L. (1986a). Carbamazepine and 10,11-epoxycarbamazepine produce use- and voltage-dependent limitation of rapidly firing action potentials of mouse central neurons in cell culture. *J. Pharmacol. Exp. Ther.* **238,** 727–738.

McLean, M. J., and Macdonald, R. L. (1986b). Sodium valproate, but not ethosuximide, produces use- and voltage-dependent limitation of high frequency repetitive firing of action potentials of mouse central neurons in cell culture. *J. Pharmacol. Exp. Ther.* **237,** 1001–1011.

McLean, M. J., Schmutz, M., Wamil, A. W., Olpe, H. R., Portet, C., and Feldmann, K. F. (1994). Oxcarbazepine: Mechanisms of action. *Epilepsia* **35**(Suppl. 3), S5–S9.

McLean, M. J., Bukhari, A. A., and Wamil, A. W. (2000). Effects of topiramate on sodium-dependent action-potential firing by mouse spinal cord neurons in cell culture. *Epilepsia* **41**(Suppl. 1), S21–S24.

Meldrum, B. (1996). Action of established and novel anticonvulsant drugs on the basic mechanisms of epilepsy. *Epilepsy Res. Suppl.* **11,** 67–77.

Michaelis, E. K. (1998). Molecular biology of glutamate receptors in the central nervous system and their role in excitotoxicity, oxidative stress and aging. *Prog. Neurobiol.* **54,** 369–415.

Micheva, K. D., Taylor, C. P., and Smith, S. J. (2006). Pregabalin reduces the release of synaptic vesicles from cultured hippocampal neurons. *Mol. Pharmacol.* **70,** 467–476.

Mimaki, T., Suzuki, Y., Tagawa, T., Tanaka, J., Itoh, N., and Yabuuchi, H. (1988). [3H]Zonisamide binding in rat brain. *Jpn. J. Psychiatry Neurol.* **42**(3), 640–642.

Mimaki, T., Suzuki, Y., Tagawa, T., Karasawa, T., and Yabuuchi, H. (1990). Interaction of zonisamide with benzodiazepine and GABA receptors in rat brain. *Med. J. Osaka Univ.* **39,** 13–17.

Ng, G. Y., Bertrand, S., Sullivan, R., Ethier, N., Wang, J., Yergey, J., Belley, M., Trimble, L., Bateman, K., Alder, L., Smith, A., McKernan, R., *et al.* (2001). Gamma-aminobutyric acid type B receptors with specific heterodimer composition and postsynaptic actions in hippocampal neurons are targets of anticonvulsant gabapentin action. *Mol. Pharmacol.* **59,** 144–152.

Niespodziany, I., Klitgaard, H., and Margineanu, D. G. (2001). Levetiracetam inhibits the high-voltage-activated Ca(2+) current in pyramidal neurones of rat hippocampal slices. *Neurosci. Lett.* **306,** 5–8.

Nieto-Barrera, M. (2002). Characteristics and indications of topiramate. *Rev. Neurol.* **35**(Suppl. 1), S88–S95.

Noyer, M., Gillard, M., Matagne, A., Henichart, J. P., and Wulfert, E. (1995). The novel antiepileptic drug levetiracetam (ucb L059) appears to act via a specific binding site in CNS membranes. *Eur. J. Pharmacol.* **286,** 137–146.

Olsen, R. W. (2002). Phenobarbital and other barbiturates: Mechanisms of action. *In* "Antiepileptic Drugs" (R. H. Levy, R. H. Mattson, B. S. Meldrum, and E. Perucca, Eds.), 5th ed., pp. 489–495. Lippincott Williams & Wilkins, Philadelphia.

Parker, D. A., Ong, J., Marino, V., and Kerr, D. I. (2004). Gabapentin activates presynaptic GABAB heteroreceptors in rat cortical slices. *Eur. J. Pharmacol.* **495,** 137–143.

Pellock, J. M., and Brodie, M. J. (1997). Felbamate: 1997 update. *Epilepsia* **38,** 1261–1264.

Pellock, J. M., Perhach, J. L., and Sofia, R. D. (2002). Felbamate. *In* "Antiepileptic Drugs" (R. H. Levy, R. H. Mattson, B. S. Meldrum, and E. Perucca, Eds.), 5th ed., pp. 301–318. Lippincott Williams & Wilkins, Philadelphia.

Petroff, O. A. C., Rothman, D. L., Behar, K. L., Lamoureax, D., and Mattson, R. H. (1995). Gabapentin increases brain gama-aminobutyric acid levels in patients with epilepsy [abstract]. *Ann. Neurol.* **38,** 295–296. Abstract M46.

Pfizer (2006). Lyrica (pregabalin) package insert. Pfizer Inc., New York, NY.

Piredda, S. G., Woodhead, J. H., and Swinyard, E. A. (1985). Effect of stimulus intensity on the profile of anticonvulsant activity of phenytoin, ethosuximide and valproate. *J. Pharmacol. Exp. Ther.* **232,** 741–745.

Pisani, A., Stefani, A., Siniscalchi, A., Mercuri, N. B., Bernardi, G., and Calabresi, P. (1995). Electrophysiological actions of felbamate on rat striatal neurones. *Br. J. Pharmacol.* **116,** 2053–2061.

Putnam, T. J., and Merritt, H. H. (1937). Experimental determination of the anticonvulsant properties of some phenyl derivatives. *Science* **85,** 525–526.

Reimann, W. (1983). Inhibition by GABA, baclofen and gabapentin of dopamine release from rabbit caudate nucleus: Are there common or different sites of action? *Eur. J. Pharmacol.* **94,** 341–344.

Reinikainen, K. J., Keranen, T., Halonen, T., Komulainen, H., and Riekkinen, P. J. (1987). Comparison of oxcarbazepine and carbamazepine: A double-blind study. *Epilepsy Res.* **1,** 284–289.

Rekling, J. C., Jahnsen, H., and Mosfeldt Laursen, A. (1990). The effect of two lipophilic gamma-aminobutyric acid uptake blockers in CA1 of the rat hippocampal slice. *Br. J. Pharmacol.* **99,** 103–106.

Reynolds, E. H., Chanarin, I., Milner, G., and Matthews, D. M. (1966). Anticonvulsant therapy, folic acid and vitamin B12 metabolism and mental symptoms. *Epilepsia* **7,** 261–270.

Rho, J. M., and Sankar, R. (1999). The pharmacologic basis of antiepileptic drug action. *Epilepsia* **40,** 1471–1483.

Rho, J. M., Donevan, S. D., and Rogawski, M. A. (1994). Mechanism of action of the anticonvulsant felbamate: Opposing effects on N-methyl-D-aspartate and gamma-aminobutyric acidA receptors. *Ann. Neurol.* **35,** 229–234.

Riddall, D. R., Clackers, M., and Leach, M. J. (1993). Correlation of inhibition of veratrine evoked [14C] guanidine uptake with inhibition of veratrine evoked release of glutamate by lamotrigine and its analogues [abstract]. *Can. J. Neurol. Sci.* **20**(Suppl. 4), S181. Abstract 6–11–17.

Rigo, J. M., Hans, G., Nguyen, L., Rocher, V., Belachew, S., Malgrange, B., Leprince, P., Moonen, G., Selak, I., Matagne, A., and Klitgaard, H. (2002). The anti-epileptic drug levetiracetam reverses the inhibition by negative allosteric modulators of neuronal GABA- and glycine-gated currents. *Br. J. Pharmacol.* **136,** 659–672.

Rock, D. M., Macdonald, R. L., and Taylor, C. P. (1989). Blockade of sustained repetitive action potentials in cultured spinal cord neurons by zonisamide (AD 810, CI 912), a novel anticonvulsant. *Epilepsy Res.* **3,** 138–143.

Roepstorff, A., and Lambert, J. D. (1992). Comparison of the effect of the GABA uptake blockers, tiagabine and nipecotic acid, on inhibitory synaptic efficacy in hippocampal CA1 neurones. *Neurosci. Lett.* **146,** 131–134.

Rogawski, M. A. (2006). Molecular targets versus models for new antiepileptic drug discovery. *Epilepsy Res.* **68,** 22–28.

Rogawski, M. A., and Porter, R. J. (1990). Antiepileptic drugs: Pharmacological mechanisms and clinical efficacy with consideration of promising developmental stage compounds. *Pharmacol. Rev.* **42,** 223–286.

Rogers, C. J., Twyman, R. E., and Macdonald, R. L. (1994). Benzodiazepine and beta-carboline regulation of single GABAA receptor channels of mouse spinal neurones in culture. *J. Physiol.* **475,** 69–82.

Sander, J. W., and Patsalos, P. N. (1992). An assessment of serum and red blood cell folate concentrations in patients with epilepsy on lamotrigine therapy. *Epilepsy Res.* **13,** 89–92.

Schauf, C. L. (1987). Zonisamide enhances slow sodium inactivation in *Myxicola*. *Brain Res.* **413**, 185–188.

Schmidt, B. (1989). Potential antiepileptic drugs: Gabapentin. *In* "Antiepileptic Drugs" (R. Levy, R. Mattson, B. Meldrum, J. K. Penry, and F. E. Dreifuss, Eds.), 3rd ed., pp. 925–935. Raven Press, New York.

Schmutz, M., Ferrat, T., Heckendorn, R., Jeker, A., Portet, C., and Olpe, H. R. (1993). GP 47779, the main human metabolite of oxcarbazepine (Trileptal), and both enantiomers have equal anticonvulsant activity [abstract]. *Epilepsia* **34**(Suppl. 2), 122.

Schwarz, J. R., and Grigat, G. (1989). Phenytoin and carbamazepine: Potential- and frequency-dependent block of Na currents in mammalian myelinated nerve fibers. *Epilepsia* **30**, 286–294.

Seino, M., Naruto, S., Ito, T., and Miyazaki, H. (1995). Other antiepileptic drugs. Zonisamide. *In* "Antiepileptic Drugs" (R. H. Levy, R. H. Mattson, and B. S. Meldrum, Eds.), 4th ed., pp. 1011–1023. Raven Press Ltd., New York.

Severt, L., Coulter, D. A., Sombati, S., and DeLorenzo, R. J. (1995). Topiramate selectively blocks kainate currents in cultured hippocampal neurons [abstract]. *Epilepsia* **36**(Suppl. 4), 38. Abstract 2.16.

Shank, R. P., Gardocki, J. F., Vaught, J. L., Davis, C. B., Schupsky, J. J., Raffa, R. B., Dodgson, S. J., Nortey, S. O., and Maryanoff, B. E. (1994). Topiramate: Preclinical evaluation of a structurally novel anticonvulsant. *Epilepsia* **35**, 450–460.

Shank, R. P., Gardocki, J. F., Streeter, A. J., and Maryanoff, B. E. (2000). An overview of the preclinical aspects of topiramate: Pharmacology, pharmacokinetics, and mechanism of action. *Epilepsia* **41**(Suppl. 1), S3–S9.

Simeone, T. A., McClellan, A. M. L., Twyman, R. E., and White, H. S. (2000). Direct activation of the recombinant $\alpha4\beta3\gamma2S$ GABAA receptor by the novel anticonvulsant topiramate. *Soc. Neurosci. Abstr.* **26**, 631.

Simeone, T. A., Wilcox, K. S., and White, H. S. (2006). Subunit selectivity of topiramate modulation of heteromeric GABA(A) receptors. *Neuropharmacology* **50**, 845–857.

Skradski, S., and White, H. S. (2000). Topiramate blocks kainate-evoked cobalt influx into cultured neurons. *Epilepsia* **41**(Suppl. 1), S45–S47.

Sofia, R. D. (1995). Felbamate. Mechanisms of action. *In* "Antiepileptic Drugs" (R. H. Levy, R. H. Mattson, and B. S. Meldrum, Eds.), 4th ed., pp. 791–797. Raven Press Ltd., New York.

Stefani, A., Pisani, A., De Murtas, M., Mercuri, N. B., Marciani, M. G., and Calabresi, P. (1995). Action of GP 47779, the active metabolite of oxcarbazepine, on the corticostriatal system. II. Modulation of high-voltage-activated calcium currents. *Epilepsia* **36**, 997–1002.

Stefani, A., Spadoni, F., Siniscalchi, A., and Bernardi, G. (1996). Lamotrigine inhibits Ca2+ currents in cortical neurons: Functional implications. *Eur. J. Pharmacol.* **307**, 113–116.

Stefani, A., Spadoni, F., and Bernardi, G. (1997). Voltage-activated calcium channels: Targets of antiepileptic drug therapy? *Epilepsia* **38**, 959–965.

Stefani, A., Spadoni, F., and Bernardi, G. (1998). Gabapentin inhibits calcium currents in isolated rat brain neurons. *Neuropharmacology* **37**, 83–91.

Stringer, J. L., and Lorenzo, N. (1999). The reduction in paired-pulse inhibition in the rat hippocampus by gabapentin is independent of GABA(B) receptor activation. *Epilepsy Res.* **33**, 169–176.

Study, R. E., and Barker, J. L. (1981). Diazepam and (−)-pentobarbital: Fluctuation analysis reveals different mechanisms for potentiation of gamma-aminobutyric acid responses in cultured central neurons. *Proc. Natl. Acad. Sci. USA* **78**, 7180–7184.

Subramaniam, S., Rho, J. M., Penix, L., Donevan, S. D., Fielding, R. P., and Rogawski, M. A. (1995). Felbamate block of the N-methyl-D-aspartate receptor. *J. Pharmacol. Exp. Ther.* **273**, 878–886.

Suman-Chauhan, N., Webdale, L., Hill, D. R., and Woodruff, G. N. (1993). Characterisation of [3H] gabapentin binding to a novel site in rat brain: Homogenate binding studies. *Eur. J. Pharmacol.* **244**, 293–301.

Suzdak, P. D., and Jansen, J. A. (1995). A review of the preclinical pharmacology of tiagabine: A potent and selective anticonvulsant GABA uptake inhibitor. *Epilepsia* **36,** 612–626.

Suzuki, S., Kawakami, K., Nishimura, S., Watanabe, Y., Yagi, K., Seino, M., and Miyamoto, K. (1992). Zonisamide blocks T-type calcium channel in cultured neurons of rat cerebral cortex. *Epilepsy Res.* **12,** 21–27.

Swinyard, E. A., Sofia, R. D., and Kupferberg, H. J. (1986). Comparative anticonvulsant activity and neurotoxicity of felbamate and four prototype antiepileptic drugs in mice and rats. *Epilepsia* **27,** 27–34.

Taglialatela, M., Ongini, E., Brown, A. M., Di Renzo, G., and Annunziato, L. (1996). Felbamate inhibits cloned voltage-dependent Na+ channels from human and rat brain. *Eur. J. Pharmacol.* **316,** 373–377.

Taverna, S., Sancini, G., Mantegazza, M., Franceschetti, S., and Avanzini, G. (1999). Inhibition of transient and persistent Na$^+$ current fractions by the new anticonvulsant topiramate. *J. Pharmacol. Exp. Ther.* **228,** 960–968.

Taylor, C. P. (2002). Gabapentin: Mechanisms of action. *In* "Antiepileptic Drugs" (R. H. Levy, R. H. Mattson, B. S. Meldrum, and E. Perucca, Eds.), 5th ed., pp. 321–334. Lippincott Williams & Wilkins, Philadelphia.

Taylor, C. P., Gee, N. S., Su, T. Z., Kocsis, J. D., Welty, D. F., Brown, J. P., Dooley, D. J., Boden, P., and Singh, L. (1998). A summary of mechanistic hypotheses of gabapentin pharmacology. *Epilepsy Res.* **29,** 233–249.

Thompson, S. M., and Gahwiler, B. H. (1992). Effects of the GABA uptake inhibitor tiagabine on inhibitory synaptic potentials in rat hippocampal slice cultures. *J. Neurophysiol.* **67,** 1698–1701.

Thurlow, R. J., Brown, J. P., Gee, N. S., Hill, D. R., and Woodruff, G. N. (1993). [3H]gabapentin may label a system-L-like neutral amino acid carrier in brain. *Eur. J. Pharmacol.* **247,** 341–345.

Ticku, M. K., Kamatchi, G. L., and Sofia, R. D. (1991). Effect of anticonvulsant felbamate on GABAA receptor system. *Epilepsia* **32,** 389–391.

Ticku, M. K., Kulkarni, S. K., and Mehta, A. K. (1992). Modulatory role of GABA receptor subtypes and glutamate receptors in the anticonvulsant effect of barbiturates. *Epilepsy Res. Suppl.* **8,** 57–62.

Tomaselli, G. F., Marban, E., and Yellen, G. (1989). Sodium channels from human brain RNA expressed in *Xenopus* oocytes. Basic electrophysiologic characteristics and their modification by diphenylhydantoin. *J. Clin. Invest.* **83,** 1724–1732.

Twyman, R. E., Rogers, C. J., and Macdonald, R. L. (1989). Differential regulation of gamma-aminobutyric acid receptor channels by diazepam and phenobarbital. *Ann. Neurol.* **25,** 213–220.

Van den Berg, R. J., Kok, P., and Voskuyl, R. A. (1993). Valproate and sodium currents in cultured hippocampal neurons. *Exp. Brain Res.* **93,** 279–287.

VanDongen, A. M., VanErp, M. G., and Voskuyl, R. A. (1986). Valproate reduces excitability by blockage of sodium and potassium conductance. *Epilepsia* **27,** 177–182.

Vartanian, M. G., Radulovic, L. L., Kinsora, J. J., Serpa, K. A., Vergnes, M., Bertram, E., and Taylor, C. P. (2006). Activity profile of pregabalin in rodent models of epilepsy and ataxia. *Epilepsy Res.* **68,** 189–205.

Vicini, S., Mienville, J. M., and Costa, E. (1987). Actions of benzodiazepine and beta-carboline derivatives on gamma-aminobutyric acid-activated Cl$^-$ channels recorded from membrane patches of neonatal rat cortical neurons in culture. *J. Pharmacol. Exp. Ther.* **243,** 1195–1201.

Wakamori, M., Kaneda, M., Oyama, Y., and Akaike, N. (1989). Effects of chlordiazepoxide, chlorpromazine, diazepam, diphenylhydantoin, flunitrazepam and haloperidol on the voltage-dependent sodium current of isolated mammalian brain neurons. *Brain Res.* **494,** 374–378.

Wallis, R. A., and Panizzon, K. L. (1993). Glycine reversal of felbamate hypoxic protection. *Neuroreport* **4,** 951–954.

Wallis, R. A., and Panizzon, K. L. (1995). Felbamate neuroprotection against CA1 traumatic neuronal injury. *Eur. J. Pharmacol.* **294,** 475–482.

Wamil, A. W. (1991). Oxcarbazepine and its monohydroxy metabolite limit action potential firing by mouse central neurons in cell culture [abstract]. *Epilepsia* **32**(Suppl. 3), 65.

Wamil, A. W., and McLean, M. J. (1994). Limitation by gabapentin of high frequency action potential firing by mouse central neurons in cell culture. *Epilepsy Res.* **17**, 1–11.

Wamsley, J. K., Sofia, R. D., Faull, R. L., Narang, N., Ary, T., and McCabe, R. T. (1994). Interaction of felbamate with [3H]DCKA-labeled strychnine-insensitive glycine receptors in human postmortem brain. *Exp. Neurol.* **129**, 244–250.

Welty, D. F., Schielke, G. P., Vartanian, M. G., and Taylor, C. P. (1993). Gabapentin anticonvulsant action in rats: Disequilibrium with peak drug concentrations in plasma and brain microdialysate. *Epilepsy Res.* **16**, 175–181.

White, H. S. (1997). Mechanisms of antiepileptic drugs. *In* "Epilepsies II" (R. Porter and D. Chadwick, Eds.), pp. 1–30. Butterworth-Heinemann, Boston.

White, H. S. (1999). Comparative anticonvulsant and mechanistic profile of the established and newer antiepileptic drugs. *Epilepsia* **40**(Suppl. 5), S2–S10.

White, H. S. (2002). Topiramate: Mechanisms of action. *In* "Antiepileptic Drugs" (R. H. Levy, R. H. Mattson, B. S. Meldrum, and E. Perucca, Eds.), 5th ed., pp. 719–726. Lippincott Williams & Wilkins, Philadelphia.

White, H. S., Wolf, H. H., Swinyard, E. A., Skeen, G. A., and Sofia, R. D. (1992). A neuropharmacological evaluation of felbamate as a novel anticonvulsant. *Epilepsia* **33**, 564–572.

White, H. S., Brown, D., Skeen, G. A., Wolf, H. H., and Twyman, R. E. (1995a). The anticonvulsant topiramate displays a unique ability to potentiate GABA-evoked chloride currents. *Epilepsia* **36** (Suppl. 3), S39–S40.

White, H. S., Harmsworth, W. L., Sofia, R. D., and Wolf, H. H. (1995b). Felbamate modulates the strychnine-insensitive glycine receptor. *Epilepsy Res.* **20**, 41–48.

White, H. S., Woodhead, J. H., Franklin, M. R., Swinyard, E. A., and Wolf, H. H. (1995c). General principles: Experimental selection, quantification, and evaluation of antiepileptic drugs. *In* "Antiepileptic Drugs" (R. H. Levy, R. H. Mattson, and B. S. Meldrum, Eds.), 4th ed., pp. 99–110. Raven Press, New York.

White, H. S., Brown, S. D., Woodhead, J. H., Skeen, G. A., and Wolf, H. H. (1997). Topiramate enhances GABA-mediated chloride flux and GABA-evoked chloride currents in murine brain neurons and increases seizure threshold. *Epilepsy Res.* **28**, 167–179.

Willow, M., and Catterall, W. A. (1982). Inhibition of binding of [3H]batrachotoxinin A 20-alpha-benzoate to sodium channels by the anticonvulsant drugs diphenylhydantoin and carbamazepine. *Mol. Pharmacol.* **22**, 627–635.

Willow, M., Kuenzel, E. A., and Catterall, W. A. (1984). Inhibition of voltage-sensitive sodium channels in neuroblastoma cells and synaptosomes by the anticonvulsant drugs diphenylhydantoin and carbamazepine. *Mol. Pharmacol.* **25**, 228–234.

Willow, M., Gonoi, T., and Catterall, W. A. (1985). Voltage clamp analysis of the inhibitory actions of diphenylhydantoin and carbamazepine on voltage-sensitive sodium channels in neuroblastoma cells. *Mol. Pharmacol.* **27**, 549–558.

Yamakura, T., Sakimura, K., Mishina, M., and Shimoji, K. (1995). The sensitivity of AMPA-selective glutamate receptor channels to pentobarbital is determined by a single amino acid residue of the alpha 2 subunit. *FEBS Lett.* **374**, 412–414.

Zhang, X., Velumian, A. A., Jones, O. T., and Carlen, P. L. (1998). Topiramate reduces high-voltage activated Ca2+ currents in CA1 pyramidal neurons *in vitro* [abstract]. *Epilepsia* **39**(Suppl. 6), 44.

Zhang, X., Velumian, A. A., Jones, O. T., and Carlen, P. L. (2000). Modulation of high-voltage-activated calcium channels in dentate granule cells by topiramate. *Epilepsia* **41**(Suppl. 1), S52–S60.

Zona, C., and Avoli, M. (1990). Effects induced by the antiepileptic drug valproic acid upon the ionic currents recorded in rat neocortical neurons in cell culture. *Exp. Brain Res.* **81**, 313–317.

Zona, C., Ciotti, M. T., and Avoli, M. (1997). Topiramate attenuates voltage-gated sodium currents in rat cerebellar granule cells. *Neurosci. Lett.* **231**, 123–126.

EPIDEMIOLOGY AND OUTCOMES OF STATUS EPILEPTICUS IN THE ELDERLY

Alan R. Towne

Department of Neurology, Virginia Commonwealth University
Richmond, Virginia 23298, USA

Status epilepticus (SE) is a serious condition of prolonged or repetitive seizures. The annual incidence (86/100,000) of SE in the elderly who are aged 60 and greater is almost twice that of the general population and is even higher in those who are 70 years and older. Either acute or remote symptomatic stroke causes approximately 60% of SE seen in the elderly. SE is associated with a high mortality in the elderly (38%), with a rate approaching 50% in patients older than 80 years of age. Etiology is a strong determinant of mortality in the elderly: mortality approaches 100% in patients with anoxia and 30% in patients with either metabolic disorders, hemorrhages, tumors, or systemic infections. Mortality is almost three times higher in SE associated with acute ischemic stroke than in stroke alone, indicating synergistic effects.

Duration of SE is also a factor in mortality. Treatment should be initiated for any convulsive seizure that lasts at least 10 min or is repetitive. An electroencephalogram (EEG) should be promptly obtained so that a diagnosis can be made without delay. Because older patients have a greater likelihood of nondiagnostic findings on routine EEGs, prolonged EEG recordings and inpatient video-EEG monitoring significantly increase the rate of establishing a definitive diagnosis. Nonconvulsive status epilepticus in the elderly is especially difficult to diagnose and should be evaluated with an EEG.

INTERNATIONAL REVIEW OF
NEUROBIOLOGY, VOL. 81
DOI: 10.1016/S0074-7742(06)81007-X

Treatment of SE is complicated by altered pharmacokinetics in the elderly. Initial treatments, usually the administration of an intravenous benzodiazepine, have overall success rates of 55% for overt convulsive SE and 14.9% for subtle SE. For refractory SE, little is gained by using additional standard drugs, and general anesthesia with continuous EEG monitoring is recommended.

I. Introduction

The elderly are the fastest growing segment of the US population. The US Department of Health and Human Services predicts that by 2030 there will be approximately 70 million adults over the age of 65 in the country (US Department of Health and Human Services, 2005). This segment of the population was 12.4% of the total population in 2000; however, by 2030 it will account for approximately 20%. The elderly have the highest incidence of seizures of any age group (Hauser et al., 1993). The increased frequency of seizures in this population can be attributed to comorbid conditions and characteristics such as an increased risk for stroke, metabolic abnormalities, and an increased use of prescription drugs. As the US population ages, physicians will increasingly face the challenge of diagnosing and effectively managing seizures in the elderly.

In this chapter, status epilepticus (SE) is described, and its epidemiology, etiology, and mortality in relation to the elderly population are discussed. The value of the electroencephalogram (EEG) in monitoring elderly patients with SE, especially those with nonconvulsive status epilepticus (NCSE), is also discussed. While there is no established protocol for SE management in the elderly patient, current treatment options are explained both in terms of initial therapy and for SE that is refractory to the initial treatment.

II. Definitions

SE has been defined as "a condition characterized by an epileptic seizure which is so frequently repeated or so prolonged as to create a fixed and lasting epileptic condition" (Gastaut, 1983). The duration of what is accepted as SE has been shrinking (Waisterlain and Chen, 2006). The Veterans Affairs (VA) cooperative trial on the treatment of SE used an operational definition of 10 min or greater to the time of treatment (Treiman et al., 1998), and it has been suggested that an operational definition should be a generalized convulsive seizure lasting 5 min or longer (Lowenstein et al., 1999b). However, the most widely used definition of SE for the older epidemiological studies is more than 30 min of either continuous seizure activity (DeLorenzo *et al.*, 1996) or intermittent seizures without full recovery of consciousness between seizures. Since most seizures last less than 2 min (Jakkampudi *et al.*, 2005), some authors have proposed that SE should be diagnosed and treated well before 30 min have elapsed (Lowenstein *et al.*, 1999b). For this reason, it is sensible to diagnose and treat any prolonged or repetitive seizures as impending SE. According to Gastaut (1983), the classification of SE contains as many types of SE as there are types of epileptic seizures.

Types of partial SE include somatosensory seizures, which can include any sensory modality (vision, taste, smell, hearing, or touch); autonomic seizures, which may include changes in visceral sensation or heart rate; and psychic seizures, during which patients report feelings of anxiety, depression, fear, or altered perception of time such as déjà vu. Complex partial SE is characterized by impairment of consciousness and may be accompanied by automatisms. Patients may describe an aura, which is a simple partial seizure, preceding the loss of consciousness. Partial seizures can secondarily generalize, and patients may also experience a complex partial seizure before secondary generalization. The major types of generalized seizures are absence, myoclonic, and tonic-clonic.

NCSE is defined as seizures lasting at least 30 min, characterized by some clinically evident alteration in mental status or behavior from baseline without overt tonic-clonic activity, with the demonstration of seizure activity on the EEG (Kaplan, 1999). NCSE is an often unrecognized cause of coma, especially in elderly patients. In a study of 236 comatose patients ($\sim$50% over the age of 60) with no overt clinical seizure activity, it was found that 8% of these patients had NCSE (Towne *et al.*, 2000). NCSE is usually categorized into two entities: complex partial SE and generalized NCSE (bilateral diffuse synchronous seizures) (Kaplan, 1999). This subject is more fully addressed in the chapter by Walker in this volume.

III. The Epidemiology of SE in the Older Patient

The age-specific incidence rates of SE have a bimodal distribution, with the highest rates in infants and the elderly (Fig. 1). The elderly, defined as greater than 60 years of age in an epidemiological study conducted in Richmond, Virginia, had an annual SE (defined as seizures $\geq$30 min) incidence rate of 86 per 100,000 (DeLorenzo *et al.*, 1996). From the same study, the highest incidence rate of SE was seen in children between the ages of 1 and 12 months, at 156 per 100,000. However, if the data are analyzed by major age groups, with the pediatric age group including ages from 1 month up to 16 years of age, adults defined as ages 16 up to 60, and the elderly group defined as age 60 and greater, the elderly group had the highest incidence rate of the three groups. The incidence of SE in the age 60 and greater

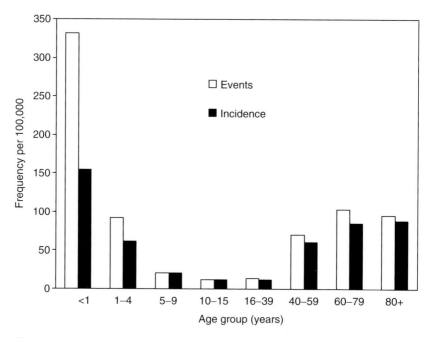

Fig. 1. Age-specific distribution of the frequency of status epilepticus (SE) events and the incidence of SE per year per 100,000 in Richmond, Virginia. The population in each age group was determined from National Census Bureau data on the demographics of Richmond, 1990. SE events included all episodes of SE per 100,000 per year. The incidence of SE represents the number of patients who developed SE per 100,000 per year in Richmond and did not include recurrent episodes of SE. Data from DeLorenzo *et al.* (1996), with permission.

group was almost twice that of the general population, and the highest incidence was seen in patients 70 years and older (DeLorenzo, 1997).

In a study of generalized convulsive status epilepticus (GCSE) in California, Wu *et al.* (2002) reviewed the International Classification of Diseases, Ninth Revision, Clinical Modification (ICD-9-CM) codes from nonfederal hospitals. In this study, the incidence of SE for all ages was 6.2 per 100,000. The rate of GCSE was highest among children under the age of 5 (7.5/100,000) and among the elderly age 75 and older (22.3/100,000). The incidence of SE was lower in this study than in other previous reports in the United States. Reasons for this lower incidence may be explained, partially, by the retrospective nature of the study, which relied on ICD-9-CM discharge coding data and the inclusion of only those patients who experienced GCSE. In a prospective study conducted in Bologna, Italy, the incidence rate was 26.2 per 100,000 in patients older than 60 years versus 5.2 per 100,000 in patients younger than 60 years (Vignatelli *et al.*, 2003). In another prospective study, undertaken in the French-speaking part of

Switzerland, the age-specific annual incidence rate again showed a bimodal distribution with a peak in children, a minimum during adolescence and young adulthood, and a progressive increase after the age of 60. In this cohort, the incidence rate was 15.1 per 100,000 in the 60–74 age group and 15.5 per 100,000 in the 75 and older age group (Coeytaux et al., 2000). Although this study revealed the same age-specific distribution, a lower specific incidence rate in the elderly was found, which could be attributed to the exclusion of patients in postanoxic coma. In Hessen, Germany, a prospective study by Knake et al. (2001) revealed an incidence of 54.5 per 100,000 in the elderly (age 60 or older) versus 4.2 per 100,000 in younger adults (18–59 years of age).

IV. Etiologies of SE in the Elderly

Etiologies of SE are important for determining the morbidity and mortality associated with this condition (Towne et al., 1994). In a retrospective study at Chang Gung Memorial Hospital in Taipei, Taiwan, of 102 patients with SE who had their first seizure after 60 years of age, cerebrovascular disease was the leading cause of seizure (35%), followed by head trauma (21%) (Sung and Chu, 1989). In patients over the age of 60 who participated in the Richmond study, 35% of SE events were caused by acute stroke and 26% were caused by remote symptomatic stroke. Thus, stroke caused SE in approximately 61% of these elderly patients (DeLorenzo et al., 1995). Other etiologies in the older age group, from the same study, included hypoxia (17%), metabolic disorder (14%), alcohol related (11%), tumor (10%), infection (6%), anoxia (6%), hemorrhage (5%), central nervous system (CNS) infection (5%), trauma (1%), idiopathic (1%), and other (1%). In a retrospective population-based study of SE from Rochester, Minnesota, another common etiology of SE was found to be dementia (Hesdorffer et al., 1998). Wu et al. (2002) found that, in elderly patients, the most common etiology of GCSE was delayed effect of stroke or brain injury. In the Coeytaux et al. (2000) study, most of the cases of SE in patients over the age of 60 were in the acute symptomatic group, which included etiologies occurring in close association with an acute cerebral insult such as cerebrovascular disease, CNS infection, alcohol-related insults, metabolic insults, and drug withdrawal.

In a prospective study of 10 patients over age 65 years with NCSE, the cause of NCSE was stroke in four patients and metabolic derangement in two others. The remaining four patients each had a different etiology: brain neoplasia, head injury, electroconvulsive therapy, and preexisting epilepsy. In comparison to reports in younger patients, elderly patients with NCSE were found to have a worse prognosis, and this was felt to be secondary to the severity of the underlying etiologies (Labar et al., 1998).

V. Mortality of SE in the Elderly

SE is associated with a high mortality in the elderly population. Previous studies have addressed this important condition in the older age groups. In the Richmond study (DeLorenzo *et al.*, 1996), the overall mortality was 22%. However, there was a considerable difference between the mortality in the pediatric population (3%) versus the adult mortality (14%). The differences were even more marked in the elderly population, with mortality at 38% (Fig. 2) (DeLorenzo *et al.*, 1995); mortality approached 50% in patients greater than 80 years of age (DeLorenzo, 1997). Along with patient age and duration of SE, etiology is a strong determinant of mortality in the elderly (Towne *et al.*, 1994). The higher mortality in elderly patients may be secondary to the increased susceptibility of these patients to systemic metabolic diseases, cerebrovascular accidents (CVAs), and progressive symptomatic conditions such as tumor and dementia. Depending on the etiologies, the mortality rates differ markedly. Etiologies with mortalities less than 9% include patients presenting with alcohol withdrawal or low levels of antiepileptic medications. Remote symptomatic cases, the majority of which consist of prior strokes, have a mortality rate of 14%. Etiologies such as metabolic disorders, hemorrhages, tumors, and systemic infections are each associated with approximately 30% mortality, and CNS infections and head trauma are associated with approximately 1% and 20% mortality, respectively. The highest mortality, approaching 100%, can be seen in patients with anoxia (Towne *et al.*, 1994).

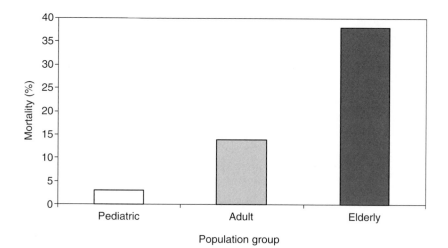

FIG. 2. Mortality of SE by age group. The pediatric group included ages from 1 month to 16 years; the adult group, 16 up to 60 years; the elderly, 60 years and greater. Data from DeLorenzo *et al.* (1995).

Because ischemic brain injury is a major cause of SE in the elderly, and since stroke and SE each have high mortalities, a study was conducted to determine whether the high mortality seen in patients with stroke and SE differs significantly from the mortality of stroke alone. Also, the question of whether the high mortality is attributable to the severity of the underlying ischemia or is a synergistic effect of the two brain insults was examined (Waterhouse *et al.*, 1998). Three groups of patients were studied: (1) patients with acute ischemic stroke without SE, (2) patients with acute ischemic stroke with SE, and (3) patients with remote ischemic stroke and SE. The results of this study demonstrated that SE with acute CVA had a 39% mortality, whereas acute CVA alone had a 14% mortality, and SE with remote CVA had the lowest mortality, at 5% (Fig. 3). No differences in radiological lesion size were seen between the groups. This study demonstrated a nearly threefold statistically significant increase in mortality in patients with SE associated with acute ischemic stroke than in patients with stroke alone ($p < 0.001$). This would indicate that the high mortality associated with concurrent SE and CVA in the elderly is a result of the synergistic effect of SE and ischemic brain injury.

NCSE also appears to be associated with high mortality. In the Labar *et al.* (1998) study of patients with NCSE, 3 of 10 died because of complications due to infections during hospitalization. Several conclusions were drawn from this study: NCSE has a worse prognosis in the elderly than in younger patients; the relatively poor outcome in the elderly is the result of more severe underlying processes and

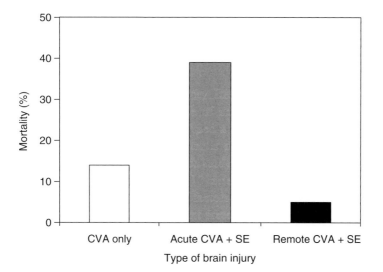

FIG. 3. Mortality of stroke and SE. CVA, cerebrovascular accident; SE, status epilepticus. Reprinted from Waterhouse *et al.* (1998), with permission from Elsevier.

is not associated with more severe NCSE, as gauged by SE duration; and NCSE in the elderly is associated with a high rate of hospital-acquired infections, which may be fatal. NCSE mortality was 52% in a study of 25 critically ill elderly patients, and death was associated with having a larger number of acute life-threatening medical problems on presentation (Litt *et al.*, 1998). In this group of critically ill patients, treatment of NCSE with benzodiazepines increased the risk of death, and aggressive anticonvulsant therapy did not improve the outcome.

VI. Electroencephalogram

Older patients with acute seizures may have a variety of EEG changes, only some of which are attributable to underlying pathology. Although benign EEG variants with epileptiform morphology occur in all age groups, three that occur with a greater frequency in the older population are subclinical rhythmic electrical discharges of adulthood, wicket spikes, and small sharp spikes (Van Cott, 2002). These patterns can potentially be misinterpreted as epileptiform abnormalities. Interictal epileptiform activity occurs less frequently in older age groups than in younger age groups (Marsan and Zivin, 1970). This indicates that older patients have a greater likelihood of nondiagnostic findings on a routine EEG. A VA cooperative study, conducted in elderly patients, found interictal epileptiform activity in approximately one-third of routine EEGs (Ramsay and Pryor, 2000). Prolonged EEG recordings and inpatient video-EEG monitoring significantly increase the diagnostic yield in elderly patients (McBride *et al.*, 2002). Although elderly patients account for approximately 25% of newly diagnosed seizures, older patients are relatively underrepresented in epilepsy monitoring units, despite the usefulness of monitoring to establish a definitive diagnosis (Kellinghaus *et al.*, 2004).

Suspected NCSE should be evaluated with an EEG. The clinical manifestations may be subtle, consisting of prolonged confusion, unusual behavior, minor motor manifestations, aphasia, or coma. Lee (1985) conducted a study of 11 patients without a history of absence seizures who presented with an acute onset of a prolonged confusional state. All patients were ambulatory with behavior ranging from mild disorientation to confusion. Some patients demonstrated prominent psychiatric manifestations. The EEGs generally showed 1–2.5 Hz generalized spike-wave or multiple spike-wave discharges. The patients were successfully treated with intravenously administered diazepam followed by orally administered phenytoin sodium and phenobarbital (Lee, 1985).

Privitera *et al.* (1994) prospectively studied hospitalized patients with altered consciousness who underwent an EEG to evaluate coma or altered mental status. In this study of 198 cases with altered consciousness but no clinical convulsions, 37% showed EEG and clinical evidence of definite or probable NCSE. In 23 cases, altered consciousness was the only clinical sign at the time of diagnosis, with subtle motor activity present in 36 others. Neither clinical signs nor prior history predicted which patients demonstrated SE on EEG. Lowenstein and Aminoff (1992) reported on 38 patients with NCSE and concluded that the majority of patients with subtle motor activity and depressed consciousness had EEG findings of SE, and that an EEG was necessary to diagnose this condition.

NCSE can be seen after the control of convulsive SE. In the Richmond study, 14% of the patients manifested NCSE after the cessation of convulsive SE (DeLorenzo et al., 1998). These patients were comatose and showed no overt clinical signs of convulsive activity. Clinical detection of NCSE in these patients would not have been possible without the use of EEG monitoring.

VII. Treatment

There is no established protocol for the management of SE in the elderly patient. Treatment is complicated by the fact that the pharmacokinetics are more complex in the elderly than in younger patients because of the altered volume of distribution, lower protein binding, decreased renal elimination, decreased hepatic metabolism, decreased enzyme inducibility, and increased use of polypharmacy in the elderly.

A. First-line Therapy

Because the goal of treatment for SE is rapid cessation of clinical and electrical seizure activity, treatment should be initiated without delay after SE diagnosis for any convulsive seizure that has lasted at least 10 min or for repetitive seizures. Emergent attention should be given to establishing an airway, monitoring oxygenation and vital signs, establishing intravenous (IV) access, measuring the blood glucose level, and sending blood for analysis of blood count, serum electrolytes, and antiepileptic drug levels. A benzodiazepine administered intravenously to patients with active convulsions is considered first-line therapy because benzodiazepines are fast acting and effective for all seizure types (Leppik et al., 1983) (Table I). Lorazepam has a longer duration of anticonvulsive action than diazepam and, when administered by paramedics to adults for out-of-hospital SE, is significantly more effective than diazepam at terminating SE (Lowenstein et al., 1999a). However, it may cause a prolonged period of depressed consciousness. Because diazepam is more lipid soluble, it enters the brain quickly but also quickly redistributes to other fatty tissues; therefore, its therapeutic effect is brief (15–30 min) compared to the 12- to 24-h therapeutic effect of lorazepam. This decreased duration of effect may lead to recurrence of SE following a single diazepam dose.

TABLE I

ACUTE DRUG THERAPY FOR CONVULSIVE SE

Initial treatment
- Lorazepam (0.1 mg/kg IV at up to 2 mg/min)[a]
- Diazepam (0.15–0.25 mg/kg IV at up to 5 mg/min)[a]

If seizures persist
- Phenytoin (20 mg/kg IV at up to 50 mg/min)[b,c]
- Fosphenytoin (150 phenytoin-equivalents/min IV)[b,c]
- Phenobarbital (20 mg/kg IV at 50–75 mg/min)[d,c]

Refractory SE treatment
- Pentobarbital (5–12 mg/kg IV loading dose, followed by continuous infusion of 1–10 mg/kg/h)[d,e]
- Midazolam (0.2 mg/kg IV loading dose, followed by an infusion of 0.1–2.0 mg/kg/h)[d,e]
- Propofol (3–5 mg/kg IV, followed by maintenance infusion of 1–15 mg/kg/h)[b,e]

---

[a]Dodson et al. (1993).
[b]Phenytoin and fosphenytoin require blood pressure and electrocardiogram monitoring.
[c]Willmore (1998).
[d]These treatments require ventilatory support, hemodynamic monitoring in an intensive care unit, and electroencephalogram monitoring.
[e]Bleck (1999).
IV, intravenous; SE, status epilepticus.
Adapted from Waterhouse and DeLorenzo (2001) (Table 1, p. 139).

When seizures persist, or in order to maintain the antiseizure effect, it is usually recommended that a longer-acting drug, such as phenytoin or fosphenytoin, be administered (Dodson et al., 1993; Lowenstein et al., 1999a). The recommended loading dose of phenytoin is 20 mg/kg intravenously in adults and 15 mg/kg in the elderly (Dodson et al., 1993; Willmore, 1998). The rate of infusion must not exceed 50 mg/min because hypotension (28–50% of patients) and cardiac arrhythmias (2% of patients) can occur at higher rates (Cranford et al., 1978; Wilder, 1983). Blood pressure and cardiac rhythm must be monitored continuously, and if these adverse effects occur, the infusion rate should be slowed. Fosphenytoin, the phosphate ester of phenytoin, is rapidly converted to phenytoin after entering the bloodstream (Browne et al., 1996); thus, its doses are expressed as phenytoin-equivalents (the amount of phenytoin delivered; 150 mg of fosphenytoin is equivalent to 100 mg of phenytoin). It can be administered at rates of up to 150 phenytoin-equivalents per minute to give a therapeutic serum phenytoin concentration within 10 min. Unlike phenytoin, it is not formulated with propylene glycol; therefore a faster infusion rate is allowed, there is better tolerability with less local irritation at the infusion site, and there is less toxicity from the drug vehicle (Willmore, 1998). Fosphenytoin can be given intramuscularly, unlike phenytoin, with therapeutic levels reached in approximately 20–30 min (Fischer et al., 2003; Leppik et al., 1990). However, as with phenytoin, hypotension and cardiac arrhythmias are possible adverse events of fosphenytoin administration. Phenobarbital has also been used for years to treat SE with an adult dosage of 20 mg/kg, given at a rate of 50–75 mg/min. Sedation and apnea as well as hypotension are possible adverse events with this regimen (Willmore, 1998). Thus, especially in the elderly, monitoring cardiac rhythm and blood pressure is critical when administering any of these drugs.

In a large VA cooperative study (Treiman et al., 1998), patients with GCSE were randomly assigned to receive initial IV treatment with lorazepam, phenobarbital, phenytoin, or diazepam followed by phenytoin. This study included many elderly patients (median age, 58.6 years in the overt convulsive SE group and 62.0 years in the subtle convulsive SE group). In patients with overt generalized SE (defined as easily visible generalized convulsion), lorazepam treatment was successful in 64.9% of patients who received it, phenytoin in 43.6%, diazepam followed by phenytoin in 55.8%, and phenobarbital in 58.2%. In a pairwise comparison, the effectiveness of lorazepam was significantly superior to that of phenytoin ($p = 0.002$). Overall in this group, the success of first drug treatment was 55%. In patients with subtle SE (defined as having ictal discharges on the EEG and being comatose), no significant differences ($p = 0.91$) were observed between the treatment groups, and the overall success rate in this group was only 14.9%. There were no differences between the four treatments in the incidence of respiratory depression, cardiac rhythm disturbances, or hypotension requiring treatment.

B. Newer Therapy

IV valproic acid is another choice for treatment of SE in the elderly. In one European study of SE, which included patients with absence, tonic-clonic, myoclonic, and partial SE, 82.6% of the patients ($n = 19$) responded within 20 min to a 15 mg/kg IV valproic acid initial injection, which was followed 30 min later by an infusion at 1 mg/kg/h for approximately 5.5 h (Giroud et al., 1993). Loading doses of IV valproic acid as high as 60–70 mg/kg, followed by maintenance infusion, have also been demonstrated to be efficacious and safe in the treatment of SE (Bleck, 1999; Knake et al., 1999; Peters and Pohlmann-Eden, 1999; Price, 1989; Short et al., 1999). Safety and tolerability of rapid infusion rates have been established (Limdi and Faught, 1999). However, because of the rapid changes of total (unbound plus bound) and unbound valproic acid levels during rapid infusion, unbound valproic acid concentrations should be monitored during IV administration (Kriel et al., 1999). The low risks of hypotension, respiratory depression, and sedation make IV valproic acid a potentially advantageous choice for the treatment of SE in the elderly. When used in this population, smaller doses may be sufficient to achieve a given serum concentration, and the potential presence of inhibiting comedications needs to be considered (Pluhar *et al.*, 1999). More information is needed before the role of valproic acid in SE treatment is established (Bleck, 1999).

Levetiracetam is chemically unrelated to existing antiepileptic drugs, and the precise mechanisms by which it exerts its antiepileptic effect are unknown. In July 2006, it became available for IV administration. Levetiracetam may be useful in treating SE, but there are few studies available addressing its effectiveness in this condition. In a study of 13 episodes of SE that occurred in 12 patients who received levetiracetam enterally, with doses ranging between 1000 and 6000 mg (median, 3000 mg), three episodes probably responded to treatment (23%), one episode responded to treatment but the patient subsequently died (8%), four episodes showed no response (31%), and five episodes had an undetermined response (38%) (Rossetti and Bromfield, 2005). This clinical outcome was comparable to outcomes of other antiepileptic drug studies. Larger prospective studies are needed to determine the efficacy of the IV levetiracetam formulation.

C. Treatment of Refractory SE

Results from the large VA cooperative study showed that when the initial treatment of SE fails, little is gained by using additional standard drugs, with only 7% of overt convulsive SE and 3% of subtle SE cases responding to treatment with a second drug (Treiman *et al.*, 1999). General anesthetic agents such as phenobarbital, pentobarbital, midazolam, or propofol are being considered as treatment options for patients who continue to have seizures after initial therapy (Bleck, 1999; Lowenstein and Alldredge, 1998; Treiman *et al.*, 1999). These treatment choices will likely result in seizure control; however, there are risks associated with these options, including respiratory depression, hypotension, and secondary complications such as infection (Bleck, 1999). All SE patients receiving general anesthesia require intubation and mechanical ventilation with hemodynamic monitoring in an intensive care unit. These patients also require continuous EEG monitoring to document electrographic cessation of seizures (Bleck, 1999; Lowenstein and Alldredge, 1998; Willmore, 1998).

Traditionally, patients who continued to seize after treatment with a benzodiazepine and phenytoin were treated with phenobarbital. However, pentobarbital is preferable to phenobarbital because it has a shorter elimination half-life, allowing shorter duration of sedation and more timely assessment of the patient's baseline clinical mental status. An IV loading dose of 5–12 mg/kg is followed by continuous infusion of 1–10 mg/kg/h. Because pentobarbital causes respiratory depression, patients must be intubated before being placed in a coma (Bleck, 1999). While the patient is in a pentobarbital coma, EEG is the only reliable method to assess cessation of convulsive seizures; after a loading dose, the

EEG pattern should change from epileptiform patterns to either burst suppression or electrocerebral inactivity (Willmore, 1998). A number of studies have confirmed that although pentobarbital is effective for treating refractory SE, it is associated with a high mortality (Krishnamurthy and Drislane, 1996; Lowenstein et al., 1988; Mirski et al., 1995; Osorio and Reed, 1989; Rashkin et al., 1987; Stecker et al., 1998; Van Ness, 1990; Yaffe and Lowenstein, 1993).

Midazolam (Versed[®]), a short-acting benzodiazepine, appears to be an effective and a safe alternative to high-dose barbiturate coma for refractory SE termination (Kumar and Bleck, 1992; Parent and Lowenstein, 1994). It has several advantages, including rapid onset, short elimination half-life (allowing the avoidance of prolonged sedation), and less hypotension than is observed with barbiturates. Disadvantages include its high cost and tachyphylaxis, which can require dramatic dose escalation after 24–48 h (Bleck, 1999). Treatment with midazolam involves a loading dose of 0.2 mg/kg followed by an infusion of 0.1–2.0 mg/kg/h, titrated to seizure suppression as monitored by EEG (Bleck, 1999).

Propofol (Diprivan[®]), a gamma-aminobutyric acid $(GABA)_A$ agonist, can effectively and quickly terminate refractory SE (Kuisma and Roine, 1995; Stecker et al., 1998). A loading dose of 1–5 mg/kg administered over 5–10 min, followed by maintenance infusion of 1–15 mg/kg/h, is sufficient for seizure suppression (Bleck, 1999; Stecker et al., 1998). Because of its rapid clearance, propofol infusions should be tapered at a rate of 5% of the maintenance infusion per hour to prevent withdrawal seizures (Stecker et al., 1998). In a retrospective study of refractory SE patients, propofol and midazolam were both equally effective in seizure control, although midazolam had fewer side effects (Prasad et al., 2001).

Topiramate administered nasogastrically has been shown to be effective in terminating refractory SE (Bensalem and Fakhoury, 2003; Towne et al., 2003). Effective doses range from 300 to 1600 mg/day. Published case series have noted no adverse events resulting from the use of topiramate for SE, except for lethargy. More studies are needed before clear guidelines for use of this agent are established.

VIII. Conclusions

With the elderly composing the fastest growing segment of the US population, it is to be expected that SE will become more common, and physicians treating this condition will need to be better informed about its semiology and treatment in this age group. SE is associated with a high mortality in the older population, partially because of the increased susceptibility of these patients to systemic metabolic diseases, strokes, and other comorbidities that may lead to increased mortality. Along with etiology and age, duration of SE is a strong determinant of mortality. Because many older patients may not present with the typical signs of convulsive SE, the diagnosis may not be made until the patient has continued to seize for a prolonged period of time. Thus, it is imperative to have a high index of suspicion of NCSE in the older patient and to obtain an EEG so that rapid treatment can be instituted. Treatment needs to be instituted as soon as the diagnosis is made, using the more recent concept of duration, also keeping in mind that the pharmacokinetics are more complex in the elderly than in younger patients.

References

Bensalem, M. K., and Fakhoury, T. A. (2003). Topiramate and status epilepticus: Report of three cases. *Epilepsy Behav.* **4,** 757–760.

Bleck, T. P. (1999). Management approaches to prolonged seizures and status epilepticus. *Epilepsia* **40**(Suppl. 1), S59–S63.

Browne, T. R., Kugler, A. R., and Eldon, M. A. (1996). Pharmacology and pharmacokinetics of fosphenytoin. *Neurology* **46,** S3–S7.

Coeytaux, A., Jallon, P., Galobardes, B., and Morabia, A. (2000). Incidence of status epilepticus in French-speaking Switzerland: (EPISTAR). *Neurology* **55,** 693–697.

Cranford, R. E., Leppik, I. E., Patrick, B., Anderson, C. B., and Kostick, B. (1978). Intravenous phenytoin: Clinical and pharmacokinetic aspects. *Neurology* **28,** 874–880.

DeLorenzo, R. J. (1997). Clinical and epidemiologic study of status epilepticus in the elderly. *In* "Seizures and Epilepsy in the Elderly" (A. J. Rowan and R. E. Ramsay, Eds.), pp. 191–205. Butterworth-Heinemann, Boston.

DeLorenzo, R. J., Pellock, J. M., Towne, A. R., and Boggs, J. G. (1995). Epidemiology of status epilepticus. *J. Clin. Neurophysiol.* **12,** 316–325.

DeLorenzo, R. J., Hauser, W. A., Towne, A. R., Boggs, J. G., Pellock, J. M., Penberthy, L., Garnett, L., Fortner, C. A., and Ko, D. (1996). A prospective, population-based epidemiologic study of status epilepticus in Richmond, Virginia. *Neurology* **46,** 1029–1035.

DeLorenzo, R. J., Waterhouse, E. J., Towne, A. R., Boggs, J. G., Ko, D., DeLorenzo, G. A., Brown, A., and Garnett, L. (1998). Persistent nonconvulsive status epilepticus after the control of convulsive status epilepticus. *Epilepsia* **39,** 833–840.

Dodson, W. E., DeLorenzo, R. J., Pedley, T. A., Shinnar, S., Treiman, D. M., and Wannamaker, B. B. (1993). Treatment of convulsive status epilepticus: Recommendations of the Epilepsy Foundation of America's Working Group on Status Epilepticus. *JAMA* **270,** 854–859.

Fischer, J. H., Patel, T. V., and Fischer, P. A. (2003). Fosphenytoin: Clinical pharmacokinetics and comparative advantages in the acute treatment of seizures. *Clin. Pharmacokinet.* **42,** 33–58.

Gastaut, H. (1983). Classification of status epilepticus. *Adv. Neurol.* **34,** 15–35.

Giroud, M., Gras, D., Escousse, A., Dumas, R., and Venaud, G. (1993). Use of injectable valproic acid in status epilepticus: A pilot study. *Drug Invest.* **5,** 154–159.

Hauser, W. A., Annegers, J. F., and Kurland, L. T. (1993). Incidence of epilepsy and unprovoked seizures in Rochester, Minnesota, 1935–1984. *Epilepsia* **34,** 453–468.

Hesdorffer, D. C., Logroscino, G., Cascino, G., Annegers, J. F., and Hauser, W. A. (1998). Incidence of status epilepticus in Rochester, Minnesota, 1965–1984. *Neurology* **50,** 735–741.

Jakkampudi, V. V., Corrie, W. S., and DeLorenzo, R. J. (2005). How long are seizures that stop without acute therapy? [abstract]. *Epilepsia* **46**(Suppl. 8), 346. Abstract 3.202.

Kaplan, P. W. (1999). Assessing the outcomes in patients with nonconvulsive status epilepticus: Nonconvulsive status epilepticus is underdiagnosed, potentially overtreated, and confounded by comorbidity. *J. Clin. Neurophysiol.* **16**, 341–352; discussion 353.

Kellinghaus, C., Loddenkemper, T., Dinner, D. S., Lachhwani, D., and Luders, H. O. (2004). Seizure semiology in the elderly: A video analysis. *Epilepsia* **45**, 263–267.

Knake, S., Vescovi, M., Hamer, H. H., Wirbatz, A., and Rosenow, F. (1999). Intravenous sodium valproate in the treatment of status epilepticus. *Epilepsia* **40**(Suppl. 7), 150. Abstract 2.269.

Knake, S., Rosenow, F., Vescovi, M., Oertel, W. H., Mueller, H. H., Wirbatz, A., Katsarou, N., and Hamer, H. M. (2001). Incidence of status epilepticus in adults in Germany: A prospective, population-based study. *Epilepsia* **42**, 714–718.

Kriel, R. L., Birnbaum, A. K., Norberg, S. K., Wical, B. S., Le, D. N., and Cloyd, J. C. (1999). Pharmacokinetics of total and unbound valproate after rapid infusion of Depacon [abstract]. *Epilepsia* **40**(Suppl. 7), 232.

Krishnamurthy, K. B., and Drislane, F. W. (1996). Relapse and survival after barbiturate anesthetic treatment of refractory status epilepticus. *Epilepsia* **37**, 863–867.

Kuisma, M., and Roine, R. O. (1995). Propofol in prehospital treatment of convulsive status epilepticus. *Epilepsia* **36**, 1241–1243.

Kumar, A., and Bleck, T. P. (1992). Intravenous midazolam for the treatment of refractory status epilepticus. *Crit. Care Med.* **20**, 483–488.

Labar, D., Barrera, J., Solomon, G., and Harden, C. (1998). Nonconvulsive status epilepticus in the elderly: A case series and a review of the literature. *J. Epilepsy* **11**, 74–78.

Lee, S. I. (1985). Nonconvulsive status epilepticus. Ictal confusion in later life. *Arch. Neurol.* **42**, 778–781.

Leppik, I. E., Derivan, A. T., Homan, R. W., Walker, J., Ramsay, R. E., and Patrick, B. (1983). Double-blind study of lorazepam and diazepam in status epilepticus. *JAMA* **249**, 1452–1454.

Leppik, I. E., Boucher, B. A., Wilder, B. J., Murthy, V. S., Watridge, C., Graves, N. M., Rangel, R. J., Rask, C. A., and Turlapaty, P. (1990). Pharmacokinetics and safety of a phenytoin prodrug given i.v. or i.m. in patients. *Neurology* **40**, 456–460.

Limdi, N. A., and Faught, E. (1999). Safety of rapid infusion of intravenous valproic acid [abstract]. *Epilepsia* **40**(Suppl. 7), 144. Abstract 2.243.

Litt, B., Wityk, R. J., Hertz, S. H., Mullen, P. D., Weiss, H., Ryan, D. D., and Henry, T. R. (1998). Nonconvulsive status epilepticus in the critically ill elderly. *Epilepsia* **39**, 1194–1202.

Lowenstein, D. H., and Alldredge, B. K. (1998). Status epilepticus. *N. Engl. J. Med.* **338**, 970–976.

Lowenstein, D. H., and Aminoff, M. J. (1992). Clinical and EEG features of status epilepticus in comatose patients. *Neurology* **42**, 100–104.

Lowenstein, D. H., Aminoff, M. J., and Simon, R. P. (1988). Barbiturate anesthesia in the treatment of status epilepticus: Clinical experience with 14 patients. *Neurology* **38**, 395–400.

Lowenstein, D. H., Alldredge, B. K., Gelb, A. M., Isaacs, S. M., Corry, M. D., Allen, F., O'Neil, N., Gottwald, M. D., Ulrich, S. K., Neuhaus, J. M., and Segal, M. R. (1999a). Results of a controlled trial of benzodiazepines for the treatment of status epilepticus in the prehospital setting [abstract]. *Epilepsia* **40**(Suppl. 7), 243.

Lowenstein, D. H., Bleck, T., and Macdonald, R. L. (1999b). It's time to revise the definition of status epilepticus. *Epilepsia* **40**, 120–122.

Marsan, C. A., and Zivin, L. S. (1970). Factors related to the occurrence of typical paroxysmal abnormalities in the EEG records of epileptic patients. *Epilepsia* **11**, 361–381.

McBride, A. E., Shih, T. T., and Hirsch, L. J. (2002). Video-EEG monitoring in the elderly: A review of 94 patients. *Epilepsia* **43**, 165–169.

Mirski, M. A., Williams, M. A., and Hanley, D. F. (1995). Prolonged pentobarbital and phenobarbital coma for refractory generalized status epilepticus. *Crit. Care Med.* **23,** 400–404.

Osorio, I., and Reed, R. C. (1989). Treatment of refractory generalized tonic-clonic status epilepticus with pentobarbital anesthesia after high-dose phenytoin. *Epilepsia* **30,** 464–471.

Parent, J. M., and Lowenstein, D. H. (1994). Treatment of refractory generalized status epilepticus with continuous infusion of midazolam. *Neurology* **44,** 1837–1840.

Peters, C. N. A., and Pohlmann-Eden, B. (1999). Efficacy and safety of intravenous valproate in status epilepticus [abstract]. *Epilepsia* **40**(Suppl. 7), 149–150. Abstract 2.267.

Pluhar, J. M., Birnbaum, A. K., Graves, N. M., Hardie, N. A., Lackner, T., Krause, S. E., and Leppik, I. E. (1999). Valproic acid serum concentrations and daily dose in elderly nursing home residents: Effects of age, gender, and concomitant medications [abstract]. *Epilepsia* **40**(Suppl. 7), 77. Abstract B.07.

Prasad, A., Worrall, B. B., Bertram, E. H., and Bleck, T. P. (2001). Propofol and midazolam in the treatment of refractory status epilepticus. *Epilepsia* **42,** 380–386.

Price, D. J. (1989). Intravenous valproate: Experience in neurosurgery. *In* "Fourth International Symposium on Sodium Valproate and Epilepsy. Royal Society of Medicine International Congress and Symposium Series No. 152" (D. Chadwick, Ed.), pp. 197–203. Royal Society of Medicine Services Limited, London.

Privitera, M., Hoffman, M., Moore, J. L., and Jester, D. (1994). EEG detection of nontonic-clonic status epilepticus in patients with altered consciousness. *Epilepsy Res.* **18,** 155–166.

Ramsay, R. E., and Pryor, F. (2000). Epilepsy in the elderly. *Neurology* **55**(5 Suppl. 1), S9–S14; discussion S54–S58.

Rashkin, M. C., Youngs, C., and Penovich, P. (1987). Pentobarbital treatment of refractory status epilepticus. *Neurology* **37,** 500–503.

Rossetti, A. O., and Bromfield, E. B. (2005). Levetiracetam in the treatment of status epilepticus in adults: A study of 13 episodes. *Eur. Neurol.* **54,** 34–38.

Short, D. W., Pack, A. M., and Bazil, C. W. (1999). Depacon administration, safety, and efficacy in a hospital setting [abstract]. *Epilepsia* **40**(Suppl. 7), 226. Abstract 3.248.

Stecker, M. M., Kramer, T. H., Raps, E. C., O'Meeghan, R., Dulaney, E., and Skaar, D. J. (1998). Treatment of refractory status epilepticus with propofol: Clinical and pharmacokinetic findings. *Epilepsia* **39,** 18–26.

Sung, C. Y., and Chu, N. S. (1989). Status epilepticus in the elderly: Etiology, seizure type and outcome. *Acta Neurol. Scand.* **80,** 51–56.

Towne, A. R., Pellock, J. M., Ko, D., and DeLorenzo, R. J. (1994). Determinants of mortality in status epilepticus. *Epilepsia* **35,** 27–34.

Towne, A. R., Waterhouse, E. J., Boggs, J. G., Garnett, L. K., Brown, A. J., Smith, J. R., Jr., and DeLorenzo, R. J. (2000). Prevalence of nonconvulsive status epilepticus in comatose patients. *Neurology* **54,** 340–345.

Towne, A. R., Garnett, L. K., Waterhouse, E. J., Morton, L. D., and DeLorenzo, R. J. (2003). The use of topiramate in refractory status epilepticus. *Neurology* **60,** 332–334.

Treiman, D. M., Meyers, P. D., Walton, N. Y., Collins, J. F., Colling, C., Rowan, A. J., Handforth, A., Faught, E., Calabrese, V. P., Uthman, B. M., Ramsay, R. E., and Mamdani, M. B. (1998). A comparison of four treatments for generalized convulsive status epilepticus. Veterans Affairs Status Epilepticus Cooperative Study Group. *N. Engl. J. Med.* **339,** 792–798.

Treiman, D. M., Walton, N. Y., and Collins, J. F. (1999). Treatment of status epilepticus if first drug fails. *Epilepsia* **40**(Suppl. 7), 243. Abstract J.02.

US Department of Health and Human Services, Administration on Aging (2005). Older population by age: 1900 to 2050. Available at: http://www.aoa.gov/prof/Statistics/online_stat_data/AgePop2050.asp.

Van Cott, A. C. (2002). Epilepsy and EEG in the elderly. *Epilepsia* **43**(Suppl. 3), 94–102.

Van Ness, P. C. (1990). Pentobarbital and EEG burst suppression in treatment of status epilepticus refractory to benzodiazepines and phenytoin. *Epilepsia* **31,** 61–67.

Vignatelli, L., Tonon, C., and D'Alessandro, R. (2003). Incidence and short-term prognosis of status epilepticus in adults in Bologna, Italy. *Epilepsia* **44,** 964–968.

Waisterlain, C. G., and Chen, J. W. (2006). Definition and classification of status epilepticus. *In* "Status Epilepticus, Mechanisms and Management" (C. G. Waisterlain and D. M. Treiman, Eds.), pp. 11–16. The MIT Press, Cambridge, MA.

Waterhouse, E. J., and DeLorenzo, R. J. (2001). Status epilepticus in older patients: Epidemiology and treatment options. *Drugs Aging* **18,** 133–142.

Waterhouse, E. J., Vaughan, J. K., Barnes, T. Y., Boggs, J. G., Towne, A. R., Kopec-Garnett, L., and DeLorenzo, R. J. (1998). Synergistic effect of status epilepticus and ischemic brain injury on mortality. *Epilepsy Res.* **29,** 175–183.

Wilder, B. J. (1983). Efficacy of phenytoin in treatment of status epilepticus. *Adv. Neurol.* **34,** 441–446.

Willmore, L. J. (1998). Epilepsy emergencies: The first seizure and status epilepticus. *Neurology* **51**(5 Suppl. 4), S34–S38.

Wu, Y. W., Shek, D. W., Garcia, P. A., Zhao, S., and Johnston, S. C. (2002). Incidence and mortality of generalized convulsive status epilepticus in California. *Neurology* **58,** 1070–1076.

Yaffe, K., and Lowenstein, D. H. (1993). Prognostic factors of pentobarbital therapy for refractory generalized status epilepticus. *Neurology* **43,** 895–900.

DIAGNOSING EPILEPSY IN THE ELDERLY

R. Eugene Ramsay,*[,†] Flavia M. Macias,[†] and A. James Rowan[‡,§]

*International Center for Epilepsy, Department of Neurology
University of Miami School of Medicine, Miami, Florida 33136, USA
[†]Department of Neurology, Miami VA Medical Center, Miami, Florida 33125, USA
[‡]Department of Neurology, Bronx VA Medical Center, Bronx, New York 10468, USA
[§]Mount Sinai School of Medicine, New York, New York 10029, USA

Elderly individuals represent the fastest-growing segment of the US population. Seizures are common among elderly persons, and the etiology, clinical presentation, and prognosis of seizure disorders can often differ between elderly patients and younger individuals. However, published information regarding the diagnosis and management of epilepsy in elderly patients is scarce. Because a number of conditions that are common in elderly patients may resemble epilepsy, diagnosis can be challenging. Cardiovascular conditions, migraines, drug effects, infections, metabolic disturbances, sleep disorders, and psychiatric disorders are all associated with signs and symptoms that may often mimic epilepsy. New paradigms must be put into practice to establish an accurate diagnosis in the elderly patient; besides an initial evaluation, the patient history and an electroencephalogram should be obtained. Proper diagnosis is essential for proper treatment in the elderly patient.

I. Introduction

The etiology, clinical presentation, and prognosis of seizure disorders differ considerably between the elderly and younger patients. Concurrent illnesses and the physiological changes that accompany aging can increase the risk of these seizure disorders and profoundly alter the response to drug therapy, often in unpredictable ways. Anticipating and recognizing these differences in the elderly are becoming increasingly important, as patients over 65 years old are the fastest-growing segment of the US population (Willmore, 1995); according to the US Census Bureau, an average of 7918 individuals will turn 60 years old every day in 2006 (http://www.sec.gov/news/press/extra/seniors/agingboomers.htm). This is the adult segment with the highest incidence of unprovoked seizures (Hauser *et al.*, 1993). The incidence of epilepsy in persons aged 65 years and over approximately doubled from the years 1935 to 1984—a trend expected to continue as the US population ages (Hauser *et al.*, 1993). This is speculated to be a result of better diagnosis in elderly patients and longer-term survival after acute neurological insults (e.g., stroke). Despite this finding, relatively little information has been published on the diagnosis and management of epilepsy in elderly individuals. Because the clinical manifestations of epilepsy can differ between older patients and younger patients, and the diagnosis may be missed, the actual incidence may be two- to threefold higher than previously reported rates.

The largest double-blind, randomized, controlled clinical trial treating new-onset geriatric seizure patients was the Veterans Affairs Cooperative Study 428 (VACS 428) (Ramsay *et al.*, 2004; Rowan *et al.*, 2005). That study was conducted in 18 centers across the United States and enrolled 593 patients. Refer to the chapter by Kraemer in this volume for details on methodology. Some of the results from that study are included in this chapter to highlight the difficulties of diagnosing epilepsy in geriatric patients.

II. Etiology

Any disorder that affects the brain can cause seizures. The etiology in approximately one-third of geriatric patients with recent-onset seizures is classified as unknown (Ramsay *et al.*, 2004), although the true figure may be far lower. Most cryptogenic seizures likely have an etiology that can be identified using better resolution magnetic resonance imaging (MRI) and further clinical evaluation. The most common cause of seizures in the elderly is cerebrovascular disease, followed by degenerative diseases and trauma; fewer seizures are attributable to neoplasms and other abnormalities (Fig. 1) (Annegers *et al.*, 1995; Hauser *et al.*, 1993; Rowan *et al.*, 2005). Results from a large population-based study showed that cerebral infarction alone raises the risk of initial late seizures in the first year after the infarct to 23 times that of the general population (So *et al.*, 1996). Common risk factors are often overlooked. For example, hypertension alone, even without clinical stroke, is a significant risk factor for unprovoked seizures [odds ratio (OR), 1.35] (Ng *et al.*, 1993). In patients with a history of hypertension, a previously occurring stroke was a seizure risk factor (adjusted OR, 4.81) (Ng *et al.*, 1993); however, many individuals with hypertension alone have had subclinical strokes (Thomas, 1997). In Hauser's epidemiological study of status epilepticus, a definitive etiology was identified in only approximately 50% of patients (Hauser, 1990). Results from VACS 428 showed that, although fewer patients were classified as having unknown etiology (24.6%), most patients had a stroke (50.9%) or had other risk factors for cerebrovascular disease such as dyslipidemia (80%), hypertension (65.9%), and diabetes (28.4%) (Ramsay *et al.*, 2004). Approximately 18% had normal imaging studies, and most were enrolled in the study before MRI scans became readily available. In the absence of overt cortical infarct, almost all the remaining patients had image (and often clinical) evidence of subcortical small vessel disease. Overall, the results indicate that vascular disease plays a significant role in the majority of elderly patients with new-onset seizures.

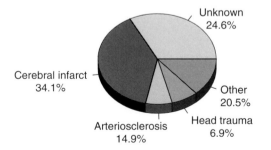

FIG. 1. Etiology of newly diagnosed epilepsy in patients ≥60 years old. From Ramsay *et al.* (2004), with permission. (See Color Insert.)

III. Clinical Manifestations

Older patients with epilepsy usually do not present with classic features of complex partial seizures (CPS) as seen in younger adults. More often, the seizure etiology is not appreciated and the patient is diagnosed as having altered mental status, memory lapses, episodes of confusion, or syncope (Ramsay *et al.*, 2004). Clinical characteristics of seizures in the general adult population and the older population are compared in Table I. Classic CPS in younger adults usually begin with an aura, followed by a disturbance of consciousness and a behavioral arrest. Subsequently, limb and oral-facial automatisms are observed. The postictal period of confusion is usually very brief, lasting 5–15 min before a return to normal. These manifestations are typical of seizures originating in the mesial-temporal structures. In contrast, an older patient's seizure focus is likely to be extratemporal, most often in the frontal lobe (Ramsay *et al.*, 2004). Strokes rarely involve the mesial-temporal areas, but rather the motor and sensory cortices, with more involvement of the anterior frontal cortical areas (Fig. 2) (Ramsay *et al.*, 2004). This accounts for the differences in the clinical presentation in older patients. In the elderly, auras are much less common than in younger patients and, when present, may be very nonspecific, such as a report of dizziness. The occurrence of secondarily generalized convulsion is lower in older patients (25.9%), compared to younger adults (65%) (Cloyd *et al.*, 2006). A disturbance of consciousness with a blank stare may be the only manifestation, as automatisms are usually not present. The lower incidence of convulsions, along with their unique clinical presentation, contributes to under-identification of epilepsy in older patients. The postictal confusional state, which may last for hours, days, or even more than a week in the older patient, is noticeably longer than in younger adults. When this lengthy confusional state

TABLE I
CLINICAL CHARACTERISTICS OF SEIZURES IN GENERAL ADULT AND OLDER PATIENT POPULATIONS

Clinical feature	General adult population	Older population
Aura	50%	Infrequent
Ictus		
Disorders of consciousness or stare	Yes	Yes
Automatisms	Oral facial	Infrequent
	Hand rubbing	
Postictal state		
Confusion	5–15 min	Prolonged
		(hours to days)

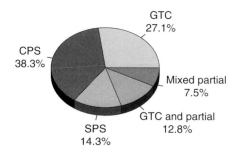

FIG. 2. Frequency of seizure types in patients ≥60 years old. CPS, complex partial seizures; GTC, generalized tonic-clonic; SPS, simple partial seizures. From Ramsay *et al.* (2004), with permission. (See Color Insert.)

happens, the patient is often felt to have a dementia and may not be diagnosed with epilepsy (Godfrey *et al.*, 1982).

IV. Differential Diagnosis

Presenting symptoms in the elderly may not strongly suggest epilepsy as a diagnosis. Only 73% of the patients in the VACS 428 who were ultimately diagnosed as having epilepsy had epilepsy as a diagnostic consideration after their initial evaluation by a primary care physician or internist (Ramsay *et al.*, 2004). The most common diagnoses were blackout spells (29.3%), syncope (16.8%), altered mental status (41.8%), and confusion (37.5%) (Ramsay *et al.*, 2004). Thus, a significant number of older patients with new-onset epilepsy are either not diagnosed or the diagnosis is significantly delayed.

Confusion is a state in which there is a global impairment of cognitive function with only minor or no alterations in the level of consciousness. The patient will have elemental responses, simple mental functions, and reactions to ordinary commands intact. There is impaired capacity to think clearly and with customary rapidity. Perception, response, and recollection of current stimuli are also impaired (DeJong, 1967). There are a multitude of underlying causes for confusion in the elderly that should be considered (Table II). When an elderly person presents with periodic or transient confusion, the possibility of seizures must be considered until proven otherwise. The ictal pattern in older patients may be only behavioral arrest, and the lack of response may often be interpreted by family members as confusion. Postictal confusion can persist for hours to days without a seizure being observed (Table I) (Godfrey *et al.*, 1982). Common diagnostic possibilities for these symptoms include metabolic disorders, toxins, migraine, sepsis, and psychiatric disorders (Table II).

TABLE II

DIFFERENTIAL DIAGNOSIS OF CONFUSION IN THE ELDERLY

Cause of confusion	Clinical presentation
Cardiovascular	Conditions with poor cardiac output
	Cardiac arrhythmias
	Congestive heart failure
	Hypotension
	Transient ischemic attack
	Broca aphasia
	Nondominant parietal lobe syndrome
	Transcortical sensory aphasia
	Wernicke aphasia
Confusional migraine	Auras >5 min
	Nausea and vomiting
	Prolonged confusion
Drug intoxication	Prescription medications
	Benzodiazepine, barbiturate, and so on
	Drugs of abuse
Infection	Encephalitis
	Human immunodeficiency virus and slow viruses
	Parasites
	Sepsis and fever
Metabolic disturbances	Hypoglycemia
	Hyperglycemia
	Thyroid storm
	Hypercapnia
	Porphyria
	Toxic metabolic encephalopathies
Sleep disorders and syncope	Abrupt loss of consciousness and tone
Psychiatric disorders	Depression
	Dissociative states
	Fugue state
	Bipolar disorder
	Anxiety disorders
	Posttraumatic stress disorder (PTSD)
	Ganser syndrome
	Nonepileptic seizures
Transient global amnesia	Abrupt memory loss
	Follows multiple-step commands
	Fluent speech
	Appears confused
Dementia	Orofacial movements
	Temper outbursts
	Wandering
	Memory lapses
Seizures	Ictal confusion
	Postictal confusion

V. Cardiovascular and Cerebrovascular Diseases

Transient ischemic attacks (TIAs) and strokes are commonly considered as possible etiologies, but a focal vascular insult cannot produce confusion. During or after a stroke, a patient may present with Wernicke or Broca aphasia and initially may appear to be confused. However, the speech and comprehension deficits clearly define these conditions as resulting from a focal cortical dysfunction and not confusional states. Conditions that result in low cardiac output with global cerebral hypoperfusion (e.g., arrhythmias, congestive heart failure, hypovolemia) or impaired oxygen delivery (marked anemia or chronic obstructive pulmonary disease) may result in confusion without focal neurological deficits.

TIAs are usually easy to differentiate from epileptic seizures. A TIA usually appears as a temporary loss of function such as hemiparesis, hemisensory deficit, or even aphasia. Seizures with such negative motor phenomena are extremely uncommon. Moreover, TIAs may have longer durations than typical seizures. One difficult situation is a TIA characterized by aphasia. Some patients with TIAs have a history of a left hemisphere cerebrovascular accident in which waxing and waning of an associated aphasia is common (Solenski, 2004). On the other hand, aphasic seizures can also occur as a result of an old infarction. In this case, the differential diagnosis is problematic. The electroencephalogram (EEG) may reveal a spike focus, making a seizure diagnosis more likely (Rosenbaum et al., 1986). A reasonable clinical diagnosis can also be based on the progress of the aphasic episode. Aphasia associated with epilepsy tends to increase over a period of perhaps several minutes (Rosenbaum et al., 1986). Aphasia resulting from a TIA tends to be sudden and without change until gradual recovery ensues (Solenski, 2004). The importance of a detailed history is again evident.

VI. Confusional Migraine

Epilepsy and migraine are highly comorbid with much symptomatic overlap. In both of these chronic conditions, patients exhibit recurrent neurological attacks with possible headache, gastrointestinal symptoms, autonomic symptoms, and psychological features (Bigal et al., 2003). In particular, confusional migraine may be difficult to differentiate from epilepsy in some cases. Patients with confusional migraine usually exhibit typical migraine aura, headache (which may or may not be prominent), and a confusional state before or after the headache (Bigal et al., 2003). The period of confusion is often followed by sleep, and thus may mimic the postical state of a patient with epilepsy (Bigal et al., 2003). However, several clues can help clinicians distinguish between migraine and epilepsy. Obtaining a patient and family history is important, as patients with confusional migraine frequently have a history of migraines without a confusional state (Evans and Gladstein, 2003), and patients with migraines often have a strong family history of migraines (Bigal et al., 2003). Migraines tend to be associated with longer auras (>5 min), more gradual onset, and longer duration than seizures (Bigal et al., 2003; Velioglu and Ozmenoglu, 1999). Nausea and vomiting are more common with migraine than with seizures, whereas periods of prolonged lethargy and confusion following the attack are seen more frequently in patients with seizures than in those with migraine (Bigal et al., 2003).

VII. Drug Use

Special attention should be given to the fact that various classes of drugs can provoke or worsen seizures (Waterhouse, 2005). These include antidepressants, antipsychotics, analgesics, anesthetics, antimicrobials, and bronchodilators, all of which are commonly prescribed in older patients (Thomas, 1997; Willmore, 1996). The aging brain is particularly vulnerable to medication side effects such as lethargy and confusion. Patients participating in the VACS 428 were prescribed an average of 6.7 medications. An estimated 10% of seizures that occur in elderly individuals are associated with the use of prescription drugs or alcohol (Waterhouse, 2005). Additionally, some sedative drugs, including benzodiazepines and barbiturates, are associated with seizures after withdrawal from chronic use (Leppik, 2000). Baclofen may also cause seizures after withdrawal (Kofler and Leis, 1992). Because patients may not have measurable amounts of these drugs in their systems at the time of seizure occurrence, a drug history is important for assessment of the patient.

Alcohol withdrawal may also cause seizures in those who abuse it (Leppik, 2000). Approximately 5% of elderly individuals are estimated to drink alcohol heavily and frequently (Hauser, 1997). Once again, a careful patient history is important in determining whether seizures may be related to alcohol withdrawal. Other drugs of abuse, including central nervous system (CNS) stimulants such as cocaine and amphetamines, can cause acute seizures (Leppik, 2000). In patients who experience seizures related to such substances, measurable concentrations of the drug(s) may be present in the blood or urine at the time of seizure occurrence (Leppik, 2000). An EEG is helpful in differentiating possible nonconvulsive seizures from encephalopathic fluctuations.

VIII. Infection

A number of infections are associated with seizures or may mimic symptoms of epilepsy. Elderly patients may be more susceptible to infection than are younger patients, in part because of increased incidence of cancer, diabetes, chronic obstructive pulmonary disease, and cardiovascular disease (Holloway, 1986). Elderly patients are often immunocompromised, and the use of steroids for chronic inflammatory conditions can contribute to infection. Sepsis and fever from infections can cause confusion (http://www.merck.com/mmpe/sec06/ch068/ch068a.html), which may be mistaken for postictal confusion, and sepsis has been associated with encephalopathy and seizures (Jackson *et al.*, 1985). Laboratory tests, including blood chemistries, complete blood count with differential, blood cultures, and urinalysis with culture, are useful in determining if an infection is the cause of a patient's confusion (http://www.merck.com/mmpe/sec06/ch068/ch068a.html).

Other systemic diseases can present with lethargy and confusion due to fever and the toxic effects of infection. In most cases, the EEG reveals a disorganized background along with bifrontal delta activity. In others, generalized epileptiform activity may be recorded, suggesting the possibility of nonconvulsive seizure activity. The seizure risk is increased if the patient has a preexisting cerebral lesion such as an infarct.

Seizures are common manifestations of encephalitis and meningitis that may occur during the acute infection stage or at a later time (Lewin *et al.*, 2005; Wang *et al.*, 2005). Because many patients with CNS infections later develop epilepsy, with estimates ranging widely from 18% to 80% (Lancman and Morris, 1996), a careful patient history is important to ascertain whether a prior CNS infection could be contributing to seizure activity. In patients who have seizures with symptoms that suggest CNS infection (e.g., fever, altered mental status, nuchal rigidity), a lumbar puncture should be performed, and the cerebrospinal fluid (CSF) should be analyzed (Shneker and Fountain, 2003).

Seizures may also be associated with parasitic infections, including cerebral malaria, cysticercosis, filariasis, toxoplasmosis, and onchocerciasis (Druet-Cabanac *et al.*, 2004; Mirdha, 2003). The gold standard for diagnosis of a parasitic infection includes isolation of the parasite or direct demonstration of the parasite from affected tissues; because this process is often unsuccessful, serological investigations may be required to establish a parasitic infection diagnosis (Mirdha, 2003).

The number of older individuals in the United States with human immunodeficiency virus (HIV) or acquired immune deficiency syndrome (AIDS) is growing. Approximately 19% of individuals with HIV or AIDS are 50 years or older (http://www.niapublications.org/agepages/aids.asp); this number was only 5–6% in 1999 (ftp://ftp.hrsa.gov/hab/hrsa2-01.pdf). Seizures in patients with HIV can occur at various stages of the infection but usually occur later in the disease process (Holtzman *et al.*, 1989; Romanelli and Ryan, 2002). Such seizures are commonly attributable to intracranial mass lesions, which may result from opportunistic infections (e.g., toxoplasmosis, herpes zoster, neurosyphilis), neoplasms, or cerebrovascular diseases (Garg, 1999). Cerebral HIV infection may also directly cause seizures (Garg, 1999). The incidence of HIV infection in the elderly is increasing; there were approximately 11,000 cases of HIV in 2004 (Centers for Disease Control and Prevention, 2005). The diagnostic workup for patients with new-onset seizures who have or are at risk for HIV infection includes a careful history and neurological examination (with special attention to changes in cognition), MRI or computed tomography, CSF analysis, and blood chemistry analysis (Garg, 1999; Holtzman *et al.*, 1989).

Seizures in elderly patients may also be attributed to slow virus infections, also known as prion diseases. The major prion diseases in humans include Creutzfeldt–Jakob disease (CJD), Gerstmann–Straussler–Scheinker disease, fatal familial insomnia, and kuru (http://professionals.epilepsy.com/page/infectious_slow.html), of which the most common is CJD (Brandel, 1999). Less than 20% of patients with CJD exhibit seizures, and seizures are even less common with the other slow virus infections (http://professionals.epilepsy.com/page/infectious_slow.html). Seizures in patients with CJD may be generalized or partial, usually occur in the later stages of the illness, and are typically resistant to treatment with antiepileptic drugs (AEDs) (Clenney, 1983; Maltete *et al.*, 2006). Most patients with CJD (82–100%) exhibit myoclonus, which may be focal or generalized, that is often asymmetric and may be precipitated by noise or other stimuli (startle myoclonus) (Maltete *et al.*, 2006). Rapidly progressive dementia and ataxia are also common in patients with CJD (Lee *et al.*, 2000). Evaluation of patients with possible CJD should include analysis of the CSF for the 14-3-3 protein and the protease-resistant prion protein (Lee *et al.*, 2000). Additionally, EEG monitoring in patients with CJD often shows slowing of background rhythms, along with periodic sharp wave complexes (1–2 Hz) (Brandel, 1999; Lee *et al.*, 2000).

IX. Metabolic Disturbances

A number of metabolic disturbances can produce seizures or mimic symptoms of epilepsy, and they may be responsible for up to 10–15% of seizures in elderly individuals (Waterhouse, 2005).

A. HYPERGLYCEMIA AND HYPOGLYCEMIA

Hyperglycemia is a common metabolic disturbance associated with seizures. Most metabolic and toxic conditions provoke generalized convulsive seizures. Nonketotic hyperglycemia is also associated with partial seizures, as well as a variety of other movement disorders (e.g., Todd's paralysis, hemiparesis, myoclonus, tonic eye deviation, nystagmus, opsoclonus, hyperreflexia, hyporeflexia, nuchal rigidity, choreoathetosis, ballism) (Morres and Dire, 1989). Seizures associated with hyperglycemia are resistant to AED therapy; the underlying metabolic disturbance must be treated instead (Morres and Dire, 1989). Simple, inexpensive testing methods (e.g., urine dip sticks, blood glucose reagent strips) can be employed to rule out hyperglycemia as a cause of seizures (Morres and Dire, 1989). Seizures can also be precipitated by hypoglycemia (blood glucose levels <40 mg/dl). As with hyperglycemia, seizures associated with hypoglycemia are unresponsive to AED therapy, and the underlying metabolic disturbance must be treated (Leppik, 2000).

B. THYROID STORM

Thyroid storm refers to a relatively rare endocrine emergency caused by life-threatening levels of thyroid hormones in the blood (Bindu *et al.*, 2005; Malchiodi, 2002). Signs and symptoms of thyroid storm include hyperthermia, tachycardia, muscle weakness/fatigue, fever, diarrhea, confusion, delirium, coma, and tremor, as well as rare cases of seizures and status epilepticus (Bindu *et al.*, 2005; Lee *et al.*, 1997; Malchiodi, 2002). Patients who develop thyroid storm typically have a history of hyperthyroidism, and patients seem to be at the greatest risk for developing thyroid storm within the first 4 years after hyperthyroidism is diagnosed (Malchiodi, 2002). The thyroid storm is often precipitated by events such as infection, surgery, discontinuation of antithyroid drug therapy, or extreme emotional stress (Malchiodi, 2002). If seizures are thought to be associated with thyroid storm, patient history and thyroid function tests are helpful in the diagnosis (Malchiodi, 2002).

C. HYPERCAPNIA

Hypercapnia (greater than normal blood carbon dioxide levels) has been proposed to account for a subset of idiopathic status epilepticus when no prior history of epilepsy exists (Legriel and Mentec, 2005). Acute hypercapnia may be associated with subtle changes in personality, headache, narcosis, and confusion (http://www.merck.com/mmpe/sec12/ch157/ch157e.html). With increased age, the ventilatory response to hypercapnia decreases (Janssens, 2005). Therefore, the risk of seizures due to hypercapnia may also increase with age.

D. PORPHYRIA

The porphyrias are a group of inherited metabolic disorders caused by defects in the enzymes involved in heme biosynthesis that result in the accumulation of porphyrin or porphyrin precursors in the blood (Solinas and Vajda, 2004).

Acute attacks of porphyria can be precipitated by alcohol or other drugs, infections, decreased caloric intake, or menstruation (Bylesjo *et al.*, 1996). Such attacks are marked by clinical features that include abdominal pain, tachycardia, hypertension, vomiting, constipation, confusion, anxiety, hallucinations, and seizures (Bylesjo *et al.*, 1996; Gordon, 1999). Up to 20% of adults with porphyria may experience associated seizures (Bylesjo *et al.*, 1996; Solinas and Vajda, 2004). CPS with or without secondary generalization are the most common seizure type associated with porphyria, but myoclonic, absence, and tonic-clonic seizures have also been reported (Solinas and Vajda, 2004). Seizures in patients with porphyria are believed to be caused by associated metabolic imbalances (e.g., hyponatremia), but a direct epileptogenic effect of δ-aminolevulinic acid (a porphyrin precursor) is also possible (Solinas and Vajda, 2004). Because the porphyrias are heritable, a confirmed family history of porphyria is extremely useful for diagnosis. Measurement of porphyrins and their precursors in the blood, urine, and feces also helps in identifying patients with porphyria (Gordon, 1999; Sies and Florkowski, 2006). Cortical damage due to porphyric attacks can lead to further seizures that do not coincide with attacks of porphyria. These seizures can occur in either patients without a history of seizures or in patients who have a history of seizures only associated with porphyric attacks. In such cases, the cause of seizures may be elucidated through video-EEG or neuroimaging techniques (Solinas and Vajda, 2004).

E. Toxic Metabolic Encephalopathies

Toxic metabolic encephalopathies are the primary examples in this group of metabolic disturbances presenting with seizures. Although there would appear to be no difficulty in diagnosing these conditions, this is not necessarily the case. These patients present with alterations in mental status ranging from confusion to coma (Harner and Katz, 1974). Fluctuating awareness is common, suggesting a paroxysmal disturbance. However, many individuals presenting with these symptoms suffer from hepatic or renal insufficiency, and the natural assumption would be that the underlying disorder is responsible. Laboratory studies may reveal typical findings (such as an elevated blood urea nitrogen/creatinine level), with little change from previous known values in older patients. In older patients, impaired water balance and electrolyte disturbances (e.g., hyponatremia) may produce confusion (Harner and Katz, 1974), particularly when diuretics are prescribed. In these cases, the EEG is of great value; during hepatic (and even renal) encephalopathy, typical EEG findings include disorganized, slow background rhythms and triphasic waves (Harner and Katz, 1974). In other cases, however, the EEG may reveal continuous or discontinuous epileptiform activity, suggesting that the mental status changes are in fact epileptic.

X. Sleep Disorders

Events occurring during sleep or related to sleep are sometimes misdiagnosed as epileptic. Myoclonic jerks commonly occur during the drowsy state preceding sleep or on awakening. The contractions can be quite vigorous and may be generalized, focal, isolated, or multiple. The physiological progression characteristic of seizures is absent, and there is no loss of consciousness (Guerrini and Genton, 2004). Sleepwalking can mimic CPS but is rare in elderly patients (Ohayon *et al.*, 1999). The individual walks about, may be unsteady, and appears confused. Evaluation in these cases should include an EEG and polysomnography. Daytime sleepiness due to sleep apnea can result in rather sudden sleep onset, even when the individual seems to be engaged in some activity. A careful history from a bed partner may elicit a pattern of snoring and even of depressed respiration preceding a sudden, loud onset of the snoring episode.

An infrequent but important condition is rapid eye movement (REM) sleep disorder (Mahowald and Schenck, 1994). This condition results from failure of the brainstem mechanism that suppresses motor activity during REM sleep (Mahowald and Schenck, 1994). The individual "acts out" dreams, sometimes resulting in disastrous outcomes. Violent kicking or striking at a bed partner may occur, along with vocalization. The person may jump out of bed, run into a wall, or even attempt to jump through a window, and be totally unaware of his or her actions (Mahowald and Schenck, 1994). Deaths and serious injury have been reported. The condition is often misdiagnosed as CPS, and polysomnography is critical for a definitive diagnosis (Mahowald and Schenck, 1994). Treatment with clonazepam has shown promise in these cases.

XI. Syncope

Syncope or drop attack is a common condition, but patients with episodic loss of consciousness are often diagnosed with syncope when a careful history would suggest otherwise. There is a prodrome, usually consisting of light-headedness, followed by collapse with a brief loss of consciousness, and then a rapid return to normal sensorium (Miller and Kruse, 2005). Although syncope is not usually followed by a confusional state, in the elderly there may be a brief period of confusion. Causes of syncope include severe anemia, cardiac arrhythmias, volume depletion, postural changes with blood pooling in the extremities, and neurally mediated syncope [neurocardiogenic syncope (NCS)]. NCS results from an abnormal brainstem reflex (http://www.ndrf.org/ParoxymalAutonomic Syncope.htm). When afferent activity from the carotid receptors signals for an increase in heart rate, the Bezold-Jarisch reflex is aberrantly activated producing bradycardia and hypotension (Stedman's, 2003). Evaluation should include a complete blood count (CBC), a Holter monitor, an echocardiogram, and a tilt test.

Drop attacks have long been perplexing and controversial. In this type of attack, the patient suddenly falls to the floor. Injury may occur, but there is little, if any, postfall confusion. Syncope may result from either vascular insufficiency in the vertebrobasilar system (disabling of the reticular-activating system and motor centers) or epilepsy. Temporal lobe syncope has been described (Jacome, 1989), but seizures of temporal lobe origin are infrequent in older patients. When awareness is altered during a seizure, the patient is also more prone to fall. A comprehensive cerebrovascular evaluation is indicated, along with an EEG.

More confounding is convulsive syncope (Aminoff *et al.*, 1988). This condition usually arises when the patient with syncope does not reach a horizontal position, for example, when a well-meaning person nearby tries to prevent the patient from falling. This prolongs the cerebral ischemia and leads to convulsive activity. There may be a few minor jerks of the limbs or a brief period of tonic posturing followed by a brief postictal state. When such an event is reported, the observer usually describes the convulsive activity (Aminoff *et al.*, 1988). This may lead to misdiagnosis and inappropriate treatment for seizures.

XII. Psychiatric Disorders

Psychiatric disorders must be considered in two ways. Psychogenic nonepileptic seizures (NES) may be secondary to the presence of a psychiatric disorder. On the other hand, there is some overlap of the signs and symptoms that are present in both psychiatric diseases and epilepsy.

Psychogenic NES were found in 13 of 94 consecutive older patients, a rate comparable to that found in younger adults (McBride *et al.*, 2002). Thus, it should be a consideration in new-onset seizures in the elderly. Symptoms may be quite subtle and consist of staring, apparent unresponsiveness, or motor activity resembling tonic-clonic convulsions (Reuber and Elger, 2003). The diagnosis may be suspected if the patient has had a recent emotional trauma, a history of seizures, or nonstereotyped events that do not conform to physiological principles. The routine EEG may be normal or may demonstrate epileptiform activity. In either case, the EEG is not in itself diagnostic. If the seizures occur with sufficient frequency (i.e., daily or several times a week), video-EEG monitoring is the procedure of choice to confirm the diagnosis (Drury *et al.*, 1999; Rowan *et al.*, 1987). However, in some cases, even video-EEG monitoring may be inconclusive (Berkhoff *et al.*, 1998). To distinguish NES from epileptic seizures, multiple data sources often must be used including interictal and ictal EEG recordings, prolonged video-EEG monitoring, brain imaging, subjective evaluation of seizures, response to AEDs, a placebo-infusion test, and induction procedures (Berkhoff *et al.*, 1998; Bowman, 1998). Also, several features may help clinicians distinguish between NES and epileptic seizures. NES tend to be longer in duration (commonly >2 min) than epileptic seizures (Mellers, 2005). Motor features also differ between NES and epileptic seizures; NES seizures tend to have a more gradual onset, and their course tends to fluctuate (Mellers, 2005). Additionally, thrashing and violent movements, side-to-side head movements, asynchronous movements, and closed eyes are common with NES but rare in epileptic seizures (Mellers, 2005). Automatisms are rare in NES but common in epileptic seizures (Mellers, 2005). Finally, recall for the unresponsive period is common with NES but rare in epileptic seizures (Mellers, 2005).

Psychiatric disorders are common among elderly individuals, many of which may be mistaken for or comorbid with epilepsy.

A. DEPRESSION

Depression is common among elderly individuals, with estimates of its prevalence ranging from 2% to 30% (Benedict and Nacoste, 1990). Symptoms, including confusion, that are manifestations of depression in elderly individuals (Benedict and Nacoste, 1990) may be mistaken for postictal confusion. To complicate matters, depression is the most common comorbid psychiatric disorder among patients with epilepsy (Benedict and Nacoste, 1990). Estimates of its prevalence range from 20% to 55% in patients with poorly controlled seizures and from 3% to 9% in patients with well-controlled epilepsy (Kanner, 2003). Additionally, a possible bidirectional relationship between depression and epilepsy has been proposed; patients with epilepsy are more likely to have a history of depression than are members of the general population (Kanner, 2003).

Depression may manifest itself in numerous ways in patients with epilepsy:

- Ictal depression refers to depression that is a clinical manifestation of a seizure. In some cases, it can be the sole expression of a simple partial seizure, making the seizure difficult to recognize as an epileptic phenomenon. However, ictal depression is often followed by alterations of consciousness, making the episode easier to recognize as a seizure (Kanner, 2003).
- Preictal depression refers to a period of dysphoria preceding a seizure, and it can last for hours, or even days, before seizure onset.
- Postictal depression refers to depression that follows a seizure. It may last as long as 2 weeks and mimic symptoms of a major depressive episode. This type of depression is common among patients whose epilepsy is poorly controlled (Kanner, 2003).

- Interictal depression refers to depression that occurs independent of seizure episodes; it is the most commonly reported presentation of depression in patients with epilepsy. Interictal depression has an intermittent course and can resemble chronic depression (Kanner, 2003).

B. DISSOCIATIVE DISORDERS AND FUGUE STATES

Dissociative disorders, such as dissociative identity disorder (DID), share a number of common symptoms with epileptic seizures, including amnesia, alterations in identity, derealization, and depersonalization (Bowman and Coons, 2000). However, some features may help distinguish between dissociative disorders and epilepsy. For example, DID is associated with amnesia of critical personal information and with consistent and recurrent alterations in identity. On the other hand, identity changes in patients with epilepsy are generally transient (Bowman and Coons, 2000). The Dissociative Experiences Scale may be useful for screening patients for dissociative disorders while the Structured Clinical Interview for Diagnostic and Statistical Manual of Mental Disorders, Fourth Edition (DSM-IV), Dissociative Disorders, may be especially useful for identification of patients with epilepsy who exhibit dissociative symptoms (Bowman and Coons, 2000). Patients with epilepsy typically have mild or moderate amnesia, as per this scale, but have low scores for depersonalization, derealization, identity confusion, and identity alteration (Bowman and Coons, 2000).

A fugue state is an altered state of consciousness with varying degrees of activity and amnesia (Akhtar and Brenner, 1979). Like dissociative disorders and epilepsy, the fugue states may be associated with organic conditions (e.g., brain tumors or head trauma), alcohol or other drugs, schizophrenia, or depression (Akhtar and Brenner, 1979; Rowan and Rosenbaum, 1991). Epileptic fugue states are most often manifestations of seizures originating in the temporal lobe, but fugue states can also occur in the postictal state. A prior or family history of seizures is helpful in determining if a fugue state is epileptic in origin. Epileptic fugue states tend to be brief, and those lasting longer than 24 h are unlikely to be attributable to epilepsy (Akhtar and Brenner, 1979).

C. BIPOLAR DISORDER

Epilepsy and bipolar disorder are frequently comorbid conditions. Manic states have been reported during ictal and postical periods following increased epileptic activity, or after suppression of seizures by AEDs in patients with epilepsy (Ettinger *et al.*, 2005; Kudo *et al.*, 2001); this only complicates the diagnosis.

Assessment of patient history is important, as it is uncommon for the symptoms of bipolar disorder to occur for the first time in an elderly individual (Bauer and Pfennig, 2005). Family history should also be evaluated. A family history of mood disorders in a patient with mania is suggestive of bipolar disorder, whereas a family history of epilepsy or convulsions is suggestive of epilepsy (Kudo *et al.*, 2001). Behavioral traits may also be used to distinguish between bipolar disorder and mania associated with epilepsy. Dependent or childish behaviors have a strong association with mania in patients with epilepsy (Kudo *et al.*, 2001). Additionally, the manic and depressive episodes seen in patients with epilepsy are often less severe than those seen in patients with bipolar disorder (Kudo *et al.*, 2001).

D. ANXIETY DISORDERS

Anxiety disorders are among the most common psychiatric disorders in elderly individuals (Vazquez and Devinsky, 2003). Some seizures, particularly those of temporal lobe origin, have features that resemble those of anxiety disorders. Ictal fear, a common emotional aura, may present as the only or predominant expression of a simple partial seizure, or it may be the first manifestation of CPS (Rosa *et al.*, 2006). Distinguishing between ictal fear and anxiety disorders (especially panic disorder) can be difficult, particularly if ictal fear is the first manifestation of epilepsy (Rosa *et al.*, 2006). Panic disorders and partial seizures share a number of common characteristics, including rapid onset, brief duration, and symptoms of impulsivity, hallucinations, and disorientation (Daly *et al.*, 2000). However, there are several differences between panic attacks and seizures. Panic attacks tend to be longer in duration than partial seizures (Rosa *et al.*, 2006; Thompson *et al.*, 2000). Automatisms are uncommon during panic attacks (Rosa *et al.*, 2006); therefore, witness reports of automatisms are suggestive of partial seizures (Thompson *et al.*, 2000). Depression and anticipatory anxiety are often associated with panic attacks but are uncommon in seizures with ictal fear (Rosa *et al.*, 2006). Ictal and interictal EEG monitoring may also be useful, since both ictal and interictal EEG readings are typically normal in patients with panic attacks but are often abnormal in patients with seizures and ictal fear (Rosa *et al.*, 2006). Additionally, MRI of temporal lobe structures frequently reveals abnormal findings in patients with seizures and ictal fear, but findings are usually normal in patients with panic attacks (Rosa *et al.*, 2006). Patient and family history are also important in assessing patients; the presentation of panic disorder for the first time in patients older than 45 years is uncommon (Thompson *et al.*, 2000). A history of febrile seizures during childhood, as well as unresponsiveness to anxiolytics, is suggestive of a seizure disorder diagnosis (Daly *et al.*, 2000; Thompson *et al.*, 2000). A family history of panic disorder is common in patients with panic disorder; a first-degree relative with epilepsy is supportive of a seizure disorder diagnosis (Daly *et al.*, 2000; Vazquez and Devinsky, 2003).

Other anxiety disorders that may mimic or be associated with epilepsy include obsessive-compulsive disorder (OCD), generalized anxiety disorder (GAD), and posttraumatic stress disorder (PTSD). Although OCD is uncommon in epilepsy, its symptoms may occur during seizures (Vazquez and Devinsky, 2003). Some patients with epilepsy may experience periods of anxiety and/or worry, which may mimic the symptoms of GAD (Vazquez and Devinsky, 2003). Patients with PTSD sometimes experience NES that may be difficult to distinguish from epileptic seizures. Assessment of patient history is critical, as a PTSD diagnosis requires a past trauma (Vazquez and Devinsky, 2003).

E. Ganser Syndrome

Ganser syndrome is a very rare condition that may resemble the abnormal mental states seen in patients with epilepsy (Dwyer and Reid, 2004). Both Ganser syndrome and seizures are commonly associated with an abrupt onset and ending of episodes, and the ictal and postical states of both may be associated with odd behavior, impairment of consciousness, and perceptual abnormalities (Dwyer and Reid, 2004). Four clinical features are required for a diagnosis of Ganser syndrome: (1) approximate answers, for example, if asked how many legs a cow has, a patient with Ganser syndrome may answer three; (2) perceptual abnormalities, such as visual or auditory hallucinations; (3) clouded consciousness; and (4) somatic conversion symptoms, such as blindness, hemiplegia, rigidity, and gait abnormalities (Dwyer and Reid, 2004). Onset of the symptoms of Ganser syndrome typically occurs in early adulthood, but onset may also occur in adolescence or childhood (Cosgray and Fawley, 1989); therefore, an accurate patient history is useful for evaluation of elderly individuals presenting with symptoms of Ganser syndrome. A previous history of epilepsy and abnormal EEG readings are suggestive of an epilepsy diagnosis in patients displaying symptoms associated with both Ganser syndrome and epilepsy (Dwyer and Reid, 2004).

XIII. Transient Global Amnesia

Transient global amnesia (TGA) is an interesting condition characterized by abrupt memory loss (Quinette *et al.*, 2006). The patient is awake and able to carry out complex motor activities, and most cognitive abilities are intact. The patient can follow multiple-step commands, has fluent speech, and appears confused. Patients repeatedly ask questions such as: Where are we now? Where are we going? What time is it? The defect is thought to be in short-term storage; registration is intact. Although TGA was initially thought to be a manifestation of epilepsy, this explanation has been discredited. TGA is most often found in elderly persons, although younger people may be affected (Quinette *et al.*, 2006). In many cases, bilateral temporal lobe ischemia is thought to be the explanation, but more recently, a migrainous mechanism has been proposed (Quinette *et al.*, 2006). Although TGA usually occurs as an isolated event, reoccurrences have been reported (Quinette *et al.*, 2006). The average duration is approximately 1–10 h (Quinette *et al.*, 2006). CPS may be confused with TGA when there is a prolonged postseizure state. The EEG in TGA is normal or shows slowing but no epileptiform activity (Quinette *et al.*, 2006). If there are temporal-lobe spike discharges, serious consideration should be given to a diagnosis of epilepsy.

XIV. Dementia

Elderly patients with dementia who exhibit episodic symptoms resembling seizures pose confounding diagnosis problems. Experience in long-term care facilities shows that demented patients often exhibit orofacial movements or other apparent automatisms, temper outbursts, wandering, fluctuating confusion, and memory lapses (Ballard *et al.*, 2001). These phenomena could all be manifestations of CPS. Unfortunately, many facilities have staff shortages; physicians may visit infrequently and may have little neurological experience. Although the episodic symptoms are probably a reflection of the dementing process, the possibility of complex seizures exists. Few of these patients have had EEG studies; an EEG study is indicated to look for an epileptiform focus if an elderly patient shows episodic behavioral symptoms.

XV. Primary Generalized Seizures (Idiopathic)

Prolonged seizures, referred to as an "epileptic twilight state," in which patients present with prolonged confusion or dementia, are being recognized more frequently in elderly patients. These patients have been reported to experience hallucinations and brief periods of improved ability to respond. In contrast, patients with temporal-lobe complex-partial status epilepticus typically display repetitive behavior such as oral-facial and limb automatisms and do not respond to commands (Belafsky *et al.*, 1978).

Older adults very infrequently present with true petit-mal (absence) status. These patients report a history of childhood seizures that have resolved with time.

Between the ages of 40–60 years, the generalized tonic-clonic seizures may recur, or the patient may also report periods of confusion during which the EEG shows continuous generalized spike-wave discharges. It is important to realize that the primary generalized epilepsies may reemerge later in life because these patients require different treatment than the treatment given for partial-onset seizures.

XVI. Summary

Overall, with the US population aging and with epilepsy being more common in the elderly than in the average adult, geriatric epilepsy is an increasing problem. Many older patients with new-onset epilepsy are either not diagnosed or have a delayed diagnosis. Many conditions present with the same symptoms as epilepsy; these symptoms include syncope, aphasia, confusion, and seizures. It is imperative that the physician make the correct diagnosis on the basis of observations, history and descriptions, and appropriate use of the EEG. Treatment of an elderly patient with AEDs can lead to serious side effects; however, lack of treatment in a geriatric patient with epilepsy can also lead to dire consequences. Proper diagnosis leads to proper treatment.

References

Akhtar, S., and Brenner, I. (1979). Differential diagnosis of fugue-like states. *J. Clin. Psychiatry* **40,** 381–385.

Aminoff, M. J., Scheinmann, M. M., Griffin, J. C., and Herre, J. M. (1988). Electrocerebral accompaniments of syncope associated with malignant ventricular arrhythmias. *Ann. Intern. Med.* **108,** 791–796.

Annegers, J. F., Hauser, W. A., Lee, J. R., and Rocca, W. A. (1995). Incidence of acute symptomatic seizures in Rochester, Minnesota, 1935–1984. *Epilepsia* **36,** 327–333.

Ballard, C. G., Margallo-Lana, M., Fossey, J., Reichelt, K., Myint, P., Potkins, D., and O'Brien, J. (2001). A 1-year follow-up study of behavioral and psychological symptoms in dementia among people in care environments. *J. Clin. Psychiatry* **62,** 631–636.

Bauer, M., and Pfennig, A. (2005). Epidemiology of bipolar disorders. *Epilepsia* **46**(Suppl. 4), 8–13.

Belafsky, M. A., Rosman, N. P., Miller, P., Waddell, G., Boxley-Johnson, J., and Delgado-Escueta, A. V. (1978). Prolonged epileptic twilight states: Continuous recordings with nasopharyngeal electrodes and viedotape analysis. *Neurology* **28,** 239–245.

Benedict, K. B., and Nacoste, D. B. (1990). Dementia and depression in the elderly: A framework for addressing difficulties in differential diagnosis. *Clin. Psychol. Rev.* **10,** 513–537.

Berkhoff, M., Briellman, R. S., Radanov, B. P., Donati, F., and Hess, C. W. (1998). Developmental background and outcome in patients with nonepileptic versus epileptic seizures: A controlled study. *Epilepsia* **39,** 463–469.

Bigal, M. E., Lipton, R. B., Cohen, J., and Silberstein, S. D. (2003). Epilepsy and migraine. *Epilepsy Behav.* **4,** S13–S24.

Bindu, M., Harinarayana, C. V., and Vengamma, B. (2005). A lady with acute confusional state and generalised tremors. *J. Ind. Acad. Clin. Med.* **6,** 76–78.

Bowman, E. S. (1998). Pseudoseizures. *Psychiatr. Clin. North Am.* **21,** 649–657.

Bowman, E. S., and Coons, P. M. (2000). The differential diagnosis of epilepsy, pseudoseizures, dissociative identity disorder, and dissociative disorder not otherwise specified. *Bull. Menninger Clin.* **64,** 164–180.

Brandel, J. P. (1999). Clinical aspects of human spongiform encephalopathies, with the exception of iatrogenic forms. *Biomed. Pharmacother.* **53,** 14–18.

Bylesjo, I., Forsgren, L., Lithner, F., and Boman, K. (1996). Epidemiology and clinical characteristics of seizures in patients with acute intermittent porphyria. *Epilepsia* **37,** 230–235.

Centers for Disease Control and Prevention (2005). HIV/AIDS Surveillance Report, 2004. US Department of Health and Human Services, Centers for Disease Control and Prevention, Atlanta.

Clenney, S. L. (1983). The clinical picture in slow virus diseases: A review. *Am. J. EEG Technol.* **23,** 205–208.

Cloyd, J., Hauser, W., Towne, A., Ramsay, R., Mattson, R., Gilliam, F., and Walczak, T. (2006). Epidemiological and medical aspects of epilepsy in the elderly. *Epilepsy Res.* **68**(Suppl. 1), S39–S48.

Cosgray, R. E., and Fawley, R. W. (1989). Could it be Ganser's syndrome? *Arch. Psychiatr. Nurs.* **3,** 241–245.

Daly, K. A., Kushner, M. G., Clayton, P. J., Crow, S., and Knopman, D. (2000). Seizure disorder is in the differential diagnosis of panic disorder. *Psychosomatics* **41,** 436–438.

DeJong, R. N. (1967). "The Neurologic Examination," 3rd ed., pp. 949–988. Harper & Row, New York.

Druet-Cabanac, M., Boussinesq, M., Dongmo, L., Farnarier, G., Bouteille, B., and Preux, P. M. (2004). Review of epidemiological studies searching for a relationship between onchocerciasis and epilepsy. *Neuroepidemiology* **23,** 144–149.

Drury, I., Selwa, L. M., Schuh, L. A., Kapur, J., Varma, N., Beydoun, A., and Henry, T. R. (1999). Value of inpatient diagnostic CCTV-EEG monitoring in the elderly. *Epilepsia* **40,** 1100–1102.

Dwyer, J., and Reid, S. (2004). Ganser's syndrome. *Lancet* **364,** 471–473.

Ettinger, A. B., Reed, M. L., Goldberg, J. F., and Hirschfeld, R. M. (2005). Prevalence of bipolar symptoms in epilepsy vs other chronic health disorders. *Neurology* **65,** 535–540.

Evans, R. W., and Gladstein, J. (2003). Confusional migraine or photoepilepsy? *Headache* **43,** 506–508.

Garg, R. K. (1999). HIV infection and seizures. *Postgrad. Med. J.* **75,** 387–390.

Godfrey, J. W., Roberts, M. A., and Caird, F. I. (1982). Epileptic seizures in the elderly: II. Diagnostic problems. *Age Ageing* **11,** 29–34.

Gordon, N. (1999). The acute porphyrias. *Brain Dev.* **21,** 373–377.

Guerrini, R., and Genton, P. (2004). Epileptic syndromes and visually induced seizures. *Epilepsia* **45** (Suppl. 1), 14–18.

Harner, R. N., and Katz, R. I. (1974). Electroencephalography in metabolic coma. *In* "Handbook of Electroencephalograpy and Clinical Neurophysiology" (A. Redmond, Ed.), pp. 47–62. Elsevier, Amsterdam.

Hauser, W. A. (1990). Status epilepticus: Epidemiologic considerations. *Neurology* **40**(5 Suppl. 2), 9–13.

Hauser, W. A. (1997). Epidemiology of seizures and epilepsy in the elderly. *In* "Seizures and Epilepsy in the Elderly" (A. J. Rowan and R. E. Ramsay, Eds.), pp. 7–20. Butterworth-Heinemann, Boston.

Hauser, W. A., Annegers, J. F., and Kurland, L. T. (1993). Incidence of epilepsy and unprovoked seizures in Rochester, Minnesota: 1935–1984. *Epilepsia* **34,** 453–468.

Holloway, W. J. (1986). Management of sepsis in the elderly. *Am. J. Med.* **80,** 143–148.

Holtzman, D. M., Kaku, D. A., and So, Y. T. (1989). New-onset seizures associated with human immunodeficiency virus infection: Causation and clinical features in 100 cases. *Am. J. Med.* **87,** 173–177.

Jackson, A. C., Gilbert, J. J., Young, G. B., and Bolton, C. F. (1985). The encephalopathy of sepsis. *Can. J. Neurol. Sci.* **12,** 303–307.

Jacome, D. E. (1989). Temporal lobe syncope: Clinical variants. *Clin. Electroencephalogr.* **20,** 58–65.

Janssens, J. P. (2005). Aging of the respiratory system: Impact on pulmonary function tests and adaptation to exertion. *Clin. Chest Med.* **26,** 469–484.

Kanner, A. M. (2003). Depression in epilepsy: Prevalence, clinical semiology, pathogenic mechanisms, and treatment. *Biol. Psychiatry* **54,** 388–398.

Kofler, M., and Leis, A. A. (1992). Prolonged seizure activity after baclofen withdrawal. *Neurology* **42,** 697–698.

Kudo, T., Ishida, S., Kubota, H., and Yagi, K. (2001). Manic episode in epilepsy and bipolar I disorder: A comparative analysis of 13 patients. *Epilepsia* **42,** 1036–1042.

Lancman, M. E., and Morris, H. H., III (1996). Epilepsy after central nervous system infection: Clinical characteristics and outcome after epilepsy surgery. *Epilepsy Res.* **25,** 285–290.

Lee, K., Haight, E., and Olejniczak, P. (2000). Epilepsia partialis continua in Creutzfeldt-Jakob disease. *Acta Neurol. Scand.* **102,** 398–402.

Lee, T. G., Ha, C. K., and Lim, B. H. (1997). Thyroid storm presenting as status epilepticus and stroke. *Postgrad. Med. J.* **73,** 61.

Legriel, S., and Mentec, H. (2005). Status epilepticus during acute hypercapnia: A case report. *Intensive Care Med.* **31,** 314.

Leppik, I. E. (2000). Evaluation of a seizure. *In* "Contemporary Diagnosis and Management of the Patient with Epilepsy," 5th ed., pp. 35–54. Handbooks in Health Care, Newtown, PA.

Lewin, J. J., III, Lapointe, M., and Ziai, W. C. (2005). Central nervous system infections in the critically ill. *J. Pharm. Pract.* **18,** 25–41.

Mahowald, M. W., and Schenck, C. H. (1994). REM sleep behavior disorder. *In* "Principles and Practice of Sleep Medicine" (M. H. Kryger, T. Roth, and W. C. Dement, Eds.), 2nd ed., pp. 574–588. W.B. Saunders, Philadelphia.

Malchiodi, L. (2002). Thyroid storm. *Am. J. Nurs.* **102,** 33–35.

Maltete, D., Guyant-Marechal, L., Mihout, B., and Hannequin, D. (2006). Movement disorders and Creutzfeldt-Jakob disease: A review. *Parkinsonism Relat. Disord.* **12,** 65–71.

McBride, A. E., Shih, T. T., and Hirsch, L. J. (2002). Video-EEG monitoring in the elderly: A review of 94 patients. *Epilepsia* **43,** 165–169.

Mellers, J. D. (2005). The approach to patients with "non-epileptic seizures." *Postgrad. Med. J.* **81,** 498–504.

Miller, T. H., and Kruse, J. E. (2005). Evaluation of syncope. *Am. Fam. Physician* **72,** 1492–1500.

Mirdha, B. R. (2003). Status of *Toxoplasma gondii* infection in the etiology of epilepsy. *J. Pediatr. Neurol.* **1,** 95–98.

Morres, C. A., and Dire, D. J. (1989). Movement disorders as a manifestation of nonketotic hyperglycemia. *J. Emerg. Med.* **7,** 359–364.

Ng, S. K., Hauser, W. A., Brust, J. C., and Susser, M. (1993). Hypertension and the risk of new-onset unprovoked seizures. *Neurology* **43,** 425–428.

Ohayon, M. M., Guilleminault, C., and Priest, R. G. (1999). Night terrors, sleepwalking, and confusional arousals in the general population: Their frequency and relationship to other sleep and mental disorders. *J. Clin. Psychiatry* **60,** 268–276.

Quinette, P., Guillery-Girard, B., Dayan, J., de la Sayette, V., Marquis, S., Viader, F., Desgranges, B., and Eustache, F. (2006). What does transient global amnesia really mean? Review of the literature and thorough study of 142 cases. *Brain* **129,** 1640–1658.

Ramsay, R. E., Rowan, A. J., and Pryor, F. M. (2004). Special considerations in treating the elderly patient with epilepsy. *Neurology* **62,** S24–S29.

Reuber, M., and Elger, C. E. (2003). Psychogenic nonepileptic seizures: Review and update. *Epilepsy Behav.* **4,** 205–216.

Romanelli, F., and Ryan, M. (2002). Seizures in HIV-seropositive individuals: Epidemiology and treatment. *CNS Drugs* **16,** 91–98.

Rosa, V. P., de Araujo Filho, G. M., Rahal, M. A., Caboclo, L. O. S. F., Sakamato, A. C., and Yacubian, E. M. T. (2006). Ictal fear: Semiologic characteristics and differential diagnosis with interictal anxiety disorders. *J. Epilepsy Clin. Neurophysiol.* **12,** 89–94.

Rosenbaum, D. H., Siegel, M., Barr, W. B., and Rowan, A. J. (1986). Epileptic aphasia. *Neurology* **36,** 822–825.

Rowan, A. J., and Rosenbaum, D. H. (1991). Ictal amnesia and fugue states. *Adv. Neurol.* **55,** 357–367.

Rowan, A. J., Siegel, M., and Rosenbaum, D. H. (1987). Daytime intensive monitoring: Comparison with prolonged intensive and ambulatory monitoring. *Neurology* **37,** 481–484.

Rowan, A. J., Ramsay, R. E., Collins, J. F., Pryor, F., Boardman, K. D., Uthman, B. M., Spitz, M., Frederick, T., Towne, A., Carter, G. S., Marks, W., Felicetta, J., *et al.* (2005). New onset geriatric epilepsy: A randomized study of gabapentin, lamotrigine, and carbamazepine. *Neurology* **64,** 1868–1873.

Shneker, B. F., and Fountain, N. B. (2003). Epilepsy. *Dis. Mon.* **49,** 426–478.

Sies, C., and Florkowski, C. (2006). A guide to the diagnosis of porphyria: Suggested methods and case examples. *N. Z. J. Med. Lab. Sci.* **60,** 7–11.

So, E. L., Annegers, J. F., Hauser, W. A., O'Brien, P. C., and Whisnant, J. P. (1996). Population-based study of seizure disorders after cerebral infarction. *Neurology* **46,** 350–355.

Solenski, N. J. (2004). Transient ischemic attacks: Part I. Diagnosis and evaluation. *Am. Fam. Physician* **69,** 1665–1674.

Solinas, C., and Vajda, F. J. (2004). Epilepsy and porphyria: New perspectives. *J. Clin. Neurosci.* **11,** 356–361.

Stedman's (2003). "Stedman's Electronic Medical Dictionary" [CD-ROM]. Version 6.0. Lippincott Williams & Wilkins, Philadelphia.

Thomas, R. J. (1997). Seizures and epilepsy in the elderly. *Arch. Intern. Med.* **157,** 605–617.

Thompson, S. A., Duncan, J. S., and Smith, S. J. (2000). Partial seizures presenting as panic attacks. *Br. Med. J.* **321,** 1002–1003.

Vazquez, B., and Devinsky, O. (2003). Epilepsy and anxiety. *Epilepsy Behav.* **4**(Suppl. 4), S20–S25.

Velioglu, S. K., and Ozmenoglu, M. (1999). Migraine-related seizures in an epileptic population. *Cephalalgia* **19,** 797–801.

Wang, K. W., Chang, W. N., Chang, H. W., Chuang, Y. C., Tsai, N. W., Wang, H. C., and Lu, C. H. (2005). The significance of seizures and other predictive factors during the acute illness for the long-term outcome after bacterial meningitis. *Seizure* **14,** 586–592.

Waterhouse, E. (2005). Seizures in the elderly: Nuances in presentation and treatment. *Cleve. Clin. J. Med.* **72**(Suppl. 3), S26–S37.

Willmore, L. J. (1995). The effect of age on pharmacokinetics of antiepileptic drugs. *Epilepsia* **36** (Suppl. 5), S14–S21.

Willmore, L. J. (1996). Management of epilepsy in the elderly. *Epilepsia* **37**(Suppl. 6), S23–S33.

PHARMACOEPIDEMIOLOGY IN COMMUNITY-DWELLING ELDERLY TAKING ANTIEPILEPTIC DRUGS

Dan R. Berlowitz*,† and Mary Jo V. Pugh‡,§

*Center for Health Quality, Outcomes, and Economic Research
Bedford VA Medical Center, Bedford, Massachusetts 01730, USA
†Department of Health Services, Boston University School of Public Health
Boston, Massachusetts 02118, USA
‡South Texas Veterans Health Care System (VERDICT), San Antonio
Texas 78229, USA
§Department of Medicine, University of Texas Health Science Center at San Antonio
San Antonio Texas 78229, USA

This study used the national inpatient, outpatient, and pharmacy databases from the US Veterans Health Administration to examine prescribing patterns for older patients with epilepsy and to determine the factors associated with receiving recommended antiepileptic drugs (AEDs) such as lamotrigine, gabapentin, or carbamazepine. Among patients with epilepsy, the AED monotherapy most prescribed was phenytoin (70%), followed by phenobarbital (17%). While the rate of phenytoin use was similar for both previously and newly diagnosed patients with epilepsy, phenobarbital was less commonly used in newly diagnosed patients. Multivariable analyses suggested that receiving outpatient neurological care was the strongest predictor of receipt of recommended AED regimens in newly diagnosed elderly patients with epilepsy. These data suggest that the challenge remains to narrow the gap between expert recommendations and actual practice if patients with epilepsy are to fully benefit from the tremendous progress that has been made in the pharmacological management of this disease.

I. Introduction

Clinical trials performed by the US Department of Veterans Affairs (VA) have demonstrated that phenobarbital and phenytoin are potentially problematic drugs for use in elderly patients with epilepsy because of the high frequency of adverse events and unfavorable cognitive effects (Mattson *et al.*, 1985; Smith *et al.*, 1987). Results from these studies have been incorporated into expert consensus recommendations that favor the use of lamotrigine, gabapentin, or carbamazepine for newly diagnosed elderly patients (Karceski *et al.*, 2001; Scottish Intercollegiate Guidelines Network, 2003). Yet experience suggests that these less problematic medications are being used relatively infrequently in actual clinical practice. However, the extent to which phenytoin and phenobarbital are still being used is uncertain, and characteristics associated with their persistent use have not been identified. Therefore, the pharmacological management of epilepsy among elderly patients in a national health care system, the VA system, was examined.

The VA is the largest integrated health care system in the country, providing comprehensive care to eligible veterans. Medications are included as part of veterans' benefits and are available either for free or for a small co-payment. The VA also maintains detailed databases describing its patients and the care they receive. This provides an ideal opportunity to study the gap in translating expert recommendations for epilepsy care into actual practice. In this chapter, profiles of elderly VA patients with epilepsy are described in terms of their comorbidities and the antiepileptic drugs (AEDs) they are receiving. Also presented are the patient and system characteristics associated with prescribing potentially inappropriate AEDs. Results presented in this chapter will offer guidance on how to improve epilepsy care for elderly patients.

II. Methods

A. DATA SOURCES

Diagnostic, demographic, and utilization data were obtained from the National Patient Care Database, including the Patient Treatment File for inpatient encounters [fiscal years (FY)96–99] and the Outpatient Clinic File (FY97–99) for outpatient encounters. Both databases contain records for each encounter that include the date, type of services used, and specific diagnoses listed as ICD-9-CM (International Classification of Diseases, Ninth Revision, Clinical Modification) codes, the official system of assigning codes to diagnoses and procedures associated with hospital utilization in the United States. Prescribed medications were obtained from the Pharmacy Benefits Management database, which contains a record for each outpatient medication dispensed by a VA pharmacy. Information used was from 1999, the first year for which national data are available.

B. Study Population

All veterans aged 65 or older receiving VA inpatient or outpatient care in FY99 were considered for inclusion in the study. Patients were considered to have epilepsy if they met two criteria. First, they must have had at least one ICD-9-CM code indicative of epilepsy (either 345 or 780.39) in the inpatient or outpatient databases. Second, they had to have received at least one prescription for an AED in FY99 that included phenobarbital, primidone, phenytoin, carbamazepine, valproate, lamotrigine, gabapentin, felbamate, topiramate, or tiagabine. Patients with epilepsy were considered newly diagnosed if their first ICD-9-CM code indicative of epilepsy appeared in FY99 and they had received VA care in prior years. Patients were considered to have previously diagnosed epilepsy if their first ICD-9-CM code was before FY99. Patients whose first ICD-9-CM code occurred in FY99 and who had not received VA care in prior years were considered to have epilepsy of undetermined onset.

C. Descriptive Data Analyses

The demographic characteristics and comorbidities of previously and newly diagnosed patients with epilepsy were examined and compared to patients without epilepsy. The AEDs prescribed for the management of epilepsy were also analyzed. Patients with epilepsy were considered to be on combination therapy if they received prescriptions for at least two different AEDs; otherwise, they were considered to be on monotherapy.

Patients with newly diagnosed epilepsy who were receiving phenobarbital/ primidone (hereafter *phenobarbital* for ease of analysis) monotherapy, phenytoin monotherapy, or combination therapy that included phenobarbital were considered to be receiving potentially inappropriate therapy. Patients receiving potentially inappropriate therapy were compared to patients receiving appropriate therapy on a range of factors that included demographics, comorbidities, disease severity, and receipt of outpatient neurological care. Comorbidities included cardiovascular disorders, neurological diseases, and psychiatric diseases, as well as a count (unweighted) of comorbidities considered in creating the Charlson

Index (Charlson *et al.*, 1987). Patients were considered to have more severe epilepsy if they received hospital or emergency room care for epilepsy, received intravenous AEDs, or had an ICD-9-CM code indicating an episode of status epilepticus. Patients were classified as having received outpatient neurological care if a visit to a neurologist was designated in the Outpatient Clinic File database.

D. STATISTICAL ANALYSIS

Chi-square tests and analysis of variance were used to compare the prevalence of different comorbidities among patients with new-onset epilepsy, chronic epilepsy, and no epilepsy. Haberman's adjusted residual was used to identify statistical differences among the three groups in chi-square analyses (Haberman, 1979). Chi-square tests were also used initially for comparing patients with and without potentially inappropriate therapy on the above-listed factors. Logistic regression models were then constructed to determine the independent effects of these factors on the receipt of each of these potentially inappropriate therapies.

III. Results

A. DEMOGRAPHICS

Among 1,130,555 veterans aged 65 or older and receiving VA care in FY99, 20,558 (1.8%) were identified as having a diagnosis of epilepsy. Of these patients with epilepsy, 16,917 had been previously diagnosed, 1893 were newly diagnosed, and for 1748, the epilepsy was of undetermined onset. Thus, new onset of epilepsy occurred in 0.17% of the population.

Patients with epilepsy generally had demographic characteristics similar to those of patients without epilepsy, although differences were significant ($p < 0.01$) because of the large sample size (Table I). While there was a difference in race, with unknown race more frequent among patients without epilepsy, this likely reflects the fact that the unknown race designation in the VA databases is associated with less frequent utilization of VA care.

B. CONCOMITANT DISEASE STATES

Neurological, cardiovascular, psychiatric, and other selective diagnoses all tended to be significantly ($p < 0.001$) more common in patients with epilepsy, particularly those with new-onset epilepsy (Table I). Cerebrovascular disease was

TABLE I

DEMOGRAPHIC CHARACTERISTICS AND COMORBIDITIES OF PATIENTS WITH NEWLY DIAGNOSED
EPILEPSY, PREVIOUSLY DIAGNOSED EPILEPSY, AND NO EPILEPSY[a]

	Epilepsy diagnosis (%)		
	New[b]	Previous[c]	None[d]
Age[e] (years)			
65–74	60	63	60
75–84	38	35	37
≥85	2	2	3
Sex[e]			
Male	98	99	98
Female	2	1	2
Race[e]			
White	70	70	67
Black	18	17	9
Hispanic	4	4	4
Other	1	1	1
Unknown	7	8	19
Neurological comorbidities[a]			
Cerebrovascular disease	41	31	9
Dementia	21	17	5
Parkinson's disease	7	5	2
Multiple sclerosis	1	1	0.3
Other neurological disorders	14	11	4
Cardiovascular comorbidities[a]			
Hypertension	79	70	67
Atrial fibrillation	14	11	10
Valvular heart disease	9	7	5
Congestive heart failure	24	18	14
Coronary artery disease	50	42	39
Peripheral vascular disease	23	20	15
Mental health disorders[a]			
Alcohol abuse/dependence	10	9	4
Depression	21	19	10
Anxiety	17	14	8
Bipolar	5	4	2
Psychosis (includes schizophrenia)	22	18	8
Other physical comorbidities[a]			
Chronic pulmonary disease	2	1	1
Diabetes	32	25	28
Peptic ulcer disease	10	8	7
Osteoarthritis	43	40	36
Unweighted Charlson Index[a]	2.79	2.24	1.53

[a]All comparisons regarding comorbidities between the epilepsy and nonepilepsy groups were significant ($p < 0.01$).

[b]$n = 1893$.

[c]$n = 16,917$.

[d]$n = 1,130,555$.

[e]Demographic information was obtained from the 1999 Veterans SF-36 Health Survey (Veterans SF-36).

noted in 41% of patients with newly diagnosed epilepsy, whereas dementia was present in 21%. The prevalence of important comorbidities is also reflected in the higher unweighted Charlson Index for patients with epilepsy (Table I).

C. MEDICATION USAGE

Among the patients with epilepsy, 79.6% were on monotherapy and 20.4% were receiving combination therapy. Phenytoin was the most frequently prescribed monotherapy; it was used in almost 70% of both previously and newly diagnosed patients with epilepsy (Fig. 1). Phenobarbital was used relatively infrequently among patients on monotherapy, but 57% of patients with previously diagnosed epilepsy and 29% of patients with newly diagnosed epilepsy who were on combination therapy received phenobarbital as one of their medications (Fig. 2). The most frequently used combination therapy was phenytoin with phenobarbital. Overall, phenobarbital was used in 17% of elderly patients with epilepsy.

D. STATISTICAL ANALYSIS

Table II presents the results of logistic regression models for patients with newly diagnosed epilepsy. The logistic regression models compared patients on the basis of treatment regimen: phenobarbital monotherapy versus other monotherapies,

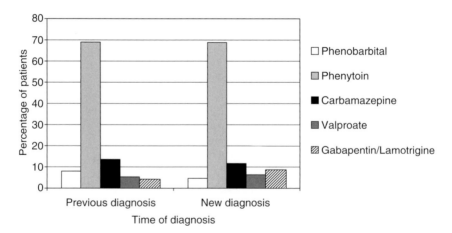

FIG. 1. Antiepileptic drug monotherapies received by patients with previously and newly diagnosed epilepsy. Adapted from Perucca et al. (2006), with permission.

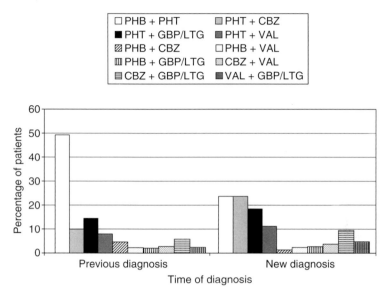

Fig. 2. Antiepileptic drug combination therapies received by patients with previously and newly diagnosed epilepsy. PHB, phenobarbital; PHT, phenytoin; GBP/LTG, gabapentin/lamotrigine; CBZ, carbamazepine; VAL, valproate. Adapted from Perucca *et al.* (2006), with permission.

phenytoin monotherapy versus other nonphenobarbital monotherapies, and phenobarbital combinations versus all other combinations. Receipt of a neurology consultation was significantly ($p < 0.05$) associated with less frequent use of phenobarbital combinations and phenytoin monotherapy; additionally, there was a trend toward less use of phenobarbital monotherapy among patients with a consultation. Nonwhite patients were significantly ($p < 0.05$) more likely to receive phenytoin monotherapy, but no racial/ethnic differences were noted for phenobarbital therapy. No consistent pattern was evident in the association between comorbidities and specific therapies prescribed, although confidence intervals for many variables were wide because of the small sample size.

IV. Discussion

Databases maintained by the VA provide a unique opportunity to study the pharmacoepidemiology of epilepsy in a large group of community-dwelling elderly patients. These databases provide comprehensive information regarding comorbidities, health care utilization, and medications for all patients using this

TABLE II

RESULTS FROM LOGISTIC REGRESSION MODELS EXAMINING POTENTIALLY INAPPROPRIATE USE OF AEDs
IN ELDERLY PATIENTS WITH NEWLY DIAGNOSED EPILEPSY

	Phenobarbital monotherapy[a]	Phenytoin monotherapy[b]	Phenobarbital combination therapy[c]
	OR[d] (95% CI)	OR[d] (95% CI)	OR[d] (95% CI)
Black race (vs white)	0.75 (0.37, 1.52)	1.68 (1.23, 2.30)[e]	0.49 (0.19, 1.29)
Hispanic race (vs white)	1.20 (0.41, 3.54)	2.38 (1.28, 4.42)[e]	0.17 (0.02, 1.67)
Other race (vs white)	N/A	2.42 (0.50, 11.69)	N/A
Unknown race (vs white)	0.78 (0.30, 2.02)	1.52 (0.93, 2.49)	0.50 (0.15, 1.46)
Aged >85 years (vs 65–74 years)	0.34 (0.04, 2.57)	0.72 (0.37, 1.41)	N/A
Aged 75–84 years (vs 65–74 years)	0.62 (0.37, 1.04)	1.22 (0.95, 1.55)	0.74 (0.40, 1.36)
Male gender	3.72 (1.19, 11.67)[e]	1.18 (0.50, 2.80)	2.28 (0.43, 12.12)
Unweighted Charlson Index	0.93 (0.78, 1.12)	1.07 (0.99, 1.16)	0.92 (0.75, 1.12)
Psychiatric comorbidity	0.75 (0.45, 1.27)	0.64 (0.51, 0.82)[e]	0.83 (0.45, 1.52)
Severe epilepsy	0.19 (0.06, 0.62)[e]	1.55 (1.13, 2.14)[e]	0.42 (0.20, 0.88)[e]
Neurology consultation	0.80 (0.48, 1.35)	0.51 (0.40, 0.65)[e]	0.27 (0.15, 0.49)[e]
Cerebrovascular disease	0.65 (0.36, 1.18)	1.37 (1.04, 1.80)[e]	1.00 (0.53, 1.90)
Dementia	1.04 (0.54, 2.00)	0.91 (0.67, 1.22)	0.43 (0.18, 1.00)
Other neurological disorders	2.57 (1.35, 4.87)[e]	0.67 (0.48, 0.94)[e]	0.83 (0.35, 1.98)
Diabetes	1.28 (0.70, 2.34)	0.70 (0.52, 0.93)[e]	0.71 (0.34, 1.50)
Hypertension	0.50 (0.30, 0.83)	0.80 (0.59, 1.08)	0.70 (0.34, 1.43)

[a]Phenobarbital monotherapy compared to other monotherapies.

[b]Phenytoin monotherapy compared to other monotherapies excluding phenobarbital.

[c]Combinations including phenobarbital compared to all other combinations.

[d]An odds ratio (OR) less than 1 shows that patients with the indicated characteristic were less likely to receive the indicated therapy.

[e]$p < 0.05$.

CI, confidence interval; N/A, not available due to small sample size.

national health care system. Moreover, the validity of these data is supported by the close agreement of VA results with those reported in other studies of elderly patients with epilepsy (Hauser, 1997; Leppik and Birnbaum, 2002).

In this study, 1.8% of elderly veterans had epilepsy, and new onset of epilepsy occurred in 0.17% of all patients. Other studies have reported a prevalence rate of 1–1.5% among patients 75 years and older (Hauser, 1997) with new onset occurring in 0.13% of patients 60 years and older (Hauser, 1997; Leppik and Birnbaum, 2002). The somewhat higher rates in the VA data could be due to the fact that elderly veterans tend to have poorer health status than the general US population (Selim et al., 2004).

In the VA system, patients with epilepsy have a large number of comorbidities, emphasizing the need for care in prescribing AEDs. Patients with epilepsy were more likely to suffer from a wide variety of neurological, cardiovascular, and psychiatric conditions; these conditions were more common in patients with newly diagnosed epilepsy than in those with previously diagnosed epilepsy. Of the conditions that are most likely to cause epilepsy, cerebrovascular disease and dementia were most common. While this raises questions as to the etiology of epilepsy in patients without one of these conditions, the results are consistent with other reports in the literature. A review of 10 different studies examining epilepsy in patients over age 60 found that stroke was the cause in 22–47% of the cases, and dementia in 2–11.5% (Loiseau, 1997).

Epilepsy treatment differed considerably from expert recommendations. These recommendations do not consider phenytoin and phenobarbital as first-line agents for epilepsy (Karceski *et al.*, 2001). Yet, the majority of patients in the study described in this chapter, with either previously or newly diagnosed epilepsy, were receiving these drugs. It should be noted that the expert recommendations are based on clinical trials that evaluated mostly newly diagnosed patients. Thus, the fact that so many previously diagnosed patients are on phenytoin or phenobarbital is understandable, as clinicians and patients may be reluctant to change a regimen that appears to be working well. Toxicities, some of which may be unrecognized or may be associated with long-term use of these medications in an aging population, need to be characterized. The possible benefits of switching therapies also need to be better defined.

It is less clear why so many newly diagnosed patients were receiving therapies considered suboptimal. Some of the clinical trials on which these recommendations are based are over a decade old, so these results cannot be considered recent. Reflecting this fact, the VA cooperative study of new-onset epilepsy in the elderly did not even consider phenytoin as a treatment option, comparing the established drug, carbamazepine, to the newer drugs gabapentin and lamotrigine (Ramsay *et al.*, 2001). The results of logistic regression models in the study reported in this chapter provide few insights into patient characteristics associated with potentially inappropriate prescribing; however, those patients seen by a neurologist tended to receive more appropriate therapy, suggesting an increased need for a neurology referral for older patients with epilepsy. It must be concluded that, in most cases, the results of clinical trials and expert recommendations have not been adequately disseminated to change the practices of those physicians most involved in treating patients with epilepsy.

Narrowing the gap between expert recommendations and actual practice remains a challenge. Experience with other diseases suggests that this will be a difficult task. Simple educational efforts such as lectures and continuing medical education programs are unlikely to be sufficient (Davis *et al.*, 1992). Rather, innovative approaches to changing physicians' practices, such as approaches that combine education with local opinion leaders and performance feedback, will be required (Bero *et al.*, 1998). Only then will patients with epilepsy fully benefit from the tremendous progress that has been made in the pharmacological management of this disease.

Acknowledgments

This study was funded by the Department of Veterans Affairs, Veterans Health Administration, Health Services Research and Development Service Merit Review Entry Program Award to M.J.V.P., PhD (MRP-02-267). We acknowledge the support of the Edith Nourse Rogers Memorial VA Medical Center/the Center for Health Quality, Outcomes, and Economic Research in Bedford, MA, and the South Texas Veterans Health Care System/VERDICT in San Antonio, TX. We also thank the Department of Veterans Affairs, Veterans Health Administration Office of Quality and Performance for providing access to data from the 1999 Large Health Survey of Veteran Enrollees.

The SF-36 is the property of the Medical Outcomes Trust.

The authors have no conflicts of interest related to the content of this chapter.

The views expressed in this chapter are those of the authors and do not necessarily represent the views of the Department of Veterans Affairs.

References

Bero, L. A., Grilli, R., Grimshaw, J. M., Harvey, E., Oxman, A. D., and Thomson, M. A. (1998). Closing the gap between research and practice: An overview of systematic reviews of interventions to promote the implementation of research findings. The Cochrane Effective Practice and Organization of Care Review Group. *BMJ* **317,** 465–468.

Charlson, M. E., Pompei, P., Ales, K. L., and MacKenzie, C. R. (1987). A new method of classifying prognostic comorbidity in longitudinal studies: Development and validation. *J. Chronic Dis.* **40,** 373–383.

Davis, D. A., Thomson, M. A., Oxman, A. D., and Haynes, R. B. (1992). Evidence for the effectiveness of CME. A review of 50 randomized controlled trials. *JAMA* **268,** 1111–1117.

Haberman, S. J. (1979). "Analysis of Qualitative Data: New Developments." Academic Press, Burlington, MA. The analysis of qualitative data series.

Hauser, W. A. (1997). Epidemiology of seizures and epilepsy in the elderly. *In* "Seizures and Epilepsy in the Elderly" (A. J. Rowan and R. E. Ramsay, Eds.), pp. 7–18. Butterworth-Heinemann, Boston.

Karceski, S., Morrell, M., and Carpenter, D. (2001). The expert consensus guideline series: Treatment of epilepsy. *Epilepsy Behav.* **2,** A1–A50.

Leppik, I. E., and Birnbaum, A. (2002). Epilepsy in the elderly. *Semin. Neurol.* **22,** 309–320.

Loiseau, P. (1997). Pathologic processes in the elderly and their association with seizures. *In* "Seizures and Epilepsy in the Elderly" (A. J. Rowan and R. E. Ramsay, Eds.), pp. 63–85. Butterworth-Heinemann, Boston.

Mattson, R. H., Cramer, J. A., Collins, J. F., Smith, D. B., Delgado-Escueta, A. V., Browne, T. R., Williamson, P. D., Treiman, D. M., McNamara, J. O., McCutchen, C. B., Homan, R. W., Crill, W. E., *et al.* (1985). Comparison of carbamazepine, phenobarbital, phenytoin, and primidone in partial and secondarily generalized tonic-clonic seizures. *N. Engl. J. Med.* **313,** 145–151.

Perucca, E., Berlowitz, D., Birnbaum, A., Cloyd, J. C., Garrard, J., Hanlon, J. T., Levy, R. H., and Pugh, M. J. (2006). Pharmacological and clinical aspects of antiepileptic drug use in the elderly. *Epilepsy Res.* **68**(Suppl. 1), S49–S63.

Ramsay, R. E., Rowan, A. J., and Pryor, F. M. (2001). Special considerations in managing epilepsy in the elderly patient. *Clin. Geriatr.* **9,** 47–56.

Scottish Intercollegiate Guidelines Network (Intercollegiate Guidelines Network 2003). *In* "Diagnosis and Management of Epilepsy in Adults," pp. 1–49. Royal College of Physicians, Edinburgh.

Selim, A. J., Berlowitz, D. R., Fincke, G., Cong, Z., Rogers, W., Haffer, S. C., Ren, X. S., Lee, A., Qian, S. X., Miller, D. R., Spiro, A., III, Selim, B. J., *et al.* (2004). The health status of elderly veteran enrollees in the veterans health administration. *J. Am. Geriatr. Soc.* **52,** 1271–1276.

Smith, D. B., Mattson, R. H., Cramer, J. A., Collins, J. F., Novelly, R. A., and Craft, B. (1987). Results of a nationwide Veterans Administration Cooperative Study comparing the efficacy and toxicity of carbamazepine, phenobarbital, phenytoin, and primidone. *Epilepsia* **28**(Suppl. 3), S50–S58.

USE OF ANTIEPILEPTIC MEDICATIONS IN NURSING HOMES

Judith Garrard,* Susan L. Harms,* Lynn E. Eberly,[†] and Ilo E. Leppik[‡]

*Division of Health Policy and Management, School of Public Health
University of Minnesota, Minneapolis, Minnesota 55455, USA
[†]Division of Biostatistics, School of Public Health, University of Minnesota
Minneapolis, Minnesota 55455, USA
[‡]Epilepsy Research and Education Program
University of Minnesota, Minneapolis, Minnesota 55455, USA

The University of Minnesota Epilepsy Research and Education Program published two studies evaluating the use of antiepileptic drugs (AEDs) among nursing home (NH) elderly. The studies used a large, nongovernmental data set for studying this population. This chapter is a summary of those two studies. In the first study, a 1-day point prevalence study, 10.5% of the NH residents had one or more AED orders, a prevalence 10 times greater than that found in the community. In a multivariate analysis of factors associated with AED treatment, seizure indication was the most important factor, and age was inversely related to AED use. Phenytoin was the most commonly used AED, followed by carbamazepine, phenobarbital, and valproic acid. The most frequently used combination was phenytoin and phenobarbital. In the second study, evaluating NH admission data, 8% of newly admitted residents were already receiving one or more

INTERNATIONAL REVIEW OF
NEUROBIOLOGY, VOL. 81
DOI: 10.1016/S0074-7742(06)81010-X

165

AEDs when they entered the NH. Factors associated with AED use in this group included epilepsy/seizure disorder, age, cognitive performance, and manic depression (bipolar disease). Among residents recently admitted who were not using an AED at entry, 3% were initiated on an AED within 3 months of admission. Among the factors associated with the initiation of AEDs during this period, the strongest association was with epilepsy/seizure disorder. Manic depression (bipolar disease) was also significantly associated with initiation of an AED after admission. In this group, there was an inverse relationship between age and initiation of an AED.

I. Introduction

This chapter summarizes research that was conducted by the University of Minnesota Epilepsy Research and Education Program to evaluate the use of antiepileptic drugs (AEDs) by elderly people in nursing homes (NHs) (Garrard *et al.*, 2000, 2003). Epilepsy, defined as two or more unprovoked seizures, was described at the population level by Hauser and colleagues in a data set from Olmsted County in Minnesota during the period 1935–1984 (Hauser, 1995; Hauser and Annegers, 1991; Hauser *et al.*, 1993, 1996). Their research concentrated on community-dwelling people of all ages and demonstrated that the incidence of epilepsy in the population forms a U-shaped curve over the life span, with one of the highest incidence rates during the first year of life, and even higher rates among the elderly, beginning at approximately 60 years of age and rising with increasing age. This finding of higher incidence among the elderly has been replicated in other populations, for example, a study of urban Danes >60 years of age (Luhdorf *et al.*, 1986). The prevalence of epilepsy, that is, the proportion of existing cases of epilepsy in the population at a specified point in time, is estimated to range between 0.68% and approximately 1.5% of the population as a whole (Hauser, 1992; Hauser and Annegers, 1991; Hauser *et al.*, 1996).

With the leading edge of the "baby boom" generation about to turn 65 years of age, the number of elderly in the US population is poised to soar, projected to increase from an estimated 40 million in 2010 to 71.5 million in 2030 (Administration on Aging, 2005). Given these statistics, the recognition and treatment of epilepsy among people 65 years and older is a concern of health practitioners and families alike.

A. The Elderly

As people age, their need for NH care increases due to greater frailty and disease. For patients aged 65 years and older, there is a lifetime risk of 43–46% of ever becoming a resident in an NH (Kemper and Murtaugh, 1991; Spillman and

Lubitz, 2002). At any one time, 4.5% of the elderly US population is residing in NHs (Hetzel and Smith, 2001).

Most existing research about epilepsy and its treatment focuses on people in the community who are below the age of 65; hence, there have been few large-scale epidemiological studies of seizures in the older population. Even less is known about the epidemiology of the treatment of epilepsy in NH elderly. Whether AEDs are prescribed differently for particular subgroups of elderly patients, for example, by age, gender, sex, racial identity, or geographic location, is not known.

1. *Demographic Factors*

In any study of elderly people, it is important not to confuse those at one end of the age spectrum, at 65 years, with those in the older subgroups. For example, to equate "elderly" who are 65 years of age with those who are 85 years of age in terms of functional or cognitive abilities is just as erroneous as equating "teenagers" who are 13 years of age with those who are 19 years of age. Those ≥ 85 years of age are more likely to develop chronic illness, be disabled, and be more dependent on others for assistance with daily activities (US Department of Commerce, 1993). Despite this, age distinctions between elderly subgroups are frequently not made in research. Another set of subgroups that has become crucial to the study of pharmacoepidemiological research is that of race/ethnicity. In 2004, 18.1% of the elderly US population were members of a minority group. This figure is projected to increase to 23.6% by the year 2020 (Administration on Aging, 2005).

2. *NH Factors*

In any research concerning NHs, the distinction must also be made between residents and admissions. A resident cohort includes all residents in the facility at a specified time, which is usually a cross-sectional sample. Such a sample consists of a mixture of newly admitted residents and others who have been in the NH for different periods. In contrast, an admission cohort includes all people admitted to a facility during a specified period. Studies of both cohorts are important in gerontological research on the treatment and outcomes of elderly people living in NHs (Coughlin *et al.*, 1990; Liu *et al.*, 1994). Because of current health care delivery practices, many patients needing temporary rehabilitation care are transferred to NHs, with the expectation by health care providers and the patients themselves that the NH stay will be temporary. Since Medicare reimburses active rehabilitation care up to the first 100 days (Liu *et al.*, 1999) and federal regulations require that Medicare-reimbursed rehabilitation patients be seen by a physician or nurse practitioner every 30 days (compared to once every 3 months for nonrehabilitation patients), an admissions cohort will have a larger proportion of residents under closer clinical scrutiny than will a cross-sectional sample of residents. Both kinds of samples, resident and admission, are important in understanding diseases and their treatment, including epilepsy and AED use.

B. Purpose

Until the publication by Garrard *et al.* (2000) on epilepsy/seizure disorder and AED use, no large-scale pharmacoepidemiology studies about the use of antiseizure medications in NH elderly, comparable to Hauser's community-based samples, had been reported in the literature. In this chapter, these NH study results are summarized, with a focus on AED use, regardless of the underlying condition being treated, although some analyses are stratified by the presence or absence of epilepsy/seizure disorder.

The first study (Garrard *et al.*, 2000), which concentrated on the use of AEDs by people already living in the facilities (i.e., NH residents), discovered a considerable difference in AED prevalence between NH residents and community-based elderly. These findings formed the basis for the next question: Do elderly people enter NHs with this level of AED use, or are they initiated on these drugs after admission? This question resulted in the second study (Garrard *et al.*, 2003), which explored AED use at the time of NH admission and at a follow-up 3 months after admission.

The results of both studies are summarized in this chapter in the context of the following three questions:

- What is the prevalence of AED use by NH residents, and what factors are associated with such use?
- What is the prevalence of AED use by elderly people on admission to the NH, and what factors distinguish between those with or without AED use at entry?
- Of newly admitted residents without AED use at NH entry, what percentage were initiated on AEDs within 3 months of admission, and what factors were associated with postadmission initiation of AEDs?

II. Methods

A. Studies Defined

1. *Residents Study*

The initial cross-sectional study (Garrard *et al.*, 2000) of AED use by NH residents was based on 21,551 residents in 346 NHs in 24 states in 1995. This was a point prevalence study with a 1-day study period.

2. *Admissions Study*

The subsequent study (Garrard *et al.*, 2003) of AED use at admission was based on a different sample: 10,318 admissions to 510 NHs in 31 states over a 3-month period in 1999. In both of the studies, all of the NHs were owned by Beverly Enterprises, the largest NH corporation in the United States (Abelson, 2002; Vickery, 2000).

B. MEDICARE-CERTIFIED NHs

To qualify for Medicare reimbursement, rehabilitative care must be delivered in a Medicare-certified facility; consequently, virtually all NHs throughout the United States have such certification. Medicare-certified NHs are a highly regulated sector of the health care delivery system. Although there are variations in the characteristics of NHs by goals, size, patient costs, and quality, the standardization brought about in recent years by Centers for Medicare and Medicaid Services (CMS) regulations has resulted in greater homogeneity across facilities. Therefore, there is the potential for increased application of these research findings to NHs not in the Beverly Enterprises system.

Secondary source data used in these two studies (Garrard *et al.*, 2000, 2003) included the physicians' orders (POs) for all medications. The second study (Garrard *et al.*, 2003) also used a federally mandated assessment form, referred to as the Minimum Data Set (MDS). The characteristics of these data sources are described below.

C. POs: MEDICATIONS DATA SET

The medication database consisted of all POs reported electronically on a daily basis to the Beverly Enterprises central office; POs were identifiable by resident ID. Regulations of the CMS require that all prescription and nonprescription drugs and any care orders (e.g., physical therapy treatment, diet, laboratory orders, and procedures administered to an NH patient) must have documentation in a PO. Beverly Enterprises requires that POs include the diagnosis or condition for which each drug is ordered. The requirement that diagnoses/conditions ascribed to medications must correspond to those in the medical record is a great advantage to researchers; rarely does a medications database include the physician's reason for a drug order.

Medication orders from the POs are available continuously from admission to the end of an NH stay. To clarify, the POs from the Beverly Enterprises system are not copies of orders written in the chart or phoned to the NH. Rather, the POs are legal medical documents that contain all orders, order changes, and discontinuations. The POs are also not the record of drugs administered, that is the Medication Administration Record (MAR), required by the CMS. The POs are used to generate the MAR forms, but POs do not record drugs that are actually administered. For example, it cannot be determined from the PO whether a resident was actually administered the drug or whether he/she was given a generic or brand name version of the ordered drug.

As part of the residents study (Garrard *et al.*, 2000), a validity study was conducted to determine the level of agreement between the electronically generated POs at the Beverly Enterprises central office and the documentation of drug orders in resident records in the NHs. The NH residents in this validity study were all aged ≥65 years, with current AED orders ($n = 42$) in seven local NHs. The percentage of agreement between the electronic POs and the gold standards (i.e., NH records for diagnostic information and the MAR for AEDs) was calculated for each of three variables: (1) presence/absence of diagnosis of epilepsy/seizure disorder, (2) presence/absence of AED orders during the study period, and (3) exact agreement in milligrams of AED dosages. These data were gathered on site in the NHs by a neurologist with a specialty in epilepsy. Agreement between the neurologist's assessment of NH records and the PO was 86% for diagnostic information (kappa $= 0.72$) and 100% for the presence/absence of AED orders (kappa $= 1.00$). Exact agreement between dosage administered (from the MAR) and dosage ordered (electronic record) ranged from 79% to 100%, depending on the AED; kappas ranged from 0.81 to 1.00. These results suggest strong agreement for each of the three variables between the electronic PO and each respective gold standard. Kappa values ≥0.75 signify excellent agreement beyond chance (Landis and Koch, 1977).

D. MINIMUM DATA SET

In 1990, the Health Care Financing Administration (HCFA, now known as the CMS) implemented the Omnibus Budget Reconciliation Act of 1987 federal legislation, which required that all NH residents in all Medicare-certified NHs be evaluated on admission, annually, and anytime there is significant change in the patient's condition using a standardized full assessment form, the MDS for Nursing Home Assessment and Care Screening. The MDS is completed on all residents, whether their care is reimbursed by Medicare or not, and the evaluator must be a nurse with MDS training. An abbreviated form of the MDS is also used for standard quarterly assessments and other reasons, but this use varies widely across states, so these forms are seldom used in research. In the admissions study (Garrard *et al.*, 2003), only the full MDS was used because it contains the complete set of required variables.

Data acquired from the MDS have been widely used for gerontological research (Brown *et al.*, 2002; Fries *et al.*, 1997; Rantz *et al.*, 1997; Won *et al.*, 1999). The validity and reliability of the MDS have been established in numerous research studies. A revision, MDS 2.0, used initially in 1990, was approved by the HCFA and implemented in 1996 (Abicht-Swensen and Debner, 1999; Frederiksen *et al.*, 1996; Hartmaier *et al.*, 1995; Hawes *et al.*, 1995; Morris *et al.*, 1997). Since MDS data for the admissions study were gathered in NHs after 1996, all data were based on MDS 2.0.

Demographic data for all subjects were obtained from the MDS, including age, gender, race, and ethnicity. Diagnoses were coded using International Classification of Diseases, Ninth Revision (ICD-9) codes extended to two places beyond the decimal point. This assured that diagnostic codes for chronic conditions, such as schizophrenia, and concomitant illnesses (e.g., acute illnesses) were captured in the database.

In the admissions study (Garrard *et al.*, 2003), cognitive performance was measured using the MDS Cognition Scale (MDS-COGS) (Hartmaier *et al.*, 1994), which was based on data collected in the MDS. This is a validated measure of cognitive performance in NH residents (Gruber-Baldini *et al.*, 2000). The instrument is not applicable to comatose patients. The MDS-COGS scores range between 0 and 10; each one-point increase indicates worsening cognitive performance. Functional ability was measured using an Activities of Daily Living (ADL) scale developed by other researchers and based on MDS items. The psychometric properties of this scale have been reported elsewhere (Abicht-Swensen and Debner, 1999).

In 1997, the HCFA required that all MDS records be transferred electronically from NHs to state-level Medicare quality assurance departments using a standardized format and that a data quality protocol of range and logic checks be established. Beverly Enterprises was in compliance with these requirements.

E. STUDY VARIABLES

The residents study (Garrard *et al.*, 2000) concentrated on four commonly used AEDs: phenytoin, carbamazepine, valproic acid, and phenobarbital, although other AEDs used by elderly NH residents were also included. This list was expanded in the admissions study (Garrard *et al.*, 2003) to include carbamazepine, ethotoin, ethosuximide, felbamate, gabapentin, lamotrigine, mephenytoin, mephobarbital, methsuximide, phenobarbital, phenytoin, primidone, phensuximide, tiagabine, topiramate, and valproic acid/valproate sodium.

In both studies, contributing factors in the analyses included age, gender, geographic location of the NH by US Census region, and indication of epilepsy/seizure disorder. The residents study also included urban versus rural NH location and mono- versus poly-AED use as associated factors. The admissions study examined a large number of other variables as possible associated factors, including:

- Patient characteristics: educational level, race/ethnicity, alcohol and tobacco use within the past year, month of admission, living situation before admission, and Medicare and Medicaid status.
- Musculoskeletal conditions: arthritis, hip fracture, amputation, osteoporosis, and pathological bone fracture.
- Cardiovascular conditions: arteriosclerotic heart disease, cardiac dysrhythmias, congestive heart failure, deep vein thrombosis, hypotension, hypertension, peripheral vascular disease (PVD), and other cardiovascular diseases.
- Neurological conditions other than epilepsy/seizure: Alzheimer's disease, aphasia, cerebral palsy, and cerebrovascular accident (stroke).
- Psychiatric/mood conditions: anxiety disorder, depression, manic depression (bipolar disease), schizophrenia, and dementia.
- Functional ability: cognitive and physical level of function as determined by the scales described above.

F. STUDY PARTICIPANTS

The residents study used a cross-sectional design with 21,551 NH residents and was a study of AED use on a single day in 1995. The dependent variables were AED treatment (yes/no) and drug type. Selected demographic variables (e.g., age, gender, race, and geographic region) were available, but information from the MDS was not (Garrard *et al.*, 2000).

The admissions study (Garrard *et al.*, 2003) used a longitudinal design to explore AED use at the time of admission, with two study groups: (1) all persons aged ≥65 years admitted between January 1 and March 31, 1999, to one of the 510 Beverly Enterprises NH facilities in 31 US states ($n = 10,318$) and (2) a follow-up cohort ($n = 9516$) of those in the admissions group who were not using an AED at NH entry. This cohort was followed up for 3 months after their individual admission dates or until NH discharge, whichever occurred first. The numbers of elderly patients in the admissions and follow-up groups are shown in Fig. 1.

G. STATISTICAL PROCEDURES

Incidence and prevalence rates were determined, and descriptive (bivariate) statistics were calculated. The significance of contributing factors was determined by means of logistic regression analysis.

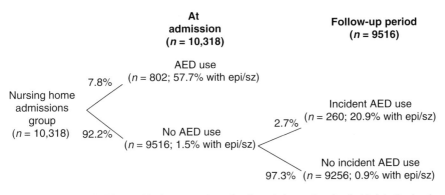

FIG. 1. Elderly residents with documentation of epilepsy/seizure disorder (epi/sz) indication by antiepileptic drug (AED) use at admission and AED use initiated during the follow-up period. Figure adapted from Garrard *et al.* (2003), with permission from John Wiley & Sons, Inc.

III. Results

A. RESIDENTS STUDY: USE OF AEDS

1. *Study Participants*

In the residents study (Garrard *et al.*, 2000), the mean age of the 21,551 residents was 83.78 years (SD, 8.13 years). Seventy-six percent were female, and the sample had the following age group distribution: 15%, 36%, and 49% for patients aged 65–74 years, 75–84 years, and ≥85 years, respectively. This distribution is similar (13%, 37%, and 50%, respectively) to that of the population of 1,557,800 people ≥65 years of age in NHs in the year 2000, on the basis of data from the US Census Bureau (Hetzel and Smith, 2001).

2. *Treatment Profile*

Of the residents in this NH sample, 10.5% had one or more AED orders on the study day and 9.2% had a seizure indication (epilepsy or seizure disorder) documented in the chart. Polypharmacy accounted for the fact that there were 2582 AED orders for the 2257 residents who were identified as AED users. Phenytoin was used by 6.2% of the residents, followed by carbamazepine (1.8%), phenobarbital (1.7%), clonazepam (1.2%), valproic acid (0.9%), and all other AEDs combined (1.2%); these percentages exceed 10.5% due to AED polytherapy.

If these results are extrapolated to all 1,557,800 elderly residents in US NHs in 2000 (Hetzel and Smith, 2001), then as many as 163,569 people were likely to have been receiving an AED. Of note, this prevalence was approximately 5 times that of AED use in the community, as reported by Hauser *et al.* in 1991 (de la Court *et al.*, 1996).

3. Factors Associated With AED Use

In the multivariate analysis of factors associated with any AED treatment (yes/no), the statistically significant factors were seizure indication ($p \leq 0.0001$), age group ($p \leq 0.0001$), and geographic region ($p \leq 0.001$). Of these, seizure indication was clearly the most important factor. Residents with a seizure indication were 87 times more likely to be taking an AED compared to those without such an indication, after taking into account the resident's age group and geographic location.

Age was inversely related to AED use. Compared to residents aged 65–74 years, those who were aged 75–84 years were half as likely to receive AED treatment, and residents ≥ 85 years were a fourth less likely. This finding was unexpected because of the upward curve in the incidence of epilepsy/seizure disorder with advancing age. Furthermore, compared to elderly people living in the community, NH residents are more likely to have experienced a stroke, which is a significant risk factor for the development of epilepsy. Thus, it could be reasoned that the percentage of those with a seizure indication and AED treatment would increase by age group. The use of AEDs as treatment for other neuropsychiatric disorders might also have been expected to increase with age. Taken together, these characteristics should have predicted a positive relationship between AED use and advancing age.

Nonetheless, the results showed a decline both in epilepsy/seizure disorder indications and in AED prevalence by age group. There may be several explanations for these findings, including the following:

1. There may have been a healthy survivor effect in which NH residents with an epileptic condition died at earlier ages, for example 65–74 years, and more of those without the condition or need for AED treatment survived to the age of 80 years.
2. Younger residents (aged 65–74 years) may have been admitted with more acute conditions associated with either epilepsy/seizure disorder or the need for AED treatment (Garrard et al., 1993).
3. Despite an epilepsy/seizure disorder indication, AED treatment was not ordered.

Both undertreatment of epilepsy/seizure disorder with advancing age and a survivor effect are plausible explanations; however, longitudinal studies are needed to determine the underlying reason for this finding.

The third factor associated with AED use was geographic region. Residents in NHs located in the South had a significantly ($p \leq 0.001$) lower prevalence of AED use than those in the Northeast. There were no other geographic differences in AED treatment rates. Whether patients with epilepsy/seizure disorder are treated differently in different geographic regions of the United States deserves more study, both for patients residing in NHs and those in the community.

4. Type of AED Used

There were also differences in the types of AEDs used. Phenytoin was the most commonly used (6% of all 21,551 residents), followed by carbamazepine (2%), phenobarbital (2%), and valproic acid (1%). Fourteen percent were treated with two or more AEDs, of which the most frequent combination was phenytoin and phenobarbital. Separate multivariate models of each of the four AEDs showed that, statistically, seizure indication was the strongest factor associated with the use of each AED. Age group was also a significant factor for three AEDs (i.e., phenytoin, carbamazepine, and valproic acid). The pattern of use by age group was consistent across the three AEDs, which replicated the overall finding of declining use with increasing age.

The use of AEDs also differed by geographic region, but the patterns were not consistent. Residents in NHs in the Northeast and South were more likely to be taking phenytoin. Valproic acid was in greater use in the Northeast, Midwest, and West, and phenobarbital was more commonly used in the Northeast compared to the other three regions. The use of carbamazepine did not differ across the four geographic regions.

Considering all the multivariate results: the older the resident, the less likely that he or she would receive any AED. However, if an AED was used, then the dosage was more likely to be lower with advancing age. Whether this was due to declining need for treatment because of lower prevalence of seizures or due to caution by the clinician cannot be determined with this data set.

B. Admissions Study: Use of AEDs

1. Study Participants

In the admissions study (Garrard et al., 2003), 66% of the 10,318 study participants were female. Persons at admission tended to be younger than residents already living in NH facilities, with an age group distribution of 20% aged 65–74 years, 44% aged 75–84 years, and 36% aged 85 years and older.

2. Treatment Profile

Approximately 8% ($n = 802$) of the admissions group used one or more AEDs at entry, and among these, greater than half (58%) had an epilepsy/seizure disorder indication. The AEDs used most frequently were phenytoin (47% of the orders among the 802 newly admitted residents with AED use), valproic acid (17%), gabapentin (15%), and carbamazepine (10%) (Table I). These four drugs constituted 89% of all AED orders at admission.

In the follow-up cohort ($n = 9516$), consisting of newly admitted residents who were not taking an AED at admission, an additional 3% ($n = 260$) were

TABLE I

Percentage and Number of Medication Orders for an AED Taken at Admission or
Initiated During the Follow-up Period by Documentation of Epilepsy/Seizure
Disorder Indication

	AED orders present at admission (882 AED orders for 802 new residents)			AEDs initiated during follow-up (312 AED orders for 260 residents)		
	Epilepsy/seizure disorder indication			Epilepsy/seizure disorder indication		
AED	Present (%)	Absent (%)	*n*	Present (%)	Absent (%)	*n*
Phenytoin	79	21	414	40	60	100
Valproic acid	40	60	154	8	92	87
Gabapentin	21	79	132	13	87	80
Carbamazepine	63	37	84	11	89	28
Phenobarbital	82	18	60	70	30	10
Primidone	35	65	34	17	83	6
Other	75	25	4	0	100	1
Any AED	60	40	882[a]	22	78	312[a]

[a]The number of medication orders for antiepileptic drugs (AEDs) at admission ($n = 882$) exceeds the total number of patients with AED use ($n = 802$) because of simultaneous use of two or more AEDs. For the same reason, the number of AEDs initiated during follow-up ($n = 312$) exceeds the number of residents ($n = 260$).

There were no users of the following AEDs at either admission or follow-up: ethotoin, ethosuximide, felbamate, mephobarbital, methsuximide, phensuximide, and tiagabine. The following other AEDs were ordered for only one or two individuals: mephenytoin, topiramate, and lamotrigine.

Table adapted from Garrard *et al.* (2003), with permission from John Wiley & Sons, Inc.

initiated on an AED after entry. Of these, approximately one-fifth (21%) had an epilepsy/seizure disorder indication. The AEDs used by this group included phenytoin (32% of the orders for the 260 follow-up subjects), valproic acid (28%), gabapentin (26%), carbamazepine (9%), and phenobarbital (3%); AED polypharmacy was 20% (Table I). The prevalence of epilepsy/seizure disorder, regardless of AED use, was 5.8% ($n = 602$) in the admissions group and 1.5% ($n = 140$) in the follow-up group.

Seven percent of the AED orders at admission were for phenobarbital compared to only 3% of all AED orders initiated during follow-up. Among the six AEDs most commonly initiated during follow-up, phenobarbital had the highest percentage of epilepsy/seizure disorder indications.

3. *Factors Associated With AED Use*

Four factors were associated with AED use in the admissions group: epilepsy/seizure disorder, age group, cognitive performance, and manic depression (bipolar disease). The epilepsy/seizure disorder variable interacted with both age group and cognitive performance, and in both interactions, epilepsy/seizure disorder was the dominant factor.

The interaction between an epilepsy/seizure disorder indication and age group showed that AED use declined significantly as age increased when epilepsy/seizure disorder was not present. Specifically, compared to individuals aged 65–74 years without epilepsy/seizure disorder, those aged 75–84 years with no epilepsy/seizure disorder were one-third less likely ($p < 0.01$) to be taking an AED at admission, and the oldest group (aged $\geq$85 years) was two-thirds less likely ($p < 0.0001$).

Patients with an epilepsy/seizure disorder indication, however, had an extremely high likelihood of AED use on admission (odds ratios of 81.57, 81.04, and 90.97 for individuals 65–74, 75–84, and 85 years or older, respectively). These odds ratios did not differ statistically across age groups among the admitted patients with epilepsy/seizure disorder.

There was also a significant interaction between an epilepsy/seizure disorder indication and cognitive performance based on the MDS-COGS. As cognitive performance worsened (with each one-point increase in the MDS-COGS), newly admitted individuals without epilepsy/seizure disorder were 9% more likely to be using an AED ($p < 0.0001$). Of those with epilepsy/seizure disorder, however, worsening cognitive performance was associated with a 2% decrease in AED use ($p < 0.0001$). The association between AED use and an epilepsy/seizure disorder indication was extremely high across the entire range of the MDS-COGS scale.

Manic depression (bipolar disease) was independently associated with AED use. Newly admitted individuals with this indication were 11 times more likely to be using an AED on NH entry than were those without manic depression (bipolar disease).

Following the analysis of any AED use, an additional multivariate analysis was performed to determine which specific AEDs were being used by individuals at admission, on the basis of the presence of either epilepsy/seizure disorder only ($n = 585$), manic depression (bipolar disease) only ($n = 105$), or both ($n = 8$). This additional analysis was conducted after the Garrard et al. (2003) paper was published and took into account diagnoses and indications from both the MDS and the POs; it was therefore more extensive than the analysis reported in the Garrard et al. (2003) paper. Results showed that the AEDs used by newly admitted individuals with epilepsy/seizure disorder ($n = 585$) included phenytoin ($n = 315$; 54%), valproic acid ($n = 57$; 10%), carbamazepine ($n = 52$; 9%), and gabapentin ($n = 27$; 5%).

Among the 105 newly admitted individuals with manic depression (bipolar disease) but not epilepsy/seizure disorder, the AEDs commonly used included valproic acid ($n = 22$; 21%), gabapentin ($n = 8$; 8%), and carbamazepine ($n = 4$; 4%). None of the individuals with only a bipolar indication received phenytoin. Among the eight individuals with both indications, AED use included valproic acid ($n = 4$), phenytoin ($n = 4$), and carbamazepine ($n = 1$).

The remainder of the admissions group with AED use had neither an epilepsy/seizure disorder nor a manic depression (bipolar disease) indication ($n = 308$; 38%).

Although other diagnoses, for example, neuropathic pain, headaches, and behavioral problems, were included as variables in the initial analyses, they did not reach statistical significance in the final analysis. The strong positive association between AED use and manic depression (bipolar disease) suggests that AEDs are being used as a treatment option, possibly as adjunctive therapy for the manic phase, which has been discussed in the psychiatric literature (Chengappa *et al.*, 1999; McElroy *et al.*, 1992, 2000).

4. *Initiation of AEDs During Follow-up*

Among residents in the follow-up cohort ($n = 9516$) who were not using an AED at admission, 260 (3%) were initiated on an AED within 3 months of admission. Factors associated with the initiation of AEDs during this period included epilepsy/seizure disorder, manic depression (bipolar disease), age group, cognitive performance (MDS-COGS), and PVD. Of these five factors, the strongest independent association was with epilepsy/seizure disorder ($p < 0.00001$). If members of the follow-up cohort developed this indication, the odds were over 25 to 1 that they would be initiated on an AED during the follow-up period.

Manic depression (bipolar disease) was also independently and significantly associated with initiation of an AED during follow-up. Cohort members with the indication were over four times more likely to be receiving an AED ($p < 0.0001$) than were those without manic depression (bipolar disease).

There was an inverse relationship between age group and initiation of an AED. Compared to the young-old (aged 65–74 years), those in the old group (aged 75–84 years) were one-third less likely to have an AED initiated ($p < 0.05$), and the oldest-old (aged ≥85 years) were only half as likely ($p < 0.0001$).

The interaction between PVD and cognitive performance was puzzling. Among cohort members without PVD, there was a 13% higher likelihood that they would use an AED as cognitive performance worsened. However, if PVD was present, then the interaction was not significantly related to AED use. The role of PVD in the use of AEDs in NHs warrants further study.

IV. Discussion

Both studies (Garrard *et al.*, 2000, 2003) had the strength of large numbers of NHs and study participants. A second strength was the use of data from up to two secondary source databases based on universally used data sources in NH research: the MDS and POs for all medications.

Several weaknesses need to be considered for both studies. The sample of NHs did not result from random sampling and was therefore not statistically representative of all NHs in the United States. For this reason, generalization of the results is not possible. A second problem concerned the definition of AED use or initiation. All medication use was based on POs, and although the order was assumed to be valid, there was the possibility that a small percentage of medications may not have been administered as ordered. This could have resulted in a slight overestimate of drug use or initiation. A third potential weakness was the assumption that if both an epilepsy/seizure disorder indication and an AED order were present, then the person was being treated for seizure control; the AED could have been used for treatment of some condition other than epilepsy or seizure disorder. Although seizure control is the primary reason for prescribing AEDs (Reynolds, 1987; Wallace et al., 1998), these drugs are also prescribed for other conditions, such as neuropathic pain (Cloyd et al., 1994; Ross, 2000), headaches (Di Trapani et al., 2000; Ross, 2000), manic depression (bipolar disease) (Calabrese et al., 1999), mood disorders (Harden et al., 1999), and behavioral problems (Cloyd et al., 1994; Gleason and Schneider, 1990).

In the residents study (Garrard et al., 2000), there were other potential design flaws. As with any point prevalence design, medication use over a 1-day study period may be problematic. Variation in treatment may be greater over a longer period, and seasonality may also be a factor. An improved study design would span 12 months. Another problem was the lack of any information in the database about the specific type of seizure disorder or the severity of the condition. In general, however, the results of both these studies provided an overview of the prevalence of AED use and an analysis of factors associated with AED use, including demographic variables, AED type, and clinical and functional variables. The conclusions from these results suggest that AED prevalence in NHs is 10 times greater than that in the community and that, among newly admitted residents, almost 8% were already taking an AED. The results also showed that 25% of the residents in the young-old age group (aged 65–74 years) were likely to receive an AED.

The residents study (Garrard et al., 2000) found that AED use decreased with age. Since publication of these results, the same finding was discovered by NH researchers in Italy (Galimberti et al., 2006). In the admissions study (Garrard et al., 2003), 63% of individuals on AEDs had an epilepsy/seizure disorder diagnosis. Subsequent studies in other countries found similar rates: 63% in a German study (Huying et al., 2006) and 67% in an Italian study (Galimberti et al., 2006).

In general, both the admissions and the residents studies were based on one of the largest nongovernmental data sets available for the study of elderly people living in NHs in the United States. Through the use of multivariate analyses, it was possible to isolate and study many of the variables that need to be considered to better understand how and why elderly NH residents and newly admitted NH patients use AEDs.

Acknowledgments

This research was supported in part by NIH grants to the Epilepsy Clinical Research Program, NIH, NINDS P50-NS16308, and NINDS 2P50-NS 16308-22A1, Ilo Leppik, MD, Principal Investigator. The University of Minnesota Epilepsy Research and Education Program includes four projects, of which one is the Project on Pharmacoepidemiology of Antiepileptic Drugs, Judith Garrard, PhD, Project Principal Investigator.

We are grateful to the leadership and staff of Beverly Enterprises, Inc., for their cooperation in making the database available for these studies. They are not responsible for the content or the opinions expressed in this chapter.

Portions of the research reported in this chapter were presented at the 2003 International Geriatric Epilepsy Symposium and at national and international meetings of the American Epilepsy Society, the American Public Health Association, and the International Society for Pharmacoepidemiology.

References

Abelson, R. (2002). Bringing discipline (and scorecards) to nursing homes. *New York Times*, New York, July 7, 2002, p. 1, 10.

Abicht-Swensen, L. M., and Debner, L. K. (1999). The Minimum Data Set 2.0: A functional assessment to predict mortality in nursing home residents. *Am. J. Hosp. Palliat. Care* **16**, 527–532.

Administration on Aging (AoA). (2005). A Profile of Older Americans: 2003. U. S. Department of Health and Human Services, Washington, DC.

Brown, M. N., Lapane, K. L., and Luisi, A. F. (2002). The management of depression in older nursing home residents. *J. Am. Geriatr. Soc.* **50**, 69–76.

Calabrese, J. R., Bowden, C. L., Sachs, G. S., Ascher, J. A., Monaghan, E., and Rudd, G. D. (1999). A double-blind placebo-controlled study of lamotrigine monotherapy in outpatients with bipolar I depression. Lamictal 602 Study Group. *J. Clin. Psychiatry* **60**, 79–88.

Chengappa, K. N., Rathore, D., Levine, J., Atzert, R., Solai, L., Parepally, H., Levin, H., Moffa, N., Delaney, J., and Brar, J. S. (1999). Topiramate as add-on treatment for patients with bipolar mania. *Bipolar Disord.* **1**, 42–53.

Cloyd, J. C., Lackner, T. E., and Leppik, I. E. (1994). Antiepileptics in the elderly. Pharmacoepidemiology and pharmacokinetics. *Arch. Fam. Med.* **3**, 589–598.

Coughlin, T. A., McBride, T. D., and Liu, K. (1990). Determinants of transitory and permanent nursing home admissions. *Med. Care* **28**, 616–631.

de la Court, A., Breteler, M. M., Meinardi, H., Hauser, W. A., and Hofman, A. (1996). Prevalence of epilepsy in the elderly: The Rotterdam Study. *Epilepsia* **37**, 141–147.

Di Trapani, G., Mei, D., Marra, C., Mazza, S., and Capuano, A. (2000). Gabapentin in the prophylaxis of migraine: A double-blind randomized placebo-controlled study. *Clin. Ter.* **151**, 145–148.

Frederiksen, K., Tariot, P., and De Jonghe, E. (1996). Minimum Data Set Plus (MDS+) scores compared with scores from five rating scales. *J. Am. Geriatr. Soc.* **44**, 305–309.

Fries, B. E., Hawes, C., Morris, J. N., Phillips, C. D., Mor, V., and Park, P. S. (1997). Effect of the National Resident Assessment Instrument on selected health conditions and problems. *J. Am. Geriatr. Soc.* **45**, 994–1001.

Galimberti, C. A., Magri, F., Magnani, B., Arbasino, C., Cravello, L., Marchioni, E., and Tartara, A. (2006). Antiepileptic drug use and epileptic seizures in elderly nursing home residents: A survey in the province of Pavia, Northern Italy. *Epilepsy Res.* **68,** 1–8.

Garrard, J., Buchanan, J. L., Ratner, E. R., Makris, L., Chan, H. C., Skay, C., and Kane, R. L. (1993). Differences between nursing home admissions and residents. *J. Gerontol.* **48,** S301–S309.

Garrard, J., Cloyd, J., Gross, C., Hardie, N., Thomas, L., Lackner, T., Graves, N., and Leppik, I. (2000). Factors associated with antiepileptic drug use among elderly nursing home residents. *J. Gerontol. A Biol. Sci. Med. Sci.* **55,** M384–M392.

Garrard, J., Harms, S., Hardie, N., Eberly, L. E., Nitz, N., Bland, P., Gross, C. R., and Leppik, I. E. (2003). Antiepileptic drug use in nursing home admissions. *Ann. Neurol.* **54,** 75–85.

Gleason, R. P., and Schneider, L. S. (1990). Carbamazepine treatment of agitation in Alzheimer's outpatients refractory to neuroleptics. *J. Clin. Psychiatry* **51,** 115–118.

Gruber-Baldini, A. L., Zimmerman, S. I., Mortimore, E., and Magaziner, J. (2000). The validity of the minimum data set in measuring the cognitive impairment of persons admitted to nursing homes. *J. Am. Geriatr. Soc.* **48,** 1601–1606.

Harden, C. L., Lazar, L. M., Pick, L. H., Nikolov, B., Goldstein, M. A., Carson, D., Ravdin, L. D., Kocsis, J. H., and Labar, D. R. (1999). A beneficial effect on mood in partial epilepsy patients treated with gabapentin. *Epilepsia* **40,** 1129–1134.

Hartmaier, S. L., Sloane, P. D., Guess, H. A., and Koch, G. G. (1994). The MDS Cognition Scale: A valid instrument for identifying and staging nursing home residents with dementia using the minimum data set. *J. Am. Geriatr. Soc.* **42,** 1173–1179.

Hartmaier, S. L., Sloane, P. D., Guess, H. A., Koch, G. G., Mitchell, C. M., and Phillips, C. D. (1995). Validation of the Minimum Data Set Cognitive Performance Scale: Agreement with the Mini-Mental State Examination. *J. Gerontol. A Biol. Sci. Med. Sci.* **50,** M128–M133.

Hauser, W. A. (1992). Seizure disorders: The changes with age. *Epilepsia* **33**(Suppl. 4), S6–S14.

Hauser, W. A. (1995). Recent developments in the epidemiology of epilepsy. *Acta Neurol. Scand. Suppl.* **162,** 17–21.

Hauser, W. A., and Annegers, J. F. (1991). Risk factors for epilepsy. *Epilepsy Res. Suppl.* **4,** 45–52.

Hauser, W. A., Annegers, J. F., and Kurland, L. T. (1993). Incidence of epilepsy and unprovoked seizures in Rochester, Minnesota: 1935–1984. *Epilepsia* **34,** 453–468.

Hauser, W. A., Annegers, J. F., and Rocca, W. A. (1996). Descriptive epidemiology of epilepsy: Contributions of population-based studies from Rochester, Minnesota. *Mayo Clin. Proc.* **71,** 576–586.

Hawes, C., Morris, J. N., Phillips, C. D., Mor, V., Fries, B. E., and Nonemaker, S. (1995). Reliability estimates for the Minimum Data Set for nursing home resident assessment and care screening (MDS). *Gerontologist* **35,** 172–178.

Hetzel, L., and Smith, A. (2001). The 65 years and over population: 2000. Census 2000 Brief: U.S. Census Bureau, C2KBR/01–10, Washington, DC.

Huying, F., Klimpe, S., and Werhahn, K. J. (2006). Antiepileptic drug use in nursing home residents: A cross-sectional, regional study. *Seizure* **15,** 194–197.

Kemper, P., and Murtaugh, C. M. (1991). Lifetime use of nursing home care. *N. Engl. J. Med.* **324,** 595–600.

Landis, J. R., and Koch, G. G. (1977). The measurement of observer agreement for categorical data. *Biometrics* **33,** 159–174.

Liu, K., McBride, T., and Coughlin, T. (1994). Risk of entering nursing homes for long versus short stays. *Med. Care* **32,** 315–327.

Liu, K., Gage, B., Harvell, J., Stevenson, D., and Brennan, N. (1999). Medicare's post-acute care benefit: Background, trends, and issues to be faced. U.S. Department of Health and Human Services, Washington, DC.

Luhdorf, K., Jensen, L. K., and Plesner, A. M. (1986). Epilepsy in the elderly: Incidence, social function, and disability. *Epilepsia* **27,** 135–141.

McElroy, S. L., Keck, P. E., Jr., Pope, H. G., Jr., and Hudson, J. I. (1992). Valproate in the treatment of bipolar disorder: Literature review and clinical guidelines. *J. Clin. Psychopharmacol.* **12,** 42S–52S.

McElroy, S. L., Suppes, T., Keck, P. E., Frye, M. A., Denicoff, K. D., Altshuler, L. L., Brown, E. S., Nolen, W. A., Kupka, R. W., Rochussen, J., Leverich, G. S., and Post, R. M. (2000). Open-label adjunctive topiramate in the treatment of bipolar disorders. *Biol. Psychiatry* **47,** 1025–1033.

Morris, J. N., Nonemaker, S., Murphy, K., Hawes, C., Fries, B. E., Mor, V., and Phillips, C. (1997). A commitment to change: Revision of HCFA's RAI. *J. Am. Geriatr. Soc.* **45,** 1011–1016.

Rantz, M. J., Popejoy, L., Mehr, D. R., Zwygart-Stauffacher, M., Hicks, L. L., Grando, V., Conn, V. S., Porter, R., Scott, J., and Maas, M. (1997). Verifying nursing home care quality using Minimum Data Set quality indicators and other quality measures. *J. Nurs. Care Qual.* **12,** 54–62.

Reynolds, E. H. (1987). Early treatment and prognosis of epilepsy. *Epilepsia* **28,** 97–106.

Ross, E. L. (2000). The evolving role of antiepileptic drugs in treating neuropathic pain. *Neurology* **55,** S41–S46; discussion S54–S58.

Spillman, B. C., and Lubitz, J. (2002). New estimates of lifetime nursing home use: Have patterns of use changed? *Med. Care* **40,** 965–975.

U.S. Department of Commerce (1993). We, the American Elderly. Economics and Statistics Administration, Bureau of the Census, Washington, DC.

Vickery, K. (2000). Top nursing facility chains. *Provider* **26,** 37–41.

Wallace, H., Shorvon, S., and Tallis, R. (1998). Age-specific incidence and prevalence rates of treated epilepsy in an unselected population of 2,052,922 and age-specific fertility rates of women with epilepsy. *Lancet* **352,** 1970–1973.

Won, A., Lapane, K., Gambassi, G., Bernabei, R., Mor, V., and Lipsitz, L. A. (1999). Correlates and management of nonmalignant pain in the nursing home. SAGE Study Group. Systematic Assessment of Geriatric drug use via Epidemiology. *J. Am. Geriatr. Soc.* **47,** 936–942.

AGE-RELATED CHANGES IN PHARMACOKINETICS: PREDICTABILITY AND ASSESSMENT METHODS

Emilio Perucca

Clinical Pharmacology Unit, Department of Internal Medicine and Therapeutics
University of Pavia, Pavia, Italy; and
Institute of Neurology IRCCS C. Mondino Foundation, Pavia, Italy

Although there have been relatively few studies of the pharmacokinetics of antiepileptic drugs (AEDs) in old age, available evidence indicates that the clearance of most old and new generation AEDs is reduced on average by about 20–40% in elderly patients compared with nonelderly adults. Depending on the pharmacokinetic characteristics of the drug, the reduction in clearance can be ascribed to a physiological reduction in rate of drug metabolism, to a decrease in renal excretion rate, or to both. Studies have consistently demonstrated that interindividual pharmacokinetic variability in old age is particularly prominent, due not only to the influence of aging-related physiological changes, but also to the impact of comorbidities and drug–drug interactions. For extensively metabolized drugs, there are no reliable tools to predict with a high degree of accuracy the pharmacokinetic behavior of an AED in an individual patient. With renally eliminated drugs, determination of creatinine clearance may provide a useful clue in predicting individual changes in drug clearance and the consequent need for dosage adjustment. In the therapeutic setting, measurement of serum AED concentrations can be valuable in individualizing dosage in an elderly person, even though it should be remembered that in the case of drugs that are highly bound to plasma proteins the total serum concentration may underestimate the level of unbound, pharmacologically active drug. Because aging is also associated with important pharmacodynamic changes that may alter the relationship between serum drug concentration and pharmacological effects, pharmacokinetic measurements alone are not a substitute for the need to monitor clinical response carefully and to adjust dosage accordingly.

I. Introduction

For almost all therapeutic agents, available information on pharmacokinetic properties refers mostly to patients aged less than 65 years, despite the fact that it is mainly elderly persons who need to receive these same drugs. Antiepileptic drugs (AEDs) are no exception: a survey of pharmacy provider organizations in the United States found that as many as 1132 of 10,168 nursing home residents

TABLE I

AGING-RELATED CHANGES THAT MAY AFFECT PHARMACOKINETICS IN THE ELDERLY

Absorption	
Gastrointestinal motility	May be decreased
Gastrointestinal secretions (including gastric acid)	May be decreased
Gastrointestinal blood flow	May be decreased
Mucosal absorptive surface	May be decreased
Distribution	
Serum albumin	Decreased
α1-acid glycoprotein	Increased or unchanged
Body fat/lean mass ratio	Increased
Total body water	Decreased
Biotransformation	
Liver mass	Decreased
Liver blood flow	Decreased
Activity of cytochrome P450 (CYP) enzymes	Decreased (most CYPs)
Activity of phase II (conjugation) enzymes	Unaffected or decreased
Renal elimination	
Renal weight	Decreased
Glomerular filtration rate	Decreased
Renal blood flow	Decreased
Filtration fraction	Increased
Tubular function	May be decreased

(11.1%) were prescribed AEDs for the treatment of seizure disorders or for other indications (Schachter *et al.*, 1998). Yet, aging-related changes in the pharmacokinetics of these drugs have been little investigated, even in the case of first generation AEDs (Bernus *et al.*, 1997; Willmore, 1995). Up to the year 2002, for example, the largest cohort of elderly patients included in various pharmacokinetic assessments of carbamazepine included only 14 individuals (Graves *et al.*, 1998), and there had not been a single study of phenobarbital pharmacokinetics in old age (Bernus *et al.*, 1997; King-Stephens, 1999).

Assessing pharmacokinetic properties in the elderly is important because each of the processes involved in drug absorption and disposition can be altered in old age (Table I), often to an extent that requires dosage adjustments (Hammerlein *et al.*, 1998; Perucca, 2006). The purpose of this chapter is to discuss how pharmacokinetics changes in old age, with special reference to AEDs; to review the tools that are available to predict such changes; and to address a number of methodological aspects in performing pharmacokinetic evaluations in this age group.

II. The Effect of Aging on Pharmacokinetics

A. Absorption

Despite the fact that several physiological processes potentially affecting drug absorption are altered in old age (Table I), most drugs appear to be adequately absorbed when administered orally in the elderly (Hammerlein *et al.*, 1998), and any absorption problem is more likely to be related to associated diseases (e.g., atrophic gastritis) than to aging per se (Russell, 2001). Although there are no known examples of AEDs whose gastrointestinal absorption is impaired in old age, most pharmacokinetic studies did not specifically investigate absorption processes, and therefore the possibility of clinically important alterations cannot be excluded. In fact, it has been speculated that day-to-day fluctuations in absorption efficiency could be responsible for the marked (up to threefold) intraindividual variability in steady-state serum phenytoin concentrations observed in elderly nursing home residents stabilized on a constant dosage of the drug (Birnbaum *et al.*, 2003).

B. Distribution

A number of AEDs, most notably phenytoin, valproic acid, and tiagabine, are extensively (about 90%) bound to plasma albumin. Albumin concentration decreases in old age, and this may result in a reduced extent of drug binding;

indeed, an increase in unbound drug fraction in the elderly has been demonstrated for both phenytoin (Perucca, 1980) and valproic acid (Bauer et al., 1985; Perucca et al., 1984).

The clinical implications of alterations in protein-binding capacity are often misunderstood. In the case of drugs such as phenytoin and valproic acid, which show restrictive (flow-independent) elimination, a change in unbound fraction should result in increased total drug clearance and decreased serum concentration of total (unbound plus protein-bound) drug, but the serum concentration of unbound, pharmacologically active drug would be expected to change only when there are concurrent changes in drug-metabolizing capacity (which, in fact, is the case for both phenytoin and valproic acid). In any case, it is important to remember that, when the unbound fraction increases, therapeutic and toxic effects will occur at total serum drug concentrations lower than usual (Grandison and Boudinot, 2000). Failure to take this into account when interpreting serum drug concentrations could have disastrous consequences. Tomson (1988), for example, described an 81-year-old woman with hypoalbuminemia in whom adjustment of phenytoin dosage based on measurement of total serum drug concentrations led to the development of severe toxicity. Subsequent measurement of unbound serum phenytoin in the same patient revealed a drug level well into the toxic range, despite the fact that the total drug concentration was within the usually accepted therapeutic window.

Because of the alterations in unbound fraction of phenytoin and valproic acid in old age, a case could be made for monitoring the unbound rather than the total drug concentrations in this age group. It is also possible to obtain a rough estimate of the change in unbound fraction by simply measuring the concentration of serum albumin (Perucca, 1980; Perucca et al., 1984). In patients with impaired renal function, however, the increase in unbound fraction may be greater than expected from the reduction in serum albumin: this is because renal insufficiency results in the accumulation of endogenous substances in plasma that displace phenytoin and valproic acid from plasma proteins (Reidenberg and Drayer, 1984).

In addition to changes in plasma protein binding, other factors may affect drug distribution in the elderly. With aging, the proportion of adipose tissue in the body increases about twofold in men and by about 50% in women; as a result, lipophilic drugs may have an increased, and hydrophilic drugs a relatively decreased, apparent volume of distribution in the elderly (Bernus et al., 1997). As most centrally active drugs are lipophilic in nature, their volume of distribution tends to increase with age to an extent that is proportional to their solubility in lipids. This, in turn, will lead to prolongation of the drug's half-life, irrespective of any change in drug-metabolizing capacity. The half-life of diazepam, for example, may increase up to fourfold between age 20 and 80, and this is largely

due to an increased volume of distribution rather than to a change in metabolizing capacity, which decreases only moderately in old age (Klotz *et al.*, 1975). This
example illustrates the pitfalls of using half-life measurements as an indicator of
the efficiency of drug elimination processes. Unlike clearance, which directly
reflects eliminating capacity and therefore represents the most useful measure in
calculating dose requirements, half-life is a hybrid parameter whose changes may
reflect alterations in volume of distribution, elimination capacity, or both.

C. BIOTRANSFORMATION

Aging is associated with various anatomical and physiological changes in the
liver, which is the major metabolizing organ for most drugs (Table I). These
changes include a decrease in liver size, a reduced hepatic blood flow, and
alterations in liver morphology (Zeeh and Platt, 2002). With respect to the activity
of cytochrome P450 (CYP)-dependent drug metabolizing mono-oxygenases, some
studies did not demonstrate deficiencies in these enzymes in the elderly, whereas
others found a decline in the activity of specific enzyme isoforms, particularly
CYP3A4 but also CYP1A2, CYP2C9, CYP2C19, CYP2D6, and CYP2E1
(Schmucker, 2001; Tanaka, 1998). Differential changes in the activity of individual
isoenzymes may explain why the ability of the aging liver to metabolize drugs does
not decline in a similar way for all pharmacological agents (Hammerlein *et al.*,
1998). Perhaps the most remarkable characteristic of liver function in the elderly is
an increase in interindividual variability, a feature that may obscure age-related
differences (Schmucker, 2001).

In general, phase I metabolic reactions (oxidation and reduction) tend to
occur at a slower rate in the elderly (Bernus *et al.*, 1997; Hammerlein *et al.*, 1998),
and AEDs that are extensively cleared by oxidation (e.g., carbamazepine, phenytoin, tiagabine, and valproic acid) show a moderately reduced drug clearance
in older patients (Table II). Although phase II reactions (conjugation) are said to
undergo little or no change in old age (Bernus *et al.*, 1997; Hammerlein *et al.*,
1998), lamotrigine and monohydroxycarbazepine, which are eliminated primarily by glucuronide conjugation, exhibit an age-related decline in clearance that
appears comparable to that observed for drugs cleared by oxidation (Table II)
(Perucca, 2006).

Because many AEDs have enzyme-inducing activity, the question arises as to
whether the sensitivity of drug-metabolizing enzymes to induction is altered in
old age. Although it has been suggested that elderly patients are less sensitive to
enzyme induction (Salem *et al.*, 1978; Twum-Barima *et al.*, 1984; Vestal *et al.*,
1975, 1979), other investigators have found that induction occurs efficiently even
in old-age groups (Battino *et al.*, 2003; Crowley *et al.*, 1988).

TABLE II

CHANGES IN CLEARANCE OF OLD AND NEW GENERATION AEDs IN ELDERLY PATIENTS[a]

Drug	Effect of old age on drug clearance	References
Carbamazepine	Decrease by 25–40%	Battino et al., 2003; Cloyd et al., 1994; Graves et al., 1998; Scheuer and McCullough, 1995
Felbamate	Decrease by 10–20%	Richens et al., 1997
Gabapentin	Decrease by 30–50%	Armijo et al., 2004; Boyd et al., 1999
Lamotrigine	Decrease by about 35%	Posner et al., 1991
Levetiracetam	Decrease by 20–40%	Patsalos, 2000; Pellock et al., 2001
Oxcarbazepine	Decrease by 25–35%[b]	Arnoldussen and Hulsman, 1991; van Heiningen et al., 1991
Phenobarbital	Decrease by 22%	Messina et al., 2005
Phenytoin	Decrease by about 25%[c]	Bachmann and Belloto, 1999; Battino et al., 2004
Pregabalin	No data	
Tiagabine	Decrease by about 30%	Snel et al., 1997
Topiramate	Decrease by 20%	Doose et al., 1998
Valproic acid	Decrease by about 40%[d]	Bauer et al., 1985; Perucca et al., 1984
Vigabatrin	Decrease by 50–90%[e]	Haegele et al., 1988
Zonisamide	No major changes[f]	Shah et al., 2002; Wallace et al., 1998

[a]In all studies, clearance was calculated after oral administration assuming complete bioavailability. Although data refer to mean changes, interindividual variation may be considerable in relation to actual age and other factors. For more information, see Perucca (2006); see also Perucca (2005).

[b]Data refer to the active metabolite monohydroxycarbazepine.

[c]Decrease in clearance of unbound drug may be greater.

[d]Decrease in unbound drug clearance. Clearance of total (unbpund plus protein-bound drug) may not change.

[e]Elderly patients had various pathologies and were preselected on the basis of impaired renal function (study group probably not representative).

[f]Studies included elderly subjects with a mean age slightly below 70 years.

Table adapted from Perucca et al. (2006), with permission.

D. RENAL EXCRETION

Aging is associated with a physiological decrease in glomerular filtration rate and other parameters of renal function (Table I). Therefore, the clearance of drugs that are eliminated primarily unchanged in urine declines with increasing age, to an extent that often requires adjustments in dosage (Muhlberg and Platt, 1999; Parker et al., 1995). Among currently used AEDs, gabapentin, levetiracetam, pregabalin, and vigabatrin are eliminated primarily by renal excretion, even though, for levetiracetam, hydrolysis to an inactive metabolite contributes to drug clearance (Patsalos, 2000). As expected, for those AEDs in this list that have been investigated, elimination was found to occur at a slower rate in elderly patients (Table II) (Perucca, 2006).

III. Predicting Aging-Associated Pharmacokinetic Changes

To some extent, pharmacokinetic changes of individual AEDs in an elderly person can be predicted on the basis of the concepts discussed above and the results of the studies summarized in Table II. However, as elegantly pointed out by Tallis (2004), "old age is less a period of predictable change than of increased variance between individuals," and the average changes listed in Table II could be grossly misleading when used to predict the pharmacokinetic pattern observed in a given individual. Apart from the fact that the expected decline in drug clearance may be very different in a 90-year-old than in a 65-year-old, many other patient-specific factors may have a profound influence on pharmacokinetics (Table III). Although a detailed discussion of these issues is beyond the purposes of this chapter, some aspects that are especially relevant to the use of AEDs in old age will be discussed below briefly.

A. DRUGS CLEARED PRIMARILY BY BIOTRANSFORMATION

There is no single test or combination of tests that can be used to predict hepatic drug metabolizing capacity in a single individual. Therefore, any attempt to predict the pharmacokinetic behavior of extensively metabolized drugs can rely only on pharmacokinetic observations made previously in a population with comparable clinical characteristics.

The interindividual variability in the pharmacokinetics of extensively metabolized drugs is usually very high, as shown in a study in which carbamazepine apparent oral clearance (CL/F) values at steady state in 157 patients aged ≥ 65 years were compared to those recorded in an equal number of younger controls matched for gender, body weight, and comedication (Battino et al., 2003).

TABLE III

PATIENT-RELATED FACTORS THAT AFFECT PHARMACOKINETICS
OF AN ANTIEPILEPTIC DRUG IN AN ELDERLY INDIVIDUAL

Genetic background
Actual chronological age
Frailty
Dietary habits
Exposure to voluptuary substances (alcohol, cigarette smoke)
Serum albumin concentration
Glomerular filtration rate and creatinine clearance
Comorbidities
Interactions caused by concomitant medications

While carbamazepine CL/F decreased with increasing age (with a reduction of almost 30% between age 35 and 80), interindividual variation was large, and in fact, age alone accounted for less than 10% of the variability in CL/F within the elderly group. Phenobarbital comedication played an additional role: specifically, carbamazepine CL/F was significantly higher in phenobarbital-comedicated patients than in patients receiving monotherapy, presumably due to enzyme induction. Multivariate analysis revealed that, in addition to age and enzyme-inducing comedication, differences in carbamazepine dosage and body weight contributed to some extent to the observed pharmacokinetic variability.

For extensively metabolized AEDs, population pharmacokinetics may provide the most accurate tool in predicting individual clearance values (e.g., by Bayesian forecasting) and, accordingly, in identifying the dosage required to attain the desired serum drug concentration in a given patient (Battino et al., 2003; Grasela et al., 1983; Graves et al., 1998). By using this approach, an elderly person's pharmacokinetic parameters can be predicted based on knowledge of age and other relevant covariates such as a gender, body weight, type of comedication, and comorbid conditions. For highly protein-bound drugs, changes in plasma protein binding can also be predicted. Even the most sophisticated models, however, usually fail to explain a considerable proportion of pharmacokinetic variability between individuals. As a result, prediction accuracy leaves much to be desired, which may explain why prediction methods based on population kinetics have not gained wide acceptance in the individualization of AED therapy.

B. Drugs Cleared Primarily by Renal Excretion

Our ability to predict changes in drug clearance is greater for renally eliminated than for extensively metabolized drugs. Indeed, the clearance of many renally excreted drugs can be predicted relatively reliably based on knowledge of the creatinine clearance (Hammerlein et al., 1998; Muhlberg and Platt, 1999).

It should be remembered that, because of reduced muscle mass, evaluation of renal function based on serum creatinine alone can be misleading in elderly persons (Landahl et al., 1981), and even the measurement of endogenous creatinine clearance (or its indirect assessment by using, e.g., the Cockroft–Gault equation) may not provide a reliable estimate of glomerular filtration rate in old age (Fliser et al., 1999). Despite these limitations, creatinine clearance is generally regarded as a useful tool for predicting changes in renal clearance in the elderly. This was shown, for example, in an elegant study in which vigabatrin pharmacokinetics was assessed in different age groups (Fig. 1) (Haegele et al., 1988). The mean clearance of the pharmacologically active S-enantiomer of vigabatrin was 1.27 ml/min/kg in 22- to 33-year-olds, 0.68 ml/min/kg in 60- to 75-year-olds, and 0.27 ml/min/kg in 76- to 97-year-olds. These correlated with creatinine

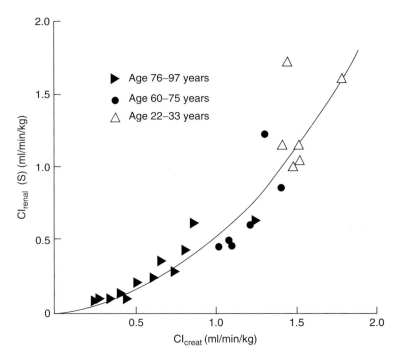

FIG. 1. Relationship between renal clearance of the active *S*-enantiomer [Cl_{renal} (S)] of vigabatrin after a single oral dose and creatinine clearance (Cl_{creat}) in 24 patients aged 22–97 years. Because elderly patients had various pathologies and were preselected based on degree of impaired renal function, clearance values in the elderly are not representative of those found in an unselected population. Reprinted from Haegele *et al.* (1988), with permission from the American Society for Clinical Pharmacology and Therapeutics, Copyright 1988.

clearance values of 97–155, 60–111, and 11–62 ml/min in the respective groups (because patients with pathologically altered renal function were included in the elderly groups, the findings of this study reflect the influence of disease-related changes in renal function rather than aging per se). Where appropriate, other variables can be incorporated into mathematical models to improve prediction power whenever creatinine clearance alone does not provide a sufficient predictive accuracy (Ducher *et al.*, 2001).

C. Comorbidities and Drug Interactions

While this chapter focuses on the pharmacokinetic implications of aging per se, an elderly person with a seizure disorder often shows comorbidities that may have a prominent influence on both pharmacokinetics and pharmacodynamics.

These persons are also more likely to receive a variety of comedications. In a survey of elderly nursing home residents in the United States, the average number of routine medications taken by AED recipients was 5.6, compared with 4.6 for residents not receiving AEDs (Lackner *et al.*, 1998). In a more recent US trial comparing carbamazepine with lamotrigine and gabapentin in elderly people with a recently diagnosed seizure disorder, the enrolled patients received an average of seven drugs as comedication (Rowan *et al.*, 2005). Use of multiple drug therapy entails a significant risk of clinically significant drug interactions, most of which involve changes in drug-metabolizing enzyme activity. In recent years, improved understanding of the isoenzymes responsible for the metabolism of AEDs and other drugs, coupled with characterization of the compounds that induce or inhibit these same enzymes, has allowed us to predict pharmacokinetic interactions at the metabolic level with remarkable accuracy (Spina *et al.*, 2005).

IV. Assessing Aging-Associated Pharmacokinetic Changes

The investigation of pharmacokinetic changes in old age entails special problems that relate partly to the difficulties in recruiting elderly persons into formal pharmacokinetic studies, and partly to the challenges involved in designing studies that address the complex factors affecting drug handling in this age group. The sections below provide a brief overview of various methodological approaches that can be used in the design of these studies.

A. Comparison of Serum Drug Levels Across Age Groups in the Therapeutic Setting

The simplest and most widely used approach to the investigation of pharmacokinetic changes in old age consists of collecting serum drug concentration data at steady state in patients receiving the drug(s) of interest in routine clinical settings (Cloyd *et al.*, 1994; Scheuer and McCullough, 1995). Serum drug concentrations can then be dose-normalized, or CL/F values can be calculated and compared with those obtained in comparable groups of younger subjects. If the age distribution of the evaluated patients is sufficiently wide, it also may be possible to determine whether a relationship exists between a specific pharmacokinetic parameter (usually CL/F) and age (Battino *et al.*, 2003, 2004; Messina *et al.*, 2005).

The obvious advantage of this approach is that collection of samples involves minimal effort, particularly with drugs that are routinely monitored in clinical practice. Moreover, the population of patients can be expected to be representative of the usual therapeutic setting, thereby allowing an estimate of the degree of interindividual variability and the factors that contribute to it. Intraindividual variability can also be explored by comparing serum drug levels at different time intervals in patients stabilized on the same dosage (Birnbaum *et al.*, 2003). Ideally, all determinations and collections of relevant clinical and pharmacological variables should be obtained prospectively. Prospective collection of information not only will improve the accuracy of the data, but may allow the important identification of individuals who may discontinue the medication early due to undesired responses related to unusual pharmacokinetic behaviors.

In studies in which different age groups are to be compared, special care should be taken when setting criteria for the selection of controls. In particular, comparing pharmacokinetic parameters between elderly patients and younger controls matched for dosage (Martines *et al.*, 1990) may lead to bias, particularly with drugs that, like many AEDs, undergo therapeutic drug monitoring. In fact, when dosages are adjusted by the physician to achieve optimal responses, or a similar target serum concentration in all patients, matching for dosage is likely to preselect patients who have similar serum drug levels (i.e., similar pharmacokinetic patterns), thereby underestimating any pharmacokinetic difference between the test group and controls (Battino *et al.*, 2003, 2004). Matching for serum drug levels would equally generate a bias if the physician had deliberately adopted different dosing policies in the two age groups.

One special problem in comparing dose-normalized drug concentration data (or pharmacokinetic parameters that require dose-normalization such as CL/F values) across age groups arises when the drugs being investigated show nonlinear (dose-dependent) pharmacokinetics. Phenytoin provides an example of nonlinear pharmacokinetics of the Michaelis–Menten type, in that the CYP2C9 enzyme contributing to its metabolism becomes progressively saturated, and therefore, within individuals, its clearance decreases with increasing dosage (Richens, 1975). If the elderly patients being assessed were prescribed, on average, lower phenytoin dosages than a randomly selected group of younger controls, any age-related decrease in metabolic rate in the elderly may be obscured by their tendency to have higher clearance values than controls as a result of the differences in dosage (Battino *et al.*, 2004). As discussed above, matching elderly persons and younger controls for dosage is not a solution because of the bias inherent in the nonrandomized dose allocation. Carbamazepine also exhibits nonlinear pharmacokinetics because its clearance increases with increasing dosage due to dose-dependent autoinduction (Kudriakova *et al.*, 1992). In this situation, comparison of clearance values between elderly patients and a group of controls receiving higher dosages may result in overestimation of any age-related decline in the metabolic clearance of the drug (Battino *et al.*, 2003). Other AEDs that exhibit nonlinear pharmacokinetics include valproic acid, which shows concentration-dependent plasma protein binding (Perucca, 2002), and gabapentin, whose oral bioavailability decreases with increasing dosage (Stewart *et al.*, 1993).

B. FORMAL PHARMACOKINETIC STUDIES

Formal pharmacokinetic studies require the frequent collection of blood and urine samples after administration of single or multiple doses in elderly volunteers or in patients who need to receive the drug for therapeutic purposes. Although pharmacokinetic parameters may be compared with those determined in historical controls, inclusion of a control group of nonelderly adults is desirable for comparison purposes. Examples of AEDs for which formal pharmacokinetic studies in the elderly have been performed include valproic acid (Bauer *et al.*, 1985; Perucca *et al.*, 1984), carbamazepine (Hockings *et al.*, 1986), oxcarbazepine (van Heiningen *et al.*, 1991), gabapentin (Boyd *et al.*, 1999), lamotrigine (Posner *et al.*, 1991), tiagabine (Snel *et al.*, 1997), and vigabatrin (Haegele *et al.*, 1988). Because these studies are labor intensive, the number of subjects investigated is usually relatively small (typically, 8–20 per group), and multiple dosing may not always be feasible.

The greatest advantage of formal pharmacokinetic studies is that they provide a precise characterization of all major pharmacokinetic parameters, including absorption rate (or peak serum drug concentration), plasma protein binding, half-life values, and a clearance estimate that is much more accurate than that obtained from sparsely collected data. Moreover, use of fixed dose allocation eliminates some of the interpretative problems discussed in the section above. Usually, inclusion criteria are designed to minimize the influence of confounding variables, with special reference to comorbidities, comedications, and differences in diet and smoking habits. Although this enables more precise characterization of the influence of aging per se, it may result in an "artificially healthy" population that may not be representative of the majority of elderly persons who will be receiving the drug in routine clinical practice. Because of the limited number of subjects and the restrictive inclusion criteria, these studies also typically fail to provide an estimate of the pharmacokinetic variability that is observed in the general therapeutic setting. Additional interpretative problems may arise when the study design is limited to the administration of single doses, which for some AEDs, such as carbamazepine and phenytoin, may not be representative of pharmacokinetic behavior at steady state.

C. POPULATION PHARMACOKINETICS APPROACHES

Population pharmacokinetics provides a more sophisticated approach to the evaluation of sparsely collected drug concentration data within the therapeutic setting. Unlike the simple comparison of relatively small groups of patients that has been discussed above, methods based on population pharmacokinetics require the collection of extensive concentration data and accurate information on associated clinical variables (Graves *et al.*, 1998; Yukawa *et al.*, 1990). As with other approaches, the highest methodological quality is achieved when the study is designed prospectively, ideally within the setting of a randomized trial involving fixed allocation to a range of dosages in a population covering a wide age range. The quality of the information may be improved by obtaining repeated samples to test for intraindividual variability and by spreading sampling times over a wide interval after the previous dose in order to obtain information on changes in serum drug concentration during a dosing interval.

The greatest advantage of population pharmacokinetics is that, with a large number of observations, an estimate can be obtained of pharmacokinetic variability in a population that is expected to be representative of the usual therapeutic setting. Careful collection of information on clinical variables should also allow detection of the major sources of variability and their relative contribution to interindividual differences in serum drug levels. As discussed above, population pharmacokinetics may also be used to predict, based on knowledge of the most relevant variables, the dosage required to achieve a desired concentration in a given individual (Graves *et al.*, 1998). Disadvantages may relate to inaccuracies inherent in the model used to analyze the data. Many physicians, additionally, find the complexities associated with the mathematical modeling unappealing.

V. Conclusions

Aging is associated with major changes in pharmacokinetics, and these may have profound clinical consequences. An understanding of the pharmacokinetic properties of individual drugs, coupled with general knowledge of the influence of aging on drug distribution and elimination processes, is essential for a rational application of pharmacological therapy in the elderly. Although studies have shown that the metabolic clearance of most AEDs is appreciably reduced in old age, interindividual variability is prominent, and it is not generally possible to predict with a high degree of accuracy the pharmacokinetic behavior of extensively metabolized drugs in an individual patient. With renally eliminated drugs, determination of creatinine clearance may provide a useful clue in predicting individual changes in drug clearance and the consequent need for dosage adjustment.

In the therapeutic setting, measurement of serum AED concentrations can be valuable in individualizing dosage in an elderly person, even though it should be remembered that, in the case of drugs that are highly bound to plasma proteins, the total serum concentration may underestimate the level of unbound, pharmacologically active drug. Because aging is also associated with important pharmacodynamic changes that may alter the relationship between serum drug concentration and pharmacological effects (Hammerlein *et al.*, 1998; Parker *et al.*, 1995), pharmacokinetic measurements alone are not a substitute for the need to monitor clinical response carefully and to adjust dosage accordingly.

References

Armijo, J. A., Pena, M. A., Adin, J., and Vega-Gil, N. (2004). Association between patient age and gabapentin serum concentration-to-dose ratio: A preliminary multivariate analysis. *Ther. Drug Monit.* **26,** 633–637.

Arnoldussen, W., and Hulsman, J. (1991). Oxcarbazepine (OCBZ) twice daily is as effective as OCBZ three times daily: A clinical and pharmacokinetic study in 24 volunteers and 6 patients. *Epilepsia* **32**(Suppl. 1), 69.

Bachmann, K. A., and Belloto, R. J., Jr. (1999). Differential kinetics of phenytoin in elderly patients. *Drugs Aging* **15,** 235–250.

Battino, D., Croci, D., Rossini, A., Messina, S., Mamoli, D., and Perucca, E. (2003). Serum carbamazepine concentrations in elderly patients: A case-matched pharmacokinetic evaluation based on therapeutic drug monitoring data. *Epilepsia* **44,** 923–929.

Battino, D., Croci, D., Mamoli, D., Messina, S., and Perucca, E. (2004). Influence of aging on serum phenytoin concentrations: A pharmacokinetic analysis based on therapeutic drug monitoring data. *Epilepsy Res.* **59,** 155–165.

Bauer, L. A., Davis, R., Wilensky, A., Raisys, V., and Levy, R. H. (1985). Valproic acid clearance: Unbound fraction and diurnal variation in young and elderly adults. *Clin. Pharmacol. Ther.* **37,** 697–700.

Bernus, I., Dickinson, R. G., Hooper, W. D., and Eadie, M. J. (1997). Anticonvulsant therapy in aged patients. Clinical pharmacokinetic considerations. *Drugs Aging* **10,** 278–289.

Birnbaum, A., Hardie, N. A., Leppik, I. E., Conway, J. M., Bowers, S. E., Lackner, T., and Graves, N. M. (2003). Variability of total phenytoin serum concentrations within elderly nursing home residents. *Neurology* **60,** 555–559.

Boyd, R. A., Turck, D., Abel, R. B., Sedman, A. J., and Bockbrader, H. N. (1999). Effects of age and gender on single-dose pharmacokinetics of gabapentin. *Epilepsia* **40,** 474–479.

Cloyd, J. C., Lackner, T. E., and Leppik, I. E. (1994). Antiepileptics in the elderly. Pharmacoepidemiology and pharmacokinetics. *Arch. Fam. Med.* **3,** 589–598.

Crowley, J. J., Cusack, B. J., Jue, S. G., Koup, J. R., Park, B. K., and Vestal, R. E. (1988). Aging and drug interactions. II. Effect of phenytoin and smoking on the oxidation of theophylline and cortisol in healthy men. *J. Pharmacol. Exp. Ther.* **245,** 513–523.

Doose, D. R., Larson, K. L., Natarajan, J., and Neto, W. (1998). Comparative single-dose pharmacokinetics of topiramate in elderly versus young men and women. *Epilepsia* **39**(Suppl. 6), 56.

Ducher, M., Maire, P., Cerutti, C., Bourhis, Y., Foltz, F., Sorensen, P., Jelliffe, R., and Fauvel, J. P. (2001). Renal elimination of amikacin and the aging process. *Clin. Pharmacokinet.* **40,** 947–953.

Fliser, D., Bischoff, I., Hanses, A., Block, S., Joest, M., Ritz, E., and Mutschler, E. (1999). Renal handling of drugs in the healthy elderly. Creatinine clearance underestimates renal function and pharmacokinetics remain virtually unchanged. *Eur. J. Clin. Pharmacol.* **55,** 205–211.

Grandison, M. K., and Boudinot, F. D. (2000). Age-related changes in protein binding of drugs: Implications for therapy. *Clin. Pharmacokinet.* **38,** 271–290.

Grasela, T. H., Sheiner, L. B., Rambeck, B., Boenigk, H. E., Dunlop, A., Mullen, P. W., Wadsworth, J., Richens, A., Ishizaki, T., Chiba, K., Miura, H., Minagawa, K., *et al.* (1983). Steady-state pharmacokinetics of phenytoin from routinely collected patient data. *Clin. Pharmacokinet.* **8,** 355–364.

Graves, N. M., Brundage, R. C., Wen, Y., Cascino, G., So, E., Ahman, P., Rarick, J., Krause, S., and Leppik, I. E. (1998). Population pharmacokinetics of carbamazepine in adults with epilepsy. *Pharmacotherapy* **18,** 273–281.

Haegele, K. D., Huebert, N. D., Ebel, M., Tell, G. P., and Schechter, P. J. (1988). Pharmacokinetics of vigabatrin: Implications of creatinine clearance. *Clin. Pharmacol. Ther.* **44,** 558–565.

Hammerlein, A., Derendorf, H., and Lowenthal, D. T. (1998). Pharmacokinetic and pharmacodynamic changes in the elderly. Clinical implications. *Clin. Pharmacokinet.* **35,** 49–64.

Hockings, N., Pall, A., Moody, J., Davidson, A. V., and Davidson, D. L. (1986). The effects of age on carbamazepine pharmacokinetics and adverse effects. *Br. J. Clin. Pharmacol.* **22,** 725–728.

King-Stephens, D. (1999). The treatment of epilepsy in the elderly. *CNS Drugs* **12,** 21–33.

Klotz, U., Avant, G. R., Hoyumpa, A., Schenker, S., and Wilkinson, G. R. (1975). The effects of age and liver disease on the disposition and elimination of diazepam in adult man. *J. Clin. Invest.* **55,** 347–359.

Kudriakova, T. B., Sirota, L. A., Rozova, G. I., and Gorkov, V. A. (1992). Autoinduction and steady-state pharmacokinetics of carbamazepine and its major metabolites. *Br. J. Clin. Pharmacol.* **33,** 611–615.

Lackner, T. E., Cloyd, J. C., Thomas, L. W., and Leppik, I. E. (1998). Antiepileptic drug use in nursing home residents: Effect of age, gender, and comedication on patterns of use. *Epilepsia* **39,** 1083–1087.

Landahl, S., Aurell, M., and Jagenburg, R. (1981). Glomerular filtration rate at the age of 70 and 75. *J. Clin. Exp. Gerontol.* **3,** 29–45.

Martines, C., Gatti, G., Sasso, E., Calzetti, S., and Perucca, E. (1990). The disposition of primidone in elderly patients. *Br. J. Clin. Pharmacol.* **30,** 607–611.

Messina, S., Battino, D., Croci, D., Mamoli, D., Ratti, S., and Perucca, E. (2005). Phenobarbital pharmacokinetics in old age: A case-matched evaluation based on therapeutic drug monitoring data. *Epilepsia* **46,** 372–377.

Muhlberg, W., and Platt, D. (1999). Age-dependent changes of the kidneys: Pharmacological implications. *Gerontology* **45,** 243–253.

Parker, B. M., Cusack, B. J., and Vestal, R. E. (1995). Pharmacokinetic optimisation of drug therapy in elderly patients. *Drugs Aging* **7,** 10–18.

Patsalos, P. N. (2000). Pharmacokinetic profile of levetiracetam: Toward ideal characteristics. *Pharmacol. Ther.* **85,** 77–85.

Pellock, J. M., Glauser, T. A., Bebin, E. M., Fountain, N. B., Ritter, F. J., Coupez, R. M., and Shields, W. D. (2001). Pharmacokinetic study of levetiracetam in children. *Epilepsia* **42,** 1574–1579.

Perucca, E. (1980). Plasma protein binding of phenytoin in health and disease: Relevance to therapeutic drug monitoring. *Ther. Drug Monit.* **2,** 331–344.

Perucca, E. (2002). Pharmacological and therapeutic properties of valproate: A summary after 35 years of clinical experience. *CNS Drugs* **16,** 695–714.

Perucca, E. (2005). Pharmacokinetic variability of new antiepileptic drugs at different ages. *Ther. Drug Monit.* **27,** 714–717.

Perucca, E. (2006). Clinical pharmacokinetics of new-generation antiepileptic drugs at the extremes of age. *Clin. Pharmacokinet.* **45,** 351–363.

Perucca, E., Grimaldi, R., Gatti, G., Pirracchio, S., Crema, F., and Frigo, G. M. (1984). Pharmacokinetics of valproic acid in the elderly. *Br. J. Clin. Pharmacol.* **17,** 665–669.

Perucca, E., Berlowitz, D., Birnbaum, A., Cloyd, J. C., Garrard, J., Hanlon, J. T., Levy, R. H., and Pugh, M. J. (2006). Pharmacological and clinical aspects of antiepileptic drug use in the elderly. *Epilepsy Res.* **68**(Suppl. 1), S49–S63.

Posner, J., Holdich, T., and Crome, P. (1991). Comparison of lamotrigine pharmacokinetics in young and elderly healthy volunteers. *J. Pharm. Med.* **1**, 121–128.

Reidenberg, M. M., and Drayer, D. E. (1984). Alteration of drug-protein binding in renal disease. *Clin. Pharmacokinet.* **9**(Suppl. 1), 18–26.

Richens, A. (1975). A study of the pharmacokinetics of phenytoin (diphenylhydantoin) in epileptic patients, and the development of a nomogram for making dose increments. *Epilepsia* **16**, 627–646.

Richens, A., Banfield, C. R., Salfi, M., Nomeir, A., Lin, C. C., Jensen, P., Affrime, M. B., and Glue, P. (1997). Single and multiple dose pharmacokinetics of felbamate in the elderly. *Br. J. Clin. Pharmacol.* **44**, 129–134.

Rowan, A. J., Ramsay, R. E., Collins, J. F., Pryor, F., Boardman, K. D., Uthman, B. M., Spitz, M., Frederick, T., Towne, A., Carter, G. S., Marks, W., Felicetta, J., *et al.* (2005). New onset geriatric epilepsy: A randomized study of gabapentin, lamotrigine, and carbamazepine. *Neurology* **64**, 1868–1873.

Russell, R. M. (2001). Factors in aging that effect the bioavailability of nutrients. *J. Nutr.* **131**(Suppl. 4), 1359S–1361S.

Salem, S. A., Rajjayabun, P., Shepherd, A. M., and Stevenson, I. H. (1978). Reduced induction of drug metabolism in the elderly. *Age Ageing* **7**, 68–73.

Schachter, S. C., Cramer, G. W., Thompson, G. D., Chaponis, R. J., Mendelson, M. A., and Lawhorne, L. (1998). An evaluation of antiepileptic drug therapy in nursing facilities. *J. Am. Geriatr. Soc.* **46**, 1137–1141.

Scheuer, M. L., and McCullough, K. (1995). Carbamazepine clearance in the elderly. *Neurology* **45**(Suppl. 4), A360.

Schmucker, D. L. (2001). Liver function and phase I drug metabolism in the elderly: A paradox. *Drugs Aging* **18**, 837–851.

Shah, J., Shellenberger, K., and Canafax, D. M. (2002). Zonisamide. Chemistry, biotransformation and pharmacokinetics. *In* "Antiepileptic Drugs" (R. H. Levy, R. H. Mattson, B. S. Meldrum, and E. Perucca, Eds.), 5th ed., pp. 873–879. Lippincott, Williams & Wilkins, Philadelphia.

Snel, S., Jansen, J. A., Mengel, H. B., Richens, A., and Larsen, S. (1997). The pharmacokinetics of tiagabine in healthy elderly volunteers and elderly patients with epilepsy. *J. Clin. Pharmacol.* **37**, 1015–1020.

Spina, E., Perucca, E., and Levy, R. (2005). Predictability of metabolic antiepileptic drug interactions. *In* "Antiepiletpic Drugs: Combination Therapy and Interactions" (J. Majkowski, B. F. D. Bourgeois, P. N. Patsalos, and R. H. Mattson, Eds.), pp. 57–92. Cambridge University Press, Cambridge.

Stewart, B. H., Kugler, A. R., Thompson, P. R., and Bockbrader, H. N. (1993). A saturable transport mechanism in the intestinal absorption of gabapentin is the underlying cause of the lack of proportionality between increasing dose and drug levels in plasma. *Pharm. Res.* **10**, 276–281.

Tallis, R. C. (2004). Management of epilepsy in the elderly person. *In* "The Treatment of Epilepsy" (S. D. Shorvon, E. Perucca, D. Fish, and W. E. Dodson, Eds.), pp. 201–214. Blackwell Science, Oxford.

Tanaka, E. (1998). *In vivo* age-related changes in hepatic drug-oxidizing capacity in humans. *J. Clin. Pharm. Ther.* **23**, 247–255.

Tomson, T. (1988). Choreoathetosis induced by ordinary phenytoin levels, explained by high free fraction?—A case report. *Ther. Drug Monit.* **10**, 239–241.

Twum-Barima, Y., Finnigan, T., Habash, A. I., Cape, R. D., and Carruthers, S. G. (1984). Impaired enzyme induction by rifampicin in the elderly. *Br. J. Clin. Pharmacol.* **17**, 595–597.

van Heiningen, P. N., Eve, M. D., Oosterhuis, B., Jonkman, J. H., de Bruin, H., Hulsman, J. A., Richens, A., and Jensen, P. K. (1991). The influence of age on the pharmacokinetics of the antiepileptic agent oxcarbazepine. *Clin. Pharmacol. Ther.* **50,** 410–419.

Vestal, R. E., Norris, A. H., Tobin, J. D., Cohen, B. H., Shock, N. W., and Andres, R. (1975). Antipyrine metabolism in man: Influence of age, alcohol, caffeine, and smoking. *Clin. Pharmacol. Ther.* **18,** 425–432.

Vestal, R. E., Wood, A. J., Branch, R. A., Shand, D. G., and Wilkinson, G. R. (1979). Effects of age and cigarette smoking on propranolol disposition. *Clin. Pharmacol. Ther.* **26,** 8–15.

Wallace, J., Shellenberger, K., and Groves, L. (1998). Pharmacokinetics of zonisamide in young and elderly subjects. *Epilepsia* **39**(Suppl. 6), 190–191.

Willmore, L. J. (1995). The effect of age on pharmacokinetics of antiepileptic drugs. *Epilepsia* **36**(Suppl. 5), S14–S21.

Yukawa, E., Higuchi, S., and Aoyama, T. (1990). Population pharmacokinetics of phenytoin from routine clinical data in Japan: An update. *Chem. Pharm. Bull. (Tokyo)* **38,** 1973–1976.

Zeeh, J., and Platt, D. (2002). The aging liver: Structural and functional changes and their consequences for drug treatment in old age. *Gerontology* **48,** 121–127.

FACTORS AFFECTING ANTIEPILEPTIC DRUG PHARMACOKINETICS IN COMMUNITY-DWELLING ELDERLY

James C. Cloyd,* Susan Marino,[†] and Angela K. Birnbaum*

*Department of Experimental and Clinical Pharmacology
College of Pharmacy, University of Minnesota, Minneapolis, Minnesota 55455, USA
[†]Department of Pharmacy Practice, College of Pharmacy, University of Florida
Gainesville, Florida 32610, USA

Because aging is associated with changes in physiological processes, it is widely believed that antiepileptic drug pharmacodynamics and pharmacokinetics in elderly patients differ from those in younger adults. In order to better characterize these differences, this chapter reports on preliminary results from an investigation of the effect of age on steady-state phenytoin (PHT) and carbamazepine (CBZ) pharmacokinetics. Parenteral formulations of stable-labeled PHT, fosphenytoin (FOS), and CBZ were administered to elderly ($\geq$65 years of age) and adult (18–64 years of age) patients on maintenance regimens of PHT or CBZ; a labeled 100-mg dose was infused over 10 min, then the remainder of the patient's AED dose was administered as unlabeled drug. Blood samples were collected just before administration of the labeled drug and for up to 192 h afterward. Samples were then assayed for the concentrations of labeled and unlabeled drug. Preliminary results from 60 patients on PHT therapy (41 elderly, mean age 76 years; 19 younger adults, mean age 41 years) indicate that PHT bioavailability did not differ between the two age groups; however, absorption and elimination half-lives were more variable in the elderly patients. The elimination half-life for the entire patient population was approximately twofold longer than the value reported in the product labeling (40–50 h vs 22 h). Preliminary results from 67 patients on CBZ therapy (14 elderly, mean age 70 years; 53 younger adults, mean age 41 years) showed no apparent difference between elderly and adult patients in any parameter; however, the mean CBZ elimination half-life for the combined groups

INTERNATIONAL REVIEW OF
NEUROBIOLOGY, VOL. 81
DOI: 10.1016/S0074-7742(06)81012-3

(21 h) was longer than previous estimates. These results indicate that the effect of age on CBZ and PHT absorption may result in greater variability in plasma concentrations in elderly patients, whereas the effect on half-life is modest.

I. Introduction

Antiepileptic drugs (AEDs) are used by a large number of elderly people ($\geq$65 years of age). Surveys of elderly persons residing in the community or in nursing homes indicate that as many as 1.4% and 10%, respectively, take one or more AEDs (Lackner *et al.*, 1998; Nitz *et al.*, 2000). Phenytoin (PHT) is the most commonly prescribed AED, regardless of resident location, representing 60–70% of AED use in this population, followed by carbamazepine (CBZ; 15–20%). There is a widely held view that AED pharmacodynamics and pharmacokinetics differ between elderly and younger patients, although data that support this view are limited. In a post hoc analysis of several large, multicenter, controlled clinical trials comparing CBZ and valproic acid (VPA), Ramsay *et al.* (1994) found that, compared to younger patients, patients $\geq$65 years of age were more often seizure free (40% of patients $\geq$65 years vs 14.7% of patients $\leq$40 years) and responded to therapy at lower serum AED concentrations (CBZ, 3.7 mg/liter vs 7.8 mg/liter; VPA, 31 mg/liter vs 43.7 mg/liter). They also found that older patients tended to experience adverse events at lower concentrations. In a retrospective study involving 117 patients aged 65–93 years with new-onset epilepsy, Stephen *et al.* (2006) found that 79% of patients were seizure free for 12 months after initiating therapy, with the majority (87 of 93 patients) attaining remission on monotherapy. In contrast, results from a prospective, randomized, double-blind, placebo-controlled study comparing CBZ, gabapentin (GBP), and lamotrigine (LTG) in elderly patients suggested that responses in the elderly were similar to responses in younger patients (Rowan *et al.*, 2005). Among 590 patients aged 60 years or older enrolled in the study, only 46.8% completed the 1-year trial. The primary reason for discontinuation was adverse events. Among those remaining in the study for 12 months, 55.3% were seizure free. The serum concentrations at the end of the study were: CBZ, 6.5 mg/liter (unbound, 0.8 mg/liter); GBP, 8.5 mg/liter; and LTG, 3.5 mg/liter. The observed AED concentrations tended to be at or below the lower ends of recommended therapeutic ranges (Gidal *et al.*, 2002). These studies generally support the assertions that the elderly require smaller doses and lower drug concentrations and are more susceptible to concentration-dependent adverse effects (Brodie and Kwan, 2005). It remains unclear whether older patients are more likely to realize seizure freedom than are younger individuals.

As shown in Table I, clinically important age-related changes can alter pharmacokinetics, further complicating the use of AEDs in the elderly (Leppik

TABLE I

PHYSIOLOGICAL CHANGES OF AGING AFFECTING AED PHARMACOKINETICS

Physiological measure[a]	Age-associated change
Gastrointestinal anatomy and physiology	Decreased blood flow, gastric pH, gastric emptying, and intestinal motility
Serum albumin	Unchanged or decreased, altered protein binding
Hepatic blood flow	Decreased
Liver mass	Decreased
Oxidative metabolism	Decreased by 1% per year after age 40
Conjugation metabolism	Unchanged or decreased
Induction of microsomal enzymes	Decreased[b]
Glomerular filtration	Decreased by 1% per year after age 40

[a]Disease may exacerbate physiological changes.
[b]Data are insufficient or inconclusive.
Adapted from Cloyd and Conway (2002), with permission from Lippincott Williams & Wilkins, Philadelphia.

and Cloyd, 2002). These changes include alterations in gastrointestinal anatomy and physiology that could cause variability in drug absorption. Drug binding to serum albumin may decrease, whereas binding to the reactive protein α-1-acid glycoprotein may increase in the elderly (Woo *et al.*, 1994). Hepatic and renal clearance decline approximately 10% per decade beginning around age 40 years. Variability in protein binding and clearance in the elderly equals or exceeds that observed in younger populations. Alterations in absorption and protein binding can lead to mistaken estimates of clearance when the estimates are based on orally administered drug doses and total drug concentrations. Information on steady-state elimination half-life is unavailable because therapy cannot be interrupted to allow adequate collection of blood samples. Nonetheless, most AED pharmacokinetic studies in the elderly have been based on oral administration of a single AED dose and typically include relatively healthy patients between ages 65 and 74 years, an age range in which pharmacokinetic changes are often minimal.

Several reports contain findings of significant age-related differences in AED pharmacokinetics. The most rigorously conducted study involving PHT characterized Michaelis–Menten parameters in 92 individuals ranging in age from 21 to 78 years who were on maintenance PHT therapy (Bauer and Blouin, 1982). Mean values for the maximum rate of metabolism (V_{max}) and the Michaelis–Menten constant (K_m) among the three age groups (20–39, 40–59, and 60–79 years) were: 7.5 mg/kg/day and 5.7 mg/liter; 6.6 mg/kg/day and 5.4 mg/liter; and 6.0 mg/kg/day and 5.6 mg/liter, respectively. The V_{max} in the oldest age group was significantly ($p < 0.05$) lower than the V_{max} in the youngest age group. The changes in V_{max} described by Bauer and Blouin could have a significant effect on PHT dosage requirements in the elderly. A simulation using estimated V_{max} and K_m values indicated that elderly patients, in comparison to younger patients, require lower PHT doses and display a greater change in plasma PHT concentration with any change in dose (Fig. 1).

Results of several studies showed decrease in CBZ clearance with advancing age. Battino *et al.* (2003) estimated apparent clearance using routinely determined plasma CBZ concentrations in 157 elderly (≥65 years) and 157 younger (20–50 years) patients matched for sex, body weight, and comedications. CBZ clearance was ~23% lower in the older group. Using a population pharmacokinetic approach, Graves *et al.* (1998) found that age was a significant covariant influencing CBZ pharmacokinetics. Apparent CBZ clearance decreased by 25% in patients 70 years of age or older.

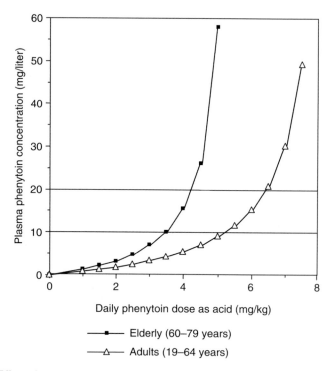

FIG. 1. Effect of age on phenytoin (PHT) dosing and plasma concentrations. Plasma PHT concentrations were calculated using estimated V_{max} and K_m values for elderly patients (60–79 years of age, $V_{max} = 6.0$ mg/kg/day, $K_m = 5.8$ mg/liter) (Bauer and Blouin, 1982) and younger adult patients (19–64 years of age, $V_{max} = 8.45$ mg/kg/day, $K_m = 6.25$ mg/liter) (Cloyd *et al.*, 1978). Results from this simulation indicated that elderly patients, in comparison to younger patients, require lower PHT doses and display a greater sensitivity to dose changes. Adapted from Perucca *et al.* (2006), with permission.

The results from these studies are consistent with the hypothesis that advancing age is associated with a decrease in AED clearance. However, both the retrospective nature of these studies and the reliance on orally administered drugs to calculate pharmacokinetic parameters limit the conclusions one can draw about the impact of age on AED pharmacokinetics.

II. Preliminary Studies of PHT and CBZ Pharmacokinetics in Community-Dwelling Elderly

A. METHODS

To rigorously investigate the effect of age on steady-state PHT and CBZ pharmacokinetics, including bioavailability and elimination half-life, parenteral formulations of stable-labeled phenytoin (SL-PHT), stable-labeled fosphenytoin (SL-FOS), and stable-labeled carbamazepine (SL-CBZ) were evaluated. These formulations were administered to young and elderly patients on maintenance regimens of PHT or CBZ. Patients participating in the study had to be at least 18 years of age; able to tolerate intravenous (IV) PHT, FOS, or CBZ, or intramuscular (IM) FOS; on a stable maintenance PHT or CBZ regimen; free of cardiac disease; and not taking medications known to interact with PHT or CBZ (Conway and Cloyd, 2005).

On the day of the study, patients were admitted for a 12- to 24-h stay at the University of Minnesota or the University of Miami, General Clinical Research Center. Following a brief physical and neurological examination, indwelling catheters were placed in both forearms of each patient, and predose blood samples were drawn. A 100-mg dose of SL-PHT, SL-FOS, or SL-CBZ was then infused over 10 min while blood pressure, heart rate, and heart rhythm were monitored. Subjects unable to tolerate an IV infusion of SL-PHT or SL-FOS were given an IM injection of SL-FOS. Subsequently, the remainder of the patient's morning AED dose (i.e., their morning dose minus 100 mg) was administered. Blood samples were collected just before drug administration and for up to 192 h afterward. Unbound PHT and CBZ were separated from bound drug using ultrafiltration. Gas chromatography/mass spectroscopy and high-performance liquid chromatography/mass spectroscopy assays were used to simultaneously determine labeled and unlabeled PHT and CBZ serum concentrations, respectively. The studies were carried out under steady-state conditions, allowing the use of a linear pharmacokinetic model to fit the data. Pharmacokinetic parameters were determined by noncompartmental analysis of plasma PHT and CBZ concentration–time data using WinNonlin® software (Pharsight Corporation, Mountain View, CA).

B. Results

1. *Phenytoin*

Preliminary PHT results were based on data from 60 enrolled patients (adults: 7 women, 12 men; elderly: 15 women, 26 men). Some patients were not included in certain analyses because of either incomplete sample collection or plasma concentrations below the lower limit of detection. The mean ages for the adult and elderly groups were 40.9 and 76.1 years, respectively (range, 21–93 years), and weights were similar in both groups.

As presented in Table II, preliminary results revealed that elderly and adult patients took similar doses and had similar total and unbound PHT concentrations. There were no statistical differences in pharmacokinetic parameters,

TABLE II

COMPARISON OF PHENYTOIN (PHT) PHARMACOKINETICS DATA OBTAINED FROM INTRAVENOUS STABLE-LABELED PHT AND ORAL PHT UNDER STEADY-STATE CONDITIONS IN YOUNGER ADULTS AND THE ELDERLY

	Younger adults	Elderly
Dose		
n	19	41
Mean (mg/kg/day)	5.3	5.2
Steady-state plasma concentration		
Total		
n	14	39
Mean (mg/liter)	13.9	11.4
Unbound		
n	13	32
Mean (mg/liter)	1.8	1.6
Bioavailability		
n	15	39
Mean (%)	0.8	1.0
Free fraction		
n	12	30
Mean	0.1	0.1
Volume of distribution		
n	16	39
Mean (liter/kg)	0.7	1.1
Clearance		
n	15	27
Mean (liter/kg/h)	0.2	0.2
Half-life		
n	15	40
Mean (h)	41.7	50.1

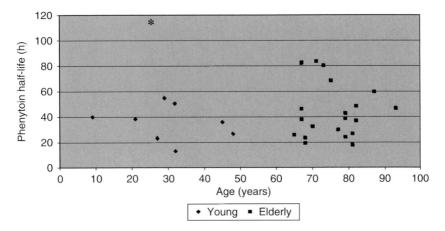

FIG. 2. Effect of age on phenytoin (PHT) half-life. PHT half-life was measured in 60 elderly (≥65 years of age; mean, 76 years) and younger adult (≤64 years of age; mean, 40 years) patients. Measurements displayed greater variability in elderly patients ($n = 41$) than in younger adults ($n = 19$). * indicates the patient with a half-life of 115 h had a genetic mutation in a cytochrome P450 enzyme.

although there were trends toward greater volume of distribution and longer elimination half-life in the elderly.

Although bioavailability (percentage of dose absorbed) did not differ significantly between the two age groups, the variability in absorption was greater in the elderly, with a range of 45–200%, whereas the range was 50–120% in the adult group. Similarly, aside from one younger subject with a prolonged PHT half-life (~115 h) due to a genetic mutation in a metabolizing enzyme, the half-life was more varied in the elderly than in younger adults, as shown in Fig. 2. For all patients combined, mean PHT half-life was approximately twofold longer (40–50 h) than that listed in product labeling (22 h; range, 7–42 h) (Pfizer Inc., 2003).

2. *Carbamazepine*

Preliminary results from the CBZ study were based on data from 67 patients who completed the study (28 women and 39 men). The mean age (±SD) for adults (i.e., aged 18–64 years) was 41 ± 12 years ($n = 53$), and the mean age for the elderly (i.e., aged ≥65 years) was 70 ± 7.3 years ($n = 14$). The daily dose of CBZ ranged from 200 to 2400 mg (both immediate- and extended-release CBZ formulations were used) with a mean dose (±SD) of 10.7 ± 6.0 mg/kg/day. The summary of CBZ pharmacokinetic results is given in Table III. There was no apparent difference between the adult and elderly groups in any parameter. The mean absolute bioavailability of CBZ was at the low end of previously reported values (Spina, 2002), but absorption was highly variable, with a threefold range (0.34–1.1) in extent of absorption. The mean (±SD) CBZ half-life for the combined adult and elderly patient groups, 21.1 ± 8.1 h (Fig. 3), was longer than previously estimated values (Spina, 2002).

TABLE III

COMPARISON OF CARBAMAZEPINE (CBZ) PHARMACOKINETICS DATA OBTAINED FROM INTRAVENOUS STABLE-LABELED CBZ AND ORAL CBZ UNDER STEADY-STATE CONDITIONS IN YOUNGER ADULTS AND THE ELDERLY

	All patients $N = 67$	Younger adults $n = 53$	Elderly $n = 14$
Age, mean (±SD), (years)	47 (16)	41 (12)	70 (7.3)
Weight, mean (±SD), (kg)	82.7 (20.7)	83.5 (21.8)	79.5 (14.3)
Dose, mean (±SD), (mg/kg/day)	10.7 (6.0)	11.2 (6.3)	8.7 (4.6)
Steady-state plasma concentration, mean (±SD), (mg/liter)	7.3 (2.3)	7.4 (2.3)	7.1 (2.7)
Bioavailability, mean (±SD), (%)	0.7 (0.2)	0.7 (0.2)	0.8 (0.2)
Volume of distribution, mean (±SD), (liter/kg)	1.1 (0.2)	1.1 (0.2)	1.1 (0.2)
Clearance, mean (±SD), (liter/kg/h)	0.04 (0.02)	0.04 (0.02)	0.04 (0.01)
Half-life, mean (±SD), (h)	21.1 (8.1)	21.1 (8.7)	20.8 (5.4)

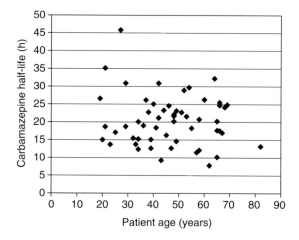

FIG. 3. Effect of age on carbamazepine (CBZ) half-life. CBZ half-life was measured in 14 elderly (≥65 years of age; mean, 70 years) and 53 younger adult (≤64 years of age; mean, 41 years) patients. Measurements in the elderly (mean 20.8 ± 5.3 h; $n = 14$) and in younger adults (mean 21.1 ± 8.7 h, $n = 53$) were similar.

III. Discussion and Conclusion

The results of these preliminary studies indicated that the effect of advancing age on CBZ and PHT elimination is modest, in contrast to previous reports and the generally accepted view that drug metabolism decreases in old age. The reasons why age had so little influence on CBZ and PHT disposition are unclear and require further investigation.

There was greater variability in PHT absorption and elimination in the elderly, which makes dose standardization difficult. This suggests that PHT absorption also varies within a patient, which could result in potentially large fluctuations in serum PHT concentrations. Prolonged elimination half-life may substantially delay the attainment of steady-state concentrations. These results also suggest that PHT therapy may require more intense monitoring in elderly patients than in younger patients. These factors could complicate assessment of therapy and limit the usefulness of PHT in selected patients.

Finally, the use of stable-labeled isotopes to study AED pharmacokinetics under steady-state conditions provides a much better understanding of drug disposition. Elimination half-life of both PHT and CBZ was found to be much longer than previously thought. A longer elimination half-life would allow many patients to take these medications using once- or twice-daily dosing regimens, which are known to enhance compliance, especially in older patients.

References

Battino, D., Croci, D., Rossini, A., Messina, S., Mamoli, D., and Perucca, E. (2003). Serum carbamazepine concentrations in elderly patients: A case-matched pharmacokinetic evaluation based on therapeutic drug monitoring data. *Epilepsia* **44,** 923–929.

Bauer, L. A., and Blouin, R. A. (1982). Age and phenytoin kinetics in adult epileptics. *Clin. Pharmacol. Ther.* **31,** 301–304.

Brodie, M. J., and Kwan, P. (2005). Epilepsy in elderly people. *BMJ* **331,** 1317–1322.

Cloyd, J. C., and Conway, J. M. (2002). Age-related changes in pharmacokinetics, drug interactions, and adverse effects. *In* "Clinical Neurology of the Older Adult" (J. I. Sirven and B. L. Malamut, Eds.), pp. 29–44. Lippincott Williams & Wilkins, Philadelphia.

Cloyd, J. C., Sawchuk, R. J., Leppik, I. E., and Pepin, S. M. (1978). The direct linear plot: Use in estimating Michaelis-Menten parameters and individualizing phenytoin dosage regimens in epileptic patients. Presented at: Epilepsy International Symposium, Vancouver, Canada.

Conway, J. M., and Cloyd, J. C. (2005). Antiepileptic drug interactions in the elderly. *In* "Antiepileptic Drugs: Combination Therapy and Interactions" (J. Majkowski, B. F. D. Bourgeois, P. N. Patsalos, and R. H. Mattson, Eds.), 1st ed., pp. 273–293. Cambridge University Press, Cambridge.

Gidal, B. E., Garnett, W. R., and Graves, N. (2002). Epilepsy. *In* "Pharmacotherapy: A Pathophysiologic Approach" (J. T. DiPiro, R. L. Talbert, G. C. Yee, G. R. Matzke, B. G. Wells, and L. M. Posey, Eds.), 5th ed., pp. 1031–1059. McGraw-Hill, New York.

Graves, N. M., Brundage, R. C., Wen, Y., Cascino, G., So, E., Ahman, P., Rarick, J., Krause, S., and Leppik, I. E. (1998). Population pharmacokinetics of carbamazepine in adults with epilepsy. *Pharmacotherapy* **18,** 273–281.

Lackner, T. E., Cloyd, J. C., Thomas, L. W., and Leppik, I. E. (1998). Antiepileptic drug use in nursing home residents: Effect of age, gender, and comedication on patterns of use. *Epilepsia* **39,** 1083–1087.

Leppik, I. E., and Cloyd, J. C. (2002). General principles: Epilepsy in the elderly. *In* "Antiepileptic Drugs" (R. H. Levy, R. H. Mattson, B. S. Meldrum, and E. Perucca, Eds.), 5th ed., pp. 149–158. Lippincott Williams & Wilkins, Philadelphia.

Nitz, N. M., Garrard, J., Harms, S. L., Hardie, N. A., Bland, P. C., Gross, C. R., and Leppik, I. E. (2000). Prevalence of antiepileptic drug use among Medicare beneficiaries [abstract]. *Epilepsia* **41** (Suppl. 7), 251.

Perucca, E., Berlowitz, D., Birnbaum, A., Cloyd, J. C., Garrard, J., Hanlon, J. T., Levy, R. H., and Pugh, M. J. (2006). Pharmacological and clinical aspects of antiepileptic drug use in the elderly. *Epilepsy Res.* **68**(Suppl. 1), S49–S63.

Pfizer Inc. (2003). Dilantin (phenytoin) package insert. Pfizer Inc., New York, NY.

Ramsay, R. E., Rowan, A. J., Slater, J. D., Collins, J., Nemire, R., and Ortiz, W. R. and the VA Cooperative Study Group (1994). Effect of age on epilepsy and its treatment: Results from the VA Cooperative Study [abstract]. *Epilepsia* **35,** 91.

Rowan, A. J., Ramsay, R. E., Collins, J. F., Pryor, F., Boardman, K. D., Uthman, B. M., Spitz, M., Frederick, T., Towne, A., Carter, G. S., Marks, W., Felicetta, J., *et al.* (2005). New onset geriatric epilepsy: A randomized study of gabapentin, lamotrigine, and carbamazepine. *Neurology* **64,** 1868–1873.

Spina, E. (2002). Carbamazepine: Chemistry, biotransformation, and pharmacokinetics. *In* "Antiepileptic Drugs" (R. H. Levy, R. H. Mattson, B. S. Meldrum, and E. Perucca, Eds.), 5th ed., pp. 236–246. Lippincott Williams & Wilkins, Philadelphia.

Stephen, L. J., Kelly, K., Mohanraj, R., and Brodie, M. J. (2006). Pharmacological outcomes in older people with newly diagnosed epilepsy. *Epilepsy Behav.* **8,** 434–437.

Woo, J., Chan, H. S., Or, K. H., and Arumanayagam, M. (1994). Effect of age and disease on two drug binding proteins: Albumin and alpha-1-acid glycoprotein. *Clin. Biochem.* **27,** 289–292.

PHARMACOKINETICS OF ANTIEPILEPTIC DRUGS IN ELDERLY NURSING HOME RESIDENTS

Angela K. Birnbaum

Department of Experimental and Clinical Pharmacology, College of Pharmacy
University of Minnesota, Minneapolis, Minnesota 55455, USA

I. Introduction
II. Older AEDs
   A. Phenytoin
   B. Valproic Acid
   C. Carbamazepine
   D. Phenobarbital
III. Newer AEDs
   References

With approximately 10% of elderly nursing home residents taking antiepileptic drugs (AEDs), it is critical to understand the pharmacokinetics, dosing, and possible adverse reactions of these AEDs. In this chapter, five AEDs commonly prescribed to nursing home residents will be discussed. Phenytoin (PHT), the most commonly used AED in this population, is extensively metabolized by the cytochrome P450 enzyme system, is highly protein bound, and interacts with many concomitant medications. Up to 45% of nursing home residents who receive PHT have concentrations below the range (subtherapeutic) used in adults (<65 years), while approximately 10% of residents have concentrations that are potentially toxic (>20 $\mu g/ml$). In addition, serum PHT concentrations can vary greatly within an individual resident and may be subtherapeutic one day and potentially toxic the next. Valproic acid is taken by approximately 9–17% of nursing home residents who are administered AEDs, with over half using it for nonseizure indications. Doses are approximately 16 mg/kg/day in elderly nursing home residents, but doses and serum concentrations are lower in the oldest age group ($\geq$85 years). A majority of residents are maintained at serum concentrations considered subtherapeutic for epilepsy, whereas relatively few ($\sim$3%) are maintained at toxic levels. The average ($\pm$SD) carbamazepine (CBZ) dose is 8.8 ± 4.7 mg/kg/day, yielding a mean serum concentration of 6.3 ± 2.2 mg/liter. Subtherapeutic concentrations are found in up to 20% of serum measurements, while 2.5% of serum measurements are in the toxic range. CBZ is highly bound to

INTERNATIONAL REVIEW OF
NEUROBIOLOGY, VOL. 81
DOI: 10.1016/S0074-7742(06)81013-5

211

serum albumin and $\alpha 1$-acid glycoprotein and is metabolized to carbamazepine-10,11-epoxide, an active metabolite thought to be responsible for some side effects. Phenobarbital (PB) is frequently combined with PHT. This combination can cause devastating side effects because both PB and PHT can produce cognitive side effects. Gabapentin is one of the newer AEDs frequently administered to nursing home residents. Its lack of both hepatic metabolism and protein binding potentially makes it a safer drug in this population.

I. Introduction

Approximately 10% of all elderly nursing home residents in the United States are prescribed an antiepileptic drug (AED) (Cloyd *et al.*, 1994; Garrard *et al.*, 2000; Lackner *et al.*, 1998; Schachter *et al.*, 1998). Since approximately 1.5 million elderly people reside in nursing homes, as many as 150,000 of those residents may be taking AEDs (Strahan, 1997). In addition to their use in epilepsy, AEDs are prescribed for a variety of conditions common among the elderly, including neuralgias, aggressive behavior disorders, essential tremor, and restless leg syndrome. Serum concentrations of AEDs are often measured to guide therapy; however, very little information exists regarding the doses and concentrations used during chronic AED therapy in nursing home residents. Further complicating treatment, elderly nursing home residents routinely take an average of five to six medications, including AEDs (Lackner *et al.*, 1998), and these AEDs are a major contributor to adverse reactions in elderly patients (Cooper, 1999; Gerety *et al.*, 1993; Gurwitz *et al.*, 2000; Moore and Jones, 1985). When prescribing AEDs for the elderly, especially those who reside in nursing homes, clinicians must consider the likelihood of concomitant disorders and the high individual variability in AED concentrations found in this population.

II. Older AEDs

A. PHENYTOIN

Phenytoin (PHT) is the most commonly used AED in nursing homes (Garrard *et al.*, 2000; Lackner *et al.*, 1998). It is extensively metabolized through the cytochrome P450 system and is a known enzyme inducer. Nursing home residents who take PHT are likely to experience interactions with other medications because of the high number of comedications these patients routinely take, age-related changes in physiology, and the tendency of PHT to exhibit nonlinear pharmacokinetics via the saturation of liver enzymes. Therefore, treatment in the elderly, particularly those who reside in nursing homes, may carry more risks than treatment in younger persons.

Few studies have evaluated the pharmacokinetics of PHT in nursing home residents. A report from one pharmacokinetic investigation of elderly patients stated that lower daily doses of PHT may be more suitable for relatively healthy, community-dwelling elderly people than the doses used in younger adults (Bauer and Blouin, 1982). However, nursing home residents may have different dosing requirements than elderly patients who live in the community. A national study of 387 nursing home residents showed that the majority of residents over the age of 65 years (56.6%) received daily PHT doses of less than 5.0 mg/kg (mean ± SD, 4.9 ± 1.8 mg/kg). Daily PHT doses did not vary according to elderly age bracket (65–74, 75–84, and 85+ years), but elderly women did require slightly higher doses than men did to maintain similar total serum PHT concentrations (Birnbaum et al., 2003b).

Up to 45% of elderly nursing home residents are being maintained at total serum PHT concentrations that are considered subtherapeutic based on the adult therapeutic range of 10–20 μg/ml (Birnbaum et al., 2003b; Schachter et al., 1998). In contrast, approximately 10% of residents who were taking PHT were maintained at potentially toxic serum levels of total PHT (>20 μg/ml; Birnbaum et al., 2003b; Schachter et al., 1998).

Community-dwelling elderly exhibit reduced PHT protein binding resulting in increases in the free fractions of PHT (unbound PHT concentration/total PHT concentration) (Hayes et al., 1975; Patterson et al., 1982), which are thought to be due to reduced albumin levels (Greenblatt, 1979). With drugs like PHT that are highly protein bound (>90%), unbound concentrations serve as good indicators of drug concentration at the active site. This may be especially true when serum albumin concentrations are lower than normal; total PHT concentrations may need to be adjusted in patients with hypoalbuminemia (Anderson et al., 1997).

Another complication is that total PHT concentrations can vary greatly within an individual nursing home resident, even while that patient is maintained on a consistent daily dose of PHT (Birnbaum et al., 2003a; Mooradian et al., 1989). In one analysis, multiple measurements of serum PHT concentration in each of 15 nursing home residents between the ages of 46 and 90 years showed twofold differences within individuals (Mooradian et al., 1989) (Fig. 1). The investigators found no explanation for this variability. Results from a second study, which measured 285 total serum PHT concentrations in 56 elderly residents of nursing homes across the United States, also showed a wide variation in total PHT concentrations within individuals (Birnbaum et al., 2003a) (Fig. 2). The high variability could not be attributed to age-related changes in metabolism, since the decreases in concentrations were sporadic throughout the study period.

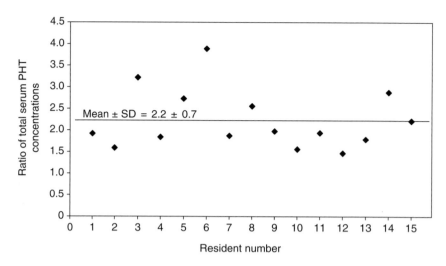

FIG. 1. Ratio of total serum PHT concentrations in 15 nursing home residents. Diamonds represent the values of the highest PHT concentration divided by the lowest PHT concentration observed in a resident. PHT indicates phenytoin. Data from Mooradian *et al.* (1989).

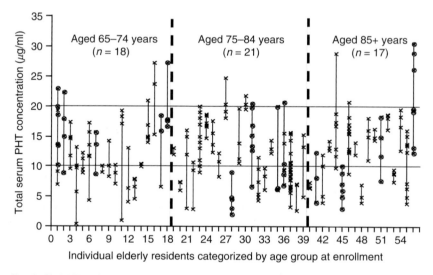

FIG. 2. Variability of total serum PHT concentrations within elderly nursing home residents. Each × denotes a single total PHT concentration, the parallel horizontal lines at 10 and 20 $\mu g/ml$ mark the standard therapeutic range of PHT used in adults, and the circles represent those total PHT concentrations resulting from a PHT dose that was given via a feeding tube. PHT indicates phenytoin. Reprinted from Birnbaum *et al.* (2003a), with permission from Lippincott Williams & Wilkins, Copyright 2003.

As in the previous study, researchers could not identify any reason that accounted for this variability.

Because of the variability in serum PHT concentrations within individual elderly residents and the standard adult therapeutic range of 10–20 μg/ml, a consistent PHT dosage could be subtherapeutic one day and toxic the next, potentially leading to breakthrough seizures or unexpected side effects. At present, there is no way to predict how or when PHT concentrations will fluctuate. The inability to anticipate toxic concentrations and subsequent adverse effects makes it difficult to provide the appropriate level of necessary nursing care. Given the side effect profile of PHT, adverse events occurring after the administration of PHT could increase falls and fractures, which could potentially have devastating consequences in the elderly population. A lower therapeutic range may be appropriate for the elderly; however, more studies are needed to determine appropriate serum PHT concentrations for the elderly nursing home population.

The only available information regarding unbound serum PHT concentrations in a nursing home population is from a group of hypoalbuminemic patients. The mean ($\pm$SD) free fraction of PHT among this group of 37 hypoalbuminemic patients was 0.11 ± 0.03; however, the total PHT concentrations ranged from 3.0 to 34.6 μg/ml (Anderson et al., 1997). It is possible that, within an individual resident, the unbound PHT concentration remains stable, while the total PHT concentration fluctuates. This scenario is plausible for a highly protein-bound drug like PHT and could explain the great variability seen in total serum PHT concentrations within residents. If this scenario is true, and patients are titrated to a daily PHT dose that optimizes both seizure control and adverse effects, then the common practice of routine measurement of total serum PHT concentration may not be useful; regular measurement of unbound serum PHT concentrations may be more appropriate.

Formulations and routes of administration in elderly nursing home residents may differ from those used in adults <65 years of age. In the larger national nursing home studies, approximately 15% of the residents receiving PHT were on feeding tubes (Birnbaum et al., 2003a,b; Schachter et al., 1998). It was noted in the Schachter study that 85% of the residents with feeding tubes continued tube feedings while PHT was administered (Schachter et al., 1998). PHT may be poorly absorbed when given with tube feedings (Holtz et al., 1987; Hooks et al., 1986), and it has been suggested that PHT should not be administered within 1–2 h of a tube feeding (Maynard et al., 1987). PHT suspension may be used more often in elderly nursing home residents; the suspension formulation accounted for 42.5% of the concentrations observed in the study by Birnbaum et al. (2003b). However, PHT suspension can settle and result in uneven drug distribution and variable patient dosing (Sarkar et al., 1989). Bottles of suspension should be thoroughly mixed before administration to patients.

Interactions between PHT and concomitant medications should be given special consideration when treating elderly nursing home residents. However, no studies currently exist that address specific drug interactions in this population. Clinicians must consider the possibility that the maintenance serum concentrations of other drugs may change when PHT is added to or withdrawn from a dosing regimen.

B. VALPROIC ACID

Of the 10% of all nursing home residents who are treated with an AED, approximately 15% take valproic acid (VPA) (Garrard et al., 2003; Schachter et al., 1998). Over half of these VPA recipients receive VPA for a nonseizure indication (Birnbaum et al., 2004; Schachter et al., 1998). Studies in healthy volunteers have shown that, on average, elderly people have higher unbound serum VPA concentrations than young adults. Unbound clearances can be as much as 40% lower in the elderly than in younger adults (Bauer et al., 1985; Perucca et al., 1984).

VPA doses of approximately 16 mg/kg/day are generally used in elderly nursing home residents (Birnbaum et al., 2004). One study (Bryson et al., 1983) compared single-dose intravenous valproate pharmacokinetics in seven young volunteers (aged 20–35 years) and six elderly hospital patients (aged 75–87 years). Total VPA clearance was similar in both groups, but the serum elimination half-life was twice as long in the elderly patients, 14.9 h compared to 7.2 h in the younger group (Bryson et al., 1983). Results from another study (Birnbaum et al., 2004) showed that the clearance of VPA did not differ among elderly age groups (65–74, 75–84, and $\geq$85 years; Table I). However, the mean doses and total serum VPA concentrations observed in the two younger elderly age groups (65–74 and 75–84 years) differed greatly from those seen for members of the oldest age group ($\geq$85 years), who were receiving a lesser amount of drug and were maintained at much lower total VPA concentrations (Birnbaum et al., 2004).

Although VPA clearance does not differ among the elderly age groups, study results have indicated that most residents ($\sim$55%) are maintained at serum VPA concentrations that are considered subtherapeutic for epilepsy ($<$50 mg/liter; Birnbaum et al., 2004). However, a report indicated that only 5 of 146 nursing home residents were maintained at a serum level considered toxic for adults ($>$100 mg/liter) (Birnbaum et al., 2004). Hence, the elderly may be more sensitive to the medication and may require less VPA to achieve seizure control. The lower observed VPA concentrations (below the adult range) may also indicate altered protein binding that has led to increased free fractions of VPA and a corresponding amplification of the therapeutic effect of VPA at lower total serum concentrations. Further studies must be conducted to address the potential need for adjusting the therapeutic range of VPA in elderly nursing home residents.

TABLE I

MEAN VALPROIC ACID DAILY DOSE, TOTAL SERUM CONCENTRATION, AND CLEARANCE BY
GENDER AND AGE

	n	Daily dose ±SD (mg/kg)	p Value	Serum total concentration ±SD (mg/liter)	p Value	Clearance ±SD (ml/h/kg)	p Value
Gender							
Men	52	18.5 ± 13.2		49.5 ± 25.6		17.0 ± 10.6	
Women	94	14.9 ± 9.8	0.058	48.0 ± 24.5	0.740	16.1 ± 17.5	0.758
Age (years)							
65–74	53	19.4 ± 11.4		56.4 ± 25.8		15.9 ± 10.5	
75–84	55	16.3 ± 12.1		47.7 ± 22.6		16.8 ± 15.8	
≥85	38	11.3 ± 7.6	0.003	38.7 ± 23.1	0.003	16.7 ± 20.1	0.948

Adapted from Birnbaum *et al.* (2004), with permission from Elsevier, Copyright 2004.

C. CARBAMAZEPINE

Carbamazepine (CBZ) is among the top four AEDs used to treat elderly nursing home residents (Garrard *et al.*, 2000; Lackner *et al.*, 1998; Schachter *et al.*, 1998). As many as 30% of nursing home residents who are receiving CBZ are doing so for nonseizure indications (Schachter *et al.*, 1998). A study of seven elderly nursing home residents (mean age, 82.3 years) showed that clearance was approximately 40% lower in the nursing home residents than in younger adults (Cloyd *et al.*, 1994). In a national nursing home study of 85 residents with an average age of 76.0 years, 47 patients were given CBZ without a concomitant inhibitor or inducer. Among these 47 patients, the average daily dose (±SD) was 8.8 ± 4.7 mg/kg, which yielded a mean total serum CBZ concentration of 6.3 ± 2.2 mg/liter (Birnbaum *et al.*, 2000). Up to 20% of total serum CBZ concentrations measured in nursing home residents are below the range used in adults (4–12 μg/ml), whereas 2.5% of measured concentrations are in the potentially toxic range (Schachter *et al.*, 1998).

Elderly nursing home residents are considered more frail than the general elderly population, and it is not uncommon for residents to have several concomitant illnesses. Both CBZ and its active metabolite, carbamazepine-10,11-epoxide (CBZ-E), bind to serum albumin and α1-acid glycoprotein (AAG) (MacKichan and Zola, 1984). Since serum AAG concentrations can increase under certain physiological conditions that are present with certain illnesses (Piafsky, 1980), it is possible that the average serum concentration of unbound CBZ could differ in this population. However, no studies have been done to validate this assumption. In addition, CBZ-E is thought to be responsible for some of the side effects seen in patients taking CBZ. No information is available regarding unbound CBZ-E concentrations in this population.

D. Phenobarbital

Phenobarbital (PB) is another commonly used AED (Cloyd *et al.*, 1994; Garrard *et al.*, 2000; Lackner *et al.*, 1998; Schachter *et al.*, 1998), and national nursing home studies have shown that PB and PHT are the most frequently used AED combination (Harms *et al.*, 2005; Schachter *et al.*, 1998). Four to five percent of measured PB concentrations were in the toxic range for elderly residents in one study (Schachter *et al.*, 1998). PB may be a metabolic inducer of PHT, and it has undesirable cognitive side effects that could be devastating in an already frail population.

III. Newer AEDs

Gabapentin (GBP) was among the four most frequently administered AEDs in a large nursing home study of 10,318 elderly residents (Garrard *et al.*, 2003). However, when the study sample was stratified according to indication, GBP was not used as frequently among residents receiving treatment for an epilepsy indication. Although there are currently no formal published studies examining the pharmacokinetics of newer AEDs in elderly nursing home residents, the pharmacokinetics of GBP make it attractive for use in this population. GBP is eliminated unchanged renally, and its clearance is expected to parallel the age-related decrease in glomerular filtration rate (Perucca, 2005). The pharmacokinetics of GBP, specifically the lack of both hepatic metabolism and protein binding, makes GBP a drug that is potentially safer than other AEDs for the elderly population. There are several other newer AEDs with very favorable pharmacokinetic profiles; however, no data exist that describe their pharmacokinetics in this population.

References

Anderson, G. D., Pak, C., Doane, K. W., Griffy, K. G., Temkin, N. R., Wilensky, A. J., and Winn, H. R. (1997). Revised Winter-Tozer equation for normalized phenytoin concentrations in trauma and elderly patients with hypoalbuminemia. *Ann. Pharmacother.* **31,** 279–284.
Bauer, L. A., and Blouin, R. A. (1982). Age and phenytoin kinetics in adult epileptics. *Clin. Pharmacol. Ther.* **31,** 301–304.

Bauer, L. A., Davis, R., Wilensky, A., Raisys, V., and Levy, R. H. (1985). Valproic acid clearance: Unbound fraction and diurnal variation in young and elderly adults. *Clin. Pharmacol. Ther.* **37,** 697–700.

Birnbaum, A. K., Hardie, N. A., Conway, J. M., Graves, N. M., Krause, S. E., Lackner, T. E., and Leppik, I. E. (2000). Serum concentration and daily doses of carbamazepine among elderly nursing home residents. *Epilepsia* **41**(Suppl. 7), 97.

Birnbaum, A., Hardie, N. A., Leppik, I. E., Conway, J. M., Bowers, S. E., Lackner, T., and Graves, N. M. (2003a). Variability of total phenytoin serum concentrations within elderly nursing home residents. *Neurology* **60,** 555–559.

Birnbaum, A. K., Hardie, N. A., Conway, J. M., Bowers, S. E., Lackner, T. E., Graves, N. M., and Leppik, I. E. (2003b). Phenytoin use in elderly nursing home residents. *Am. J. Geriatr. Pharmacother.* **1,** 90–95.

Birnbaum, A. K., Hardie, N. A., Conway, J. M., Bowers, S. E., Lackner, T. E., Graves, N. M., and Leppik, I. E. (2004). Valproic acid doses, concentrations, and clearances in elderly nursing home residents. *Epilepsy Res.* **62,** 157–162.

Bryson, S. M., Verma, N., Scott, P. J., and Rubin, P. C. (1983). Pharmacokinetics of valproic acid in young and elderly subjects. *Br. J. Clin. Pharmacol.* **16,** 104–105.

Cloyd, J. C., Lackner, T. E., and Leppik, I. E. (1994). Antiepileptics in the elderly. Pharmacoepidemiology and pharmacokinetics. *Arch. Fam. Med.* **3,** 589–598.

Cooper, J. W. (1999). Adverse drug reaction-related hospitalizations of nursing facility patients: A 4-year study. *South. Med. J.* **92,** 485–490.

Garrard, J., Cloyd, J., Gross, C., Hardie, N., Thomas, L., Lackner, T., Graves, N., and Leppik, I. (2000). Factors associated with antiepileptic drug use among elderly nursing home residents. *J. Gerontol. A Biol. Sci. Med. Sci.* **55,** M384–M392.

Garrard, J., Harms, S., Hardie, N., Eberly, L. E., Nitz, N., Bland, P., Gross, C. R., and Leppik, I. E. (2003). Antiepileptic drug use in nursing home admissions. *Ann. Neurol.* **54,** 75–85.

Gerety, M. B., Cornell, J. E., Plichta, D. T., and Eimer, M. (1993). Adverse events related to drugs and drug withdrawal in nursing home residents. *J. Am. Geriatr. Soc.* **41,** 1326–1332.

Greenblatt, D. J. (1979). Reduced serum albumin concentration in the elderly: A report from the Boston Collaborative Drug Surveillance Program. *J. Am. Geriatr. Soc.* **27,** 20–22.

Gurwitz, J. H., Field, T. S., Avorn, J., McCormick, D., Jain, S., Eckler, M., Benser, M., Edmondson, A. C., and Bates, D. W. (2000). Incidence and preventability of adverse drug events in nursing homes. *Am. J. Med.* **109,** 87–94.

Harms, S. L., Eberly, L. E., Garrard, J. M., Hardie, N. A., Bland, P. C., and Leppik, I. E. (2005). Prevalence of appropriate and problematic antiepileptic combination therapy in older people in the nursing home. *J. Am. Geriatr. Soc.* **53,** 1023–1028.

Hayes, M. J., Langman, M. J., and Short, A. H. (1975). Changes in drug metabolism with increasing age. 2. Phenytoin clearance and protein binding. *Br. J. Clin. Pharmacol.* **2,** 73–79.

Holtz, L., Milton, J., and Sturek, J. K. (1987). Compatibility of medications with enteral feedings. *J. Parenter. Enteral Nutr.* **11,** 183–186.

Hooks, M. A., Longe, R. L., Taylor, A. T., and Francisco, G. E. (1986). Recovery of phenytoin from an enteral nutrient formula. *Am. J. Hosp. Pharm.* **43,** 685–688.

Lackner, T. E., Cloyd, J. C., Thomas, L. W., and Leppik, I. E. (1998). Antiepileptic drug use in nursing home residents: Effect of age, gender, and comedication on patterns of use. *Epilepsia* **39,** 1083–1087.

MacKichan, J. J., and Zola, E. M. (1984). Determinants of carbamazepine and carbamazepine 10,11-epoxide binding to serum protein, albumin and alpha-1-acid glycoprotein. *Br. J. Clin. Pharmacol.* **18,** 487–493.

Maynard, G. A., Jones, K. M., and Guidry, J. R. (1987). Phenytoin absorption from tube feedings. *Arch. Intern. Med.* **147,** 1821.

Mooradian, A. D., Hernandez, L., Tamai, I. C., and Marshall, C. (1989). Variability of serum phenytoin concentrations in nursing home patients. *Arch. Intern. Med.* **149,** 890–892.

Moore, S. R., and Jones, J. K. (1985). Adverse drug reaction surveillance in the geriatric population: A preliminary view. *In* "Geriatric Drug Use: Clinical & Social Perspectives" (S. R. Moore and T. W. Teal, Eds.), pp. 70–77. Pergamon Press, New York.

Patterson, M., Heazelwood, R., Smithurst, B., and Eadie, M. J. (1982). Plasma protein binding of phenytoin in the aged: *In vivo* studies. *Br. J. Clin. Pharmacol.* **13,** 423–425.

Perucca, E. (2005). Pharmacokinetic variability of new antiepileptic drugs at different ages. *Ther. Drug Monit.* **27,** 714–717.

Perucca, E., Grimaldi, R., Gatti, G., Pirracchio, S., Crema, F., and Frigo, G. M. (1984). Pharmacokinetics of valproic acid in the elderly. *Br. J. Clin. Pharmacol.* **17,** 665–669.

Piafsky, K. M. (1980). Disease-induced changes in the plasma binding of basic drugs. *Clin. Pharmacokinet.* **5,** 246–262.

Sarkar, M. A., Garnett, W. R., and Karnes, H. T. (1989). The effects of storage and shaking on the settling properties of phenytoin suspension. *Neurology* **39,** 207–209.

Schachter, S. C., Cramer, G. W., Thompson, G. D., Chaponis, R. J., Mendelson, M. A., and Lawhorne, L. (1998). An evaluation of antiepileptic drug therapy in nursing facilities. *J. Am. Geriatr. Soc.* **46,** 1137–1141.

Strahan, G. W. (1997). An overview of nursing homes and their current residents: Data from the 1995 National Nursing Home Survey. *Adv. Data* **280,** 1–12.

THE IMPACT OF EPILEPSY ON OLDER VETERANS

Mary Jo V. Pugh,*,† Dan R. Berlowitz,‡,§ and Lewis Kazis‡,§

*South Texas Veterans Health Care System (VERDICT), San Antonio, Texas 78229, USA
†Department of Medicine, University of Texas Health Science Center at San Antonio
San Antonio, Texas 78229, USA
‡Center for Health Quality, Outcomes, and Economic Research, Bedford VA Medical Center
Bedford, Massachusetts 01730, USA
§Department of Health Services, Boston University School of Public Health, Boston
Massachusetts 02118, USA

Despite the fact that old age is the time with the highest incidence of epilepsy, little is known specifically about the impact of epilepsy on the daily lives of the elderly. Previous studies have explored the impact of epilepsy on health status in a general population, but typically have not included enough older individuals to adequately describe this population. The study on which this chapter is based used a general survey instrument to begin exploration of this issue in a population of older veterans with epilepsy.

Older patients (≥65 years of age) were identified who had both International Classification of Diseases, Ninth Revision, Clinical Modification (ICD-9-CM), codes indicating epilepsy and prescriptions of antiepileptic drugs in national Veterans Affairs (VA) administrative and pharmacy databases during fiscal year 1999. Using these databases, patients were further identified as newly or previously diagnosed. Diagnostic data were then linked with data from the 1999 Large Health Survey of Veteran Enrollees, using encrypted identifiers, and the impact of epilepsy on patients of different ages was assessed using individual scales and component summaries of the Veterans SF-36.

Results showed that older individuals with epilepsy had lower scores on measures of both physical and mental health than did their counterparts with no

221

epilepsy. Further, scores associated with mental health functioning were significantly lower for those with newly diagnosed epilepsy than for those with chronic epilepsy, but differences associated with scores on physical functioning were not significant. Thus, while previous studies suggest that the effects of chronic neurological disorders such as epilepsy are most obvious on measures of mental health, these data suggest that older patients experience difficulties in both physical and mental health.

I. Introduction

Epilepsy is a common neurological disorder affecting approximately 1.5% of Americans over the age of 65 (Leppik and Birnbaum, 2002), and it is believed that this number will rise with the aging of society. Yet, little is known about the impact of epilepsy on the health status of elderly patients. Previous studies have demonstrated that patients with epilepsy tend to have lower scores on measures of health-related quality of life, including physical and mental health status, than the general population (Baker, 1998; Baker *et al.*, 1998; Donker *et al.*, 1997; Fisher, 2000; Fisher *et al.*, 2000). However, the impact of epilepsy is greater for patients with frequent and more severe seizures; those whose seizures are controlled tend to have health status scores that are similar to the scores of individuals in the general population (Abetz *et al.*, 2000; Baker *et al.*, 1998; Birbeck *et al.*, 2002; Devinsky *et al.*, 1995; Raty *et al.*, 1999). Study results further suggest that patients newly diagnosed with epilepsy experience lower levels of physical and mental health than patients who were previously diagnosed with epilepsy, but that this postdiagnosis decline in health status is transient (Petersen *et al.*, 1998; Raty *et al.*, 1999).

Although these studies provide insight into the general population, the extent to which they describe the elderly population is unknown. The average age of individuals in published studies generally falls in the fourth decade of life (Abetz *et al.*, 2000; Baker *et al.*, 1998; Birbeck *et al.*, 2002; Devinsky *et al.*, 1995; Donker *et al.*, 1997; Fisher, 2000; Petersen *et al.*, 1998; Raty *et al.*, 1999), and study reports have suggested that elderly individuals are less likely to respond to surveys than are younger individuals (Wagner *et al.*, 1996). The developmental needs of older patients are different from those of young adults and adults in middle age, and physiological changes that occur with aging complicate this picture.

Successful aging occurs when patients are able to remain engaged in life, have a low risk of disease, and maintain high cognitive and physical function (Rowe and Kahn, 1998). Patients with epilepsy cannot avoid disease, so it is even more important to maintain active mental, physical, and social functioning. The purpose of the research presented in this chapter was to begin to describe the impact of epilepsy on the health status of older veterans receiving care for epilepsy within the US Department of Veterans Affairs (VA).

II. Methods

A. DATA SOURCES AND POPULATION

The VA National Patient Care Database and the national Pharmacy Benefits Management database (PBM v.3) were used to identify the cohort and describe the type of care received. Veterans ≥65 years of age were first identified and then those with diagnoses for epilepsy were selected [International Classification of Diseases, Ninth Revision, Clinical Modification (ICD-9-CM), codes 345.XX (epilepsy) or 780.3 (convulsion)] in VA inpatient [fiscal year 1996 (FY96) to FY99] or outpatient (FY97 to FY99) databases (Pugh *et al.*, 2004). Those who received at least one prescription for phenobarbital, phenytoin, carbamazepine, valproate, lamotrigine, or gabapentin in FY99 were identified using pharmacy data ($n = 20,558$). Because previous research indicates that patients newly diagnosed with epilepsy may have a qualitatively different experience than patients who have had the disease for one year or more, older veterans were further classified as being newly or previously diagnosed with epilepsy. Those diagnosed between FY96 and FY98 were classified as previously diagnosed ($n = 16,917$). Those who had a first diagnosis in FY99 and who had received care in the VA system during FY97 to FY99 were classified as newly diagnosed ($n = 1893$). Those first diagnosed with epilepsy in FY99 without prior VA care were identified as new to the VA and were excluded from the study ($n = 1748$). Approximately 1.1 million older veterans had no history of epilepsy diagnosis between FY96 and FY99.

Basic demographic information (age, sex, race) was obtained from VA administrative data (inpatient and outpatient). Additional demographic information (age, sex, race, marital status, education) was obtained from the 1999 Veterans SF-36 Health Survey (VSHS). In that survey, 1.5 million VA enrollees were randomly selected from the VA enrollee database and asked to describe their activity level, health habits, and health status (Veterans SF-36). Data collection took place between July 1, 1999, and January 1, 2000, with a response rate of approximately 75% for those ≥65 years of age. From this cohort, 718 individuals identified as newly diagnosed with epilepsy, 7806 identified as previously diagnosed with epilepsy, and 436,903 identified as having no history of epilepsy responded to the survey.

B. MEASURES

1. *Primary Outcomes*

The selected outcomes described the experiences of epilepsy from the patients' point of view using data from the VSHS. The primary outcome measures were obtained via the Veterans SF-36, a modification of the well-known Medical

Outcomes Study SF-36 (Ware and Sherbourne, 1992; Ware *et al.*, 1994). The SF-36 measures eight constructs of health: physical functioning, role limitations due to physical problems, bodily pain, general health perceptions, energy/vitality, social functioning, role limitations due to emotional problems, and mental health (Ware *et al.*, 1994). The Veterans SF-36 expanded the response choices of role limitations due to emotional and physical problems from a dichotomous response to a five-point scale, enhancing discriminant validity and reducing floor and ceiling effects (Kazis, 2000). From these eight scales, two component summaries—physical (PCS) and mental (MCS) summaries—are calculated. These component summaries are important because they are standardized to the US population and are norm based (mean $= 50$; $\pm SD = 10$), allowing for comparisons between different groups. Previous research using generic health status assessments such as the SF-36 in a population with epilepsy indicates that individual scales are more specific than component summaries in distinguishing between patients with various levels of symptoms and those who have no symptoms (Wagner *et al.*, 1996). Therefore, component summaries and individual scale scores were both examined, focusing on specific scales that may have a more serious impact on the ability of older individuals to remain engaged in their lives and to maintain physical, social, and role functioning.

In addition, individuals' perceptions of change in mental and physical health status (worse, the same, or better) over the past year and rates of participation per week in regular physical exercise (0, <1, 1–2, 3–4, or ≥5 times per week) were described.

2. *Demographic Information*

Inpatient and outpatient VA data were used in conjunction with information from the VSHS to identify demographic information. Age, sex, race, and marital status information was integrated from both data sources. Previous research has found that information about race in VA administrative data is quite accurate (Kressin *et al.*, 2003). Thus, information on sex, race, and marital status was supplemented when values were missing in the administrative data. Information on living situation (alone or with someone) and education (less than high school, high school, some college, or complete college) was obtained from the VSHS.

3. *Comorbidity*

Because health status is also related to level of disease burden, other disease burden was controlled for using the Physical and Mental Comorbidity Indices (CIs) developed by Selim *et al.*, which include counts of diseases for 30 physical and 6 mental comorbidities. These indices were developed to control for comorbidity in analyses of health status and were validated using the SF-36 (Selim *et al.*, 2004b). Findings indicated that the Physical and Mental CI scores are significantly associated with all SF-36 scales, explain 24% and 36% of the variance in the PCS and MCS, respectively, and are significant predictors of both health care utilization and mortality.

C. ANALYSES

Analyses included descriptions of older veterans' scores for individual scales of the Veterans SF-36, the PCS and MCS, change in physical and mental health in the past year, and participation in regular exercise. In order to assess differential impact, health status scores were compared between older veterans without epilepsy, those with previously diagnosed epilepsy, and those with newly diagnosed epilepsy. Further, analysis of covariance (ANCOVA) models was used to determine whether differences were significant when demographic characteristics and comorbidity were controlled. Chi-square analyses were used to assess differences between these groups on categorical variables.

III. Results

The majority of older veterans in this cohort were male (98%), under the age of 75 (~60%), and Caucasian (70%). Table I provides specific demographic and comorbidity information for those with no epilepsy diagnosis, those with newly diagnosed epilepsy, and those with previously diagnosed epilepsy. The only significant differences between the groups were that African Americans were more likely to have a diagnosis of epilepsy ($p < 0.01$) and those with unknown race were less likely to be diagnosed with epilepsy ($p < 0.01$).

A. VETERANS SF-36 SCORES

Scores on individual scales and component summaries for the Veterans SF-36 are presented in Table II. Those without epilepsy had higher scores than those with an epilepsy diagnosis on individual scales, indicating better physical and mental health on all measures. Moreover, individuals newly diagnosed with epilepsy consistently had lower levels of health status than did those who were diagnosed at least one year previously. These differences were not only statistically significant, they were also clinically significant. Differences between those with no epilepsy and those who were newly diagnosed were approximately one-half of a standard deviation. Differences both between those with no epilepsy versus those previously diagnosed and between those previously diagnosed versus those newly diagnosed averaged approximately one-third of a standard deviation.

TABLE I

DEMOGRAPHIC CHARACTERISTICS OF OLDER VETERANS BY EPILEPSY STATUS

	Epilepsy diagnosis		
	None ($n = 1,109,997$)	Previous ($n = 16,917$)	New ($n = 1893$)
Age (%)			
65–74	60	63	60
75–84	37	35	38
≥85	3	2	2
Sex (%)			
Male	98	99	98
Female	2	1	2
Race (%)			
Caucasian	67	70	70
African American	9	17	19
Hispanic	4	5	5
Other	1	1	1
Unknown	19	8	7
Marital status (%)			
Married	66	61	62
Unmarried	34	39	38
Education[a] (%)			
<High school	41	45	41
High school	29	30	30
Some college	19	16	19
Completed college	11	10	10
Living situation[b] (%)			
Alone	23	20	14
With someone	77	80	86

[a]For those responding to the 1999 Veterans SF-36 Health Survey: no epilepsy, $n = 308,701$; previous diagnosis, $n = 4882$; new diagnosis, $n = 431$.

[b]For those responding to the 1999 Veterans SF-36 Health Survey: no epilepsy, $n = 307,285$; previous diagnosis, $n = 4886$; new diagnosis, $n = 416$.

These differences on individual scales were also reflected in the PCS and MCS, with differences between those with no epilepsy and those with newly diagnosed epilepsy being about one-half of a standard deviation for the PCS (32.95 vs 28.69) and three-quarters of a standard deviation for the MCS (46.07 vs 38.68). Consistent with previous research, scores for all three groups were lower than were scores for respondents to the 1998 Medicare Health Outcomes Survey (Selim *et al.*, 2004a). However, there were also variations between the groups on both the PCS and MCS (Table II).

Next, when differences in age, sex, race, marital status, living situation, education, and comorbidities were controlled among all groups, the adjusted scores reported in Table III indicate that, although the differences were attenuated,

TABLE II

VETERANS SF-36 SCORES FOR INDIVIDUALS WITH NO EPILEPSY, NEW EPILEPSY, OR PREVIOUS EPILEPSY[a]

Veterans SF-36 scale	Observed score, mean ($\pm$SD)
Physical function (PF)	
No epilepsy	46.14 (29.35)
Previous epilepsy	37.18 (30.13)
New epilepsy	29.07 (28.92)
Role physical (RP)	
No epilepsy	27.21 (37.34)
Previous epilepsy	18.23 (34.21)
New epilepsy	10.34 (29.45)
General health (GH)	
No epilepsy	45.45 (23.82)
Previous epilepsy	37.39 (22.92)
New epilepsy	31.48 (21.65)
Bodily pain (BP)	
No epilepsy	47.92 (26.78)
Previous epilepsy	43.25 (27.45)
New epilepsy	38.54 (26.77)
Mental health (MH)	
No epilepsy	67.22 (22.29)
Previous epilepsy	59.91 (23.49)
New epilepsy	54.00 (24.04)
Role emotional (RE)	
No epilepsy	49.27 (48.69)
Previous epilepsy	36.33 (46.71)
New epilepsy	25.70 (43.33)
Vitality (VT)	
No epilepsy	41.85 (24.13)
Previous epilepsy	35.58 (23.20)
New epilepsy	28.98 (21.97)
Social functioning (SF)	
No epilepsy	60.59 (30.93)
Previous epilepsy	49.05 (32.04)
New epilepsy	41.27 (31.37)
Physical component summary (PCS)	
No epilepsy	32.95 (3.99)
Previous epilepsy	30.27 (4.48)
New epilepsy	28.69 (4.71)
Mental component summary (MCS)	
No epilepsy	46.07 (5.12)
Previous epilepsy	41.76 (5.95)
New epilepsy	38.68 (6.71)

[a]Because diagnoses could happen at any point in fiscal year 1999 for those newly diagnosed, we included only those newly diagnosed patients who were diagnosed before completing the Veterans SF-36.

TABLE III

Adjusted[a] Veterans SF-36 Scores for Individuals With No Epilepsy, New Epilepsy, or Previous Epilepsy

Veterans SF-36 scale	Adjusted score, mean (±SE)	Comparisons
Physical function (PF)		
No epilepsy	43.85 (0.25)	No vs New[b]
Previous epilepsy	39.32 (0.48)	New vs Previous[c]
New epilepsy	34.80 (1.41)	No vs Previous[b]
Role physical (RP)		
No epilepsy	29.80 (0.32)	No vs New[b]
Previous epilepsy	25.61 (0.62)	New vs Previous
New epilepsy	21.47 (1.84)	No vs Previous[b]
General health (GH)		
No epilepsy	49.24 (0.20)	No vs New[b]
Previous epilepsy	45.20 (0.39)	New vs Previous[c]
New epilepsy	42.08 (1.15)	No vs Previous[b]
Bodily pain (BP)		
No epilepsy	48.20 (0.23)	No vs New
Previous epilepsy	47.45 (0.44)	New vs Previous
New epilepsy	45.83 (1.30)	No vs Previous
Mental health (MH)		
No epilepsy	68.77 (0.19)	No vs New[b]
Previous epilepsy	64.79 (0.36)	New vs Previous[b]
New epilepsy	61.50 (1.05)	No vs Previous[c]
Role emotional (RE)		
No epilepsy	50.43 (0.42)	No vs New[b]
Previous epilepsy	43.94 (0.81)	New vs Previous
New epilepsy	37.89 (2.41)	No vs Previous[b]
Vitality (VT)		
No epilepsy	44.97 (0.21)	No vs New[b]
Previous epilepsy	42.23 (0.39)	New vs Previous[c]
New epilepsy	38.34 (1.16)	No vs Previous[b]
Social functioning (SF)		
No epilepsy	59.38 (0.26)	No vs New[b]
Previous epilepsy	52.67 (0.50)	New vs Previous
New epilepsy	48.88 (1.48)	No vs Previous[b]
Physical component summary (PCS)		
No epilepsy	33.04 (0.10)	No vs New[b]
Previous epilepsy	32.01 (0.19)	New vs Previous
New epilepsy	30.66 (0.55)	No vs Previous[b]
Mental component summary (MCS)		
No epilepsy	46.90 (0.11)	No vs New[b]
Previous epilepsy	44.51 (0.21)	New vs Previous[c]
New epilepsy	42.82 (0.63)	No vs Previous[b]

[a]Adjusted scores represent differences between groups after controlling for age, sex, race, marital status, living situation, education, and comorbidities.

[b]$p < 0.0001$.

[c]$p < 0.01$.

differences remained both statistically and clinically important, and that the same pattern of differences prevailed. However, in several cases, differences were not statistically different between new and previously diagnosed patients (i.e., bodily pain, social functioning, role physical, role emotional, and the PCS).

B. OTHER INDICATORS OF HEALTH STATUS

Other aspects providing insight into the impact of epilepsy on older individuals include changes in mental and physical health over the past year and the extent to which individuals participate in regular exercise. Results presented in Table IV indicate that those with no epilepsy diagnosis were more likely to report that their mental and physical health were about the same, and were less likely to report a decrement in mental or physical health over the past year than were individuals with epilepsy. Furthermore, for both mental and physical health, those newly diagnosed were more likely to say they were worse and were less likely to say they were about the same over the past year than those who were previously diagnosed. These differences remained significant despite controlling for age, sex, race, marital status, and comorbidities using multinomial logistic regression models.

Results for participation in regular exercise are presented in Table V. Patterns were similar to those seen in previous analyses, with individuals with epilepsy being less likely to participate in regular exercise (no regular exercise: 44.57% for those with no epilepsy, 59.02% for those previously diagnosed, and 68.42% for those newly diagnosed). Because nonparticipation in exercise may be due in part to comorbidities, multinomial logistic regression was used to determine if

TABLE IV
CHANGE IN HEALTH DURING THE PAST YEAR

	Epilepsy diagnosis		
	None	Previous	New
Physical health (%)			
Worse	40.68	48.21	62.05
About the same	45.76	39.69	28.79
Better	13.56	12.10	9.15
Mental health (%)			
Worse	23.51	31.89	43.68
About the same	62.78	55.19	44.57
Better	13.71	12.92	11.75

All comparisons are significant: $p < 0.01$.

TABLE V

PARTICIPATION IN REGULAR PHYSICAL EXERCISE

	Epilepsy diagnosis		
Times per week	None (%)	Previous (%)	New (%)
0	44.57[a]	59.02[a]	68.42[a]
<1	13.19[a]	10.60	10.53
1–2	15.04[a]	11.53[a]	5.95[a]
3–4	15.10[a]	10.32[a]	7.09[a]
≥5	12.10[a]	8.53	8.01

[a] $p < 0.01$.

these findings were consistent when age, sex, race, marital status, and comorbidities were controlled. Findings indicated that those with epilepsy were about 1.5 times more likely to have no regular participation in exercise ($p < 0.01$; 99% confidence intervals: 1.08–2.52 for those newly diagnosed and 1.36–1.81 for those previously diagnosed) than those without epilepsy.

IV. Discussion

Despite the fact that old age is the time with the highest incidence of epilepsy, little is known specifically about the impact of epilepsy on the daily lives of the elderly. Previous studies have explored the impact of epilepsy on health status in a general population, but typically have not included enough older individuals to adequately describe this population. The study on which this chapter is based used a general survey instrument to begin exploration of this issue. Even with such a general instrument, this study found that the impact of epilepsy on the elderly is substantial and that the impact was different for those newly and previously diagnosed.

First, results from this study showed that older individuals with epilepsy had lower scores on measures of both physical and mental health. Thus, while previous studies suggest that the effects of chronic neurological disorders such as epilepsy are most obvious on scales measuring mental health (Hermann *et al.*, 1996), these data suggest that older patients experience difficulties in both physical and mental health. Older patients are more susceptible to the adverse effects of drugs (Ramsay *et al.*, 1994), and they may be more likely to experience somnolence, ataxia, disturbance of gait, and so on, which may lead to falls and decreased physical activity. This latter hypothesis is supported by the findings of lower rates of participation in regular physical activity by older veterans with epilepsy. Lower levels of physical activity leading to decreased muscle mass and falls are thought to be the cause of most hip fracture injuries (Marks *et al.*, 2003). The association of decreased activity, decreased muscle mass, and a negative spiral toward frailty also suggests that the physical impact of epilepsy may be more serious for the elderly than the nonelderly adult population.

In addition, older individuals with epilepsy were more likely to report decrements in both physical and mental health over the past year than were those without epilepsy. Whereas results from previous studies suggest that, after adjustment to epilepsy, health status returns to previous levels (Stavem *et al.*, 2000), this study indicates that this may not be so for the elderly. These data suggest that health status is not as high for individuals who have had epilepsy for some time as it is for individuals without epilepsy. Rather, measures of both physical and mental health, participation in activities, and change in mental and physical health status for those previously diagnosed with epilepsy were between the levels of newly diagnosed patients and those who did not have epilepsy diagnoses. Although this may be due to the fact that some individuals have more frequent seizures, these data point to the importance of prospective research to more fully explicate the changes in health status experienced by older patients with epilepsy over time.

These data indicate that the impact of epilepsy on older individuals is substantial and that differences are not only significant, but also clinically important. While analyses were limited to VA data, the epilepsy cohort had demographic characteristics and comorbidity profiles that were similar to those of the general geriatric epilepsy population, and thus the results may be generalizable (Hauser, 1997). These data suggest that further research is needed to improve understanding of the impact of epilepsy on older patients. Because recent clinical recommendations suggest that there is variation in the side-effect profiles of antiepileptic drugs (AEDs) in the elderly (Karceski *et al.*, 2001), research examining the impact of long-term AED use by the type of AED may help providers understand the effectiveness of these drugs in a geriatric population.

Acknowledgments

The study on which this chapter is based was funded by the Department of Veterans Affairs, Veterans Health Administration, Health Services Research and Development Service Merit Review Entry Program Award to M.J.V.P., Ph.D. (MRP-02-267). We acknowledge the support of the South Texas Veterans Health Care System/VERDICT in San Antonio, TX, and the Edith Nourse Rogers Memorial VA Medical Center/The Center for Health Quality, Outcomes, and Economic Research in Bedford, MA. We also thank the Department of Veterans Affairs, Veterans Health Administration Office of Quality and Performance for providing access to data from the 1999 Large Health Survey of Veteran Enrollees. The SF-36 is the property of the Medical Outcomes Trust.

The authors have no conflicts of interest related to the content of this chapter.

The views expressed in this article are those of the authors and do not necessarily represent the views of the Department of Veterans Affairs.

References

Abetz, L., Jacoby, A., Baker, G. A., and McNulty, P. (2000). Patient-based assessments of quality of life in newly diagnosed epilepsy patients: Alidation of the NEWQOL. *Epilepsia* **41,** 1119–1128.

Baker, G. A. (1998). Quality of life and epilepsy: The Liverpool experience. *Clin. Ther.* **20**(Suppl. A), A2–A12.

Baker, G. A., Gagnon, D., and McNulty, P. (1998). The relationship between seizure frequency, seizure type and quality of life: Findings from three European countries. *Epilepsy Res.* **30,** 231–240.

Birbeck, G. L., Hays, R. D., Cui, X., and Vickrey, B. G. (2002). Seizure reduction and quality of life improvements in people with epilepsy. *Epilepsia* **43,** 535–538.

Devinsky, O., Vickrey, B. G., Cramer, J., Perrine, K., Hermann, B., Meador, K., and Hays, R. D. (1995). Development of the quality of life in epilepsy inventory. *Epilepsia* **36,** 1089–1104.

Donker, G. A., Foets, M., and Spreeuwenberg, P. (1997). Epilepsy patients: Health status and medical consumption. *J. Neurol.* **244,** 365–370.

Fisher, R. S. (2000). Epilepsy from the patient's perspective: Review of results of a community-based survey. *Epilepsy Behav.* **1,** S9–S14.

Fisher, R. S., Vickrey, B. G., Gibson, P., Hermann, B., Penovich, P., Scherer, A., and Walker, S. (2000). The impact of epilepsy from the patient's perspective I. Descriptions and subjective perceptions. *Epilepsy Res.* **41,** 39–51.

Hauser, W. A. (1997). Epidemiology of seizures and epilepsy in the elderly. *In* "Seizures and Epilepsy in the Elderly" (A. J. Rowan and R. E. Ramsay, Eds.), pp. 7–18. Butterworth-Heinemann, Newton, MA.

Hermann, B. P., Vickrey, B., Hays, R. D., Cramer, J., Devinsky, O., Meador, K., Perrine, K., Myers, L. W., and Ellison, G. W. (1996). A comparison of health-related quality of life in patients with epilepsy, diabetes and multiple sclerosis. *Epilepsy Res.* **25,** 113–118.

Karceski, S., Morrell, M., and Carpenter, D. (2001). The expert consensus guideline series: Treatment of epilepsy. *Epilepsy Behav.* **2,** A1–A50.

Kazis, L. E. (2000). The Veterans SF-36 Health Status Questionnaire: Development and application in the Veterans Health Administration. *Med. Outcomes Trust Monitor* **5,** 1–2 and 13–14.

Kressin, N. R., Chang, B. H., Hendricks, A., and Kazis, L. E. (2003). Agreement between administrative data and patients' self-reports of race/ethnicity. *Am. J. Public Health* **93,** 1734–1739.

Leppik, I. E., and Birnbaum, A. (2002). Epilepsy in the elderly. *Semin. Neurol.* **22,** 309–320.

Marks, R., Allegrante, J. P., Ronald MacKenzie, C., and Lane, J. M. (2003). Hip fractures among the elderly: Causes, consequences and control. *Ageing Res. Rev.* **2,** 57–93.

Petersen, B., Walker, M. L., Runge, U., and Kessler, C. (1998). Quality of life in patients with idiopathic, generalized epilepsy. *J. Epilepsy* **11,** 306–313.

Pugh, M. J., Cramer, J., Knoefel, J., Charbonneau, A., Mandell, A., Kazis, L., and Berlowitz, D. (2004). Potentially inappropriate antiepileptic drugs for elderly patients with epilepsy. *J. Am. Geriatr. Soc.* **52,** 417–422.

Ramsay, R. E., Rowan, A. J., Slater, J. D., Collins, J., Nemire, R., Ortiz, W. R. and the VA Cooperative Study Group (1994). Effect of age on epilepsy and its treatment: Results from the VA Cooperative Study (abstract). *Epilepsia* **35,** 91.

Raty, L., Hamrin, E., and Soderfeldt, B. (1999). Quality of life in newly-debuted epilepsy. An empirical study. *Acta Neurol. Scand.* **100,** 221–226.

Rowe, J. W., and Kahn, R. L. (1998). "Successful Aging." Pantheon Books, New York.

Selim, A. J., Berlowitz, D. R., Fincke, G., Cong, Z., Rogers, W., Haffer, S. C., Ren, X. S., Lee, A., Qian, S. X., Miller, D. R., Spiro, A., III, Selim, B. J., *et al.* (2004a). The health status of elderly veteran enrollees in the Veterans Health Administration. *J. Am. Geriatr. Soc.* **52,** 1271–1276.

Selim, A. J., Fincke, G., Ren, X. S., Lee, A., Rogers, W. H., Miller, D. R., Skinner, K. M., Linzer, M., and Kazis, L. E. (2004b). Comorbidity assessments based on patient report: Results from the Veterans Health Study. *J. Ambul. Care Manage.* **27,** 281–295.

Stavem, K., Loge, J. H., and Kaasa, S. (2000). Health status of people with epilepsy compared with a general reference population. *Epilepsia* **41,** 85–90.

Wagner, A. K., Bungay, K. M., Kosinski, M., Bromfield, E. B., and Ehrenberg, B. L. (1996). The health status of adults with epilepsy compared with that of people without chronic conditions. *Pharmacotherapy* **16,** 1–9.

Ware, J. E., Jr., and Sherbourne, C. D. (1992). The MOS 36-item short-form health survey (SF-36). I. Conceptual framework and item selection. *Med. Care* **30,** 473–483.

Ware, J. E., Jr., Kosinski, M., and Keller, S. D. (1994). "SF-36 Physical and Mental Health Summary Scales: A Users Manual." The Health Institute, New England Medical Center, Boston.

RISK AND PREDICTABILITY OF DRUG INTERACTIONS IN THE ELDERLY

René H. Levy*,† and Carol Collins*

*Department of Pharmaceutics, University of Washington
Seattle, Washington 98195, USA
†Department of Neurological Surgery, University of Washington
Seattle, Washington 98195, USA

The issue of drug–drug interactions is particularly relevant for geriatric patients with epilepsy because they are often treated with multiple medications for concurrent diseases such as cardiovascular disease and psychiatric disorders (e.g., dementia and depression). The antidepressants with the least potential for altering antiepileptic drug (AED) metabolism are citalopram, escitalopram, venlafaxine, duloxetine, and mirtazapine. The use of established AEDs with enzyme-inducing properties, such as carbamazepine, phenytoin, and phenobarbital, may be associated with reductions in the levels of drugs such as donepezil, galantamine, and particularly warfarin. Carbamazepine, phenytoin, and phenobarbital have been reported to decrease prothrombin time in patients taking oral anticoagulants, although with phenytoin, an increase in prothrombin time has also been reported. Drugs associated with increased risk of bleeding in patients taking oral anticoagulants include selective serotonin reuptake inhibitors

INTERNATIONAL REVIEW OF
NEUROBIOLOGY, VOL. 81
DOI: 10.1016/S0074-7742(06)81015-9

235

(especially fluoxetine), gemfibrozil, fluvastatin, and lovastatin. Other drugs affected by enzyme inducers include cytochrome P450 3A4 substrates, such as calcium channel blockers (e.g., nimodipine, nilvadipine, nisoldipine, and felodipine) and the 3-hydroxy-3-methylglutaryl coenzyme A (HMG-CoA) reductase inhibitors atorvastatin, lovastatin, and simvastatin.

Although there have been no reports of AEDs altering ticlopidine metabolism, ticlopidine coadministration can result in carbamazepine and phenytoin toxicity. Also, there is a significant risk of elevated levels of carbamazepine when diltiazem and verapamil are administered. In addition, there are case reports of phenytoin toxicity when administered with diltiazem. Drugs with a lower potential for metabolic drug interactions include (1) cholinesterase inhibitors (although the theoretical possibility of a reduction in donepezil and galantamine levels by enzyme-inducing AEDs should be considered) and the N-methyl-D-aspartate receptor antagonist memantine and (2) antihypertensives such as angiotensin-converting enzyme inhibitors, angiotensin receptor blockers, hydrophilic beta-blockers, and thiazide diuretics. There is a moderate risk that enzyme-inducing AEDs will decrease levels of lipophilic beta-blockers. Newer AEDs have a lower potential for drug interactions. In particular, levetiracetam and gabapentin have not been reported to alter enzyme activity. In summary, there is a significant potential for drug interactions between AEDs and drugs commonly prescribed in geriatric patients with epilepsy.

I. Introduction

Geriatric patients with epilepsy often receive multiple medications for concurrent diseases (Lackner *et al.*, 1998). Therefore, these patients are particularly vulnerable to drug–drug interactions. Elderly patients with epilepsy are also more sensitive to side effects of medications, particularly cognitive impairment and cardiac conduction abnormalities (Turnheim, 2003). Two areas of particular concern for comorbid diseases are cardiovascular disease and psychiatric disorders, such as dementia and depression. The most common known etiology of seizures in the elderly is stroke (DeToledo, 1999; Hauser, 1992; Kramer, 2001; Lossius *et al.*, 2002). Furthermore, the most common type of stroke is ischemic stroke, which is associated with multiple cardiovascular risk factors, including atrial fibrillation, carotid artery stenosis, hypertension, and coronary heart disease (Gorelick *et al.*, 1999). Therefore, the clinician should have a high index of suspicion for the presence of other types of cardiovascular disease in stroke patients, and drugs with known cardiovascular toxicity should be used with caution. The associations between stroke, epilepsy, and Alzheimer's disease are complicated by an increased incidence of Alzheimer's disease among individuals with a history of stroke (Honig *et al.*, 2003), and conversely, an increased incidence of seizures among patients with dementia (Hauser, 1992; Stephen and Brodie, 2000). The associations between stroke, epilepsy, and depression are similarly complicated. The incidence of depression is increased in individuals after a stroke (Robinson, 2003) and in individuals with epilepsy (Kanner, 2003), and increased incidences of stroke and cardiovascular disease are seen among patients with depression (Larson *et al.*, 2001). Epidemiological data were used to focus on medications that would be frequently used in geriatric patients with epilepsy and to evaluate the evidence for interactions between antiepileptic drugs (AEDs) and such medications.

II. Antiepileptic Drugs Used in the Elderly

Antiepileptic drugs reviewed in this chapter include carbamazepine, gabapentin, lamotrigine, levetiracetam, oxcarbazepine, phenobarbital, phenytoin, topiramate, valproic acid, and zonisamide. Some AEDs are known to alter the metabolism of other drugs through effects on cytochrome P450 (CYP450) enzymes. Carbamazepine, phenytoin, and phenobarbital are inducers of CYP450 enzymes, including CYP1A2, CYP2C9, CYP2C19, and CYP3A4. Carbamazepine, phenytoin, and phenobarbital also induce uridine diphosphate glucuronosyltransferases (UGTs) 1A4, 2B7, and 2B15 (Anderson, 2004). Oxcarbazepine and topiramate are weak inducers of CYP3A4. In addition, oxcarbazepine is a weak inhibitor of CYP2C19. Lamotrigine has been reported to decrease valproate levels (Anderson *et al.*, 1996), and administration of lamotrigine 300 mg/day has been found to cause a modest decrease in the serum levels of levonorgestrel (Sidhu *et al.*, 2006). Levetiracetam and gabapentin have not been reported to alter enzyme activity (Anderson, 2004).

The CYP450 enzymes involved in the metabolism of AEDs have been established. Carbamazepine and zonisamide are predominantly metabolized by CYP3A4. Phenytoin and phenobarbital are metabolized by both CYP2C9 and CYP2C19. Valproic acid is metabolized by β-oxidation and glucuronidated by UGT isoforms UGT1A6, UGT1A9, and UGT2B7; UGT1A4 catalyzes the metabolism of lamotrigine. Oxcarbazepine is converted to its active monohydroxy derivative by cytosol arylketone reductase, and the monohydroxy derivative is primarily metabolized by glucuronidation. Levetiracetam undergoes minimal metabolism by CYP450 enzymes, and its metabolism is not altered to a major extent by known CYP450 inducers or inhibitors. Gabapentin is not metabolized in humans, and topiramate is primarily excreted by renal excretion, although a fraction of its dose undergoes inducible oxidative metabolism (Hachad *et al.*, 2002).

III. Methods for Pharmacokinetic Evaluation

The literature of reported drug–drug interactions involving AEDs with some cardiovascular and psychiatric drugs used frequently in the geriatric population was analyzed using the Metabolism and Transport Drug Interaction Database (http://www.druginteractioninfo.org/) and the Medical Literature Analysis and Retrieval System Online (MEDLINE). When no interactions had been reported, the potential for interactions was evaluated using established principles of pharmacology and available literature on the metabolic behavior of compounds.

Inhibition of metabolic enzymes is isoenzyme specific and substrate independent. That is, a selective inhibitor of a specific CYP450 isoenzyme would be expected to inhibit all substrates of that enzyme but should not inhibit substrates metabolized by other enzymes. *In vitro* data can be used to establish the enzymes involved in catalyzing the metabolism of specific substrates and to evaluate the effects of drugs on CYP450 enzyme activity. *In vivo* studies can be used to evaluate the potential for enzyme induction and to confirm whether inhibition occurs. Specific *in vivo* interaction studies are not available for many drug pairs; in these cases, predictions were made from available information.

IV. Antidepressants and Drugs Used to Treat Dementia

A. ANTIDEPRESSANTS

Several different antidepressants are used in clinical practice. This section focuses on some antidepressants recommended for or commonly used in the elderly. Characteristics of these antidepressants include wider therapeutic indices, low risk of cardiotoxicity, and relatively lower tendencies to reduce seizure thresholds compared with other antidepressants. These antidepressant drugs include fluoxetine, fluvoxamine, paroxetine, sertraline, citalopram, escitalopram, venlafaxine, duloxetine, and mirtazapine.

Fluoxetine is primarily metabolized by CYP2D6, with lesser contributions by CYP2C9 and CYP3A4 (Margolis *et al.*, 2000). No inhibition or induction of fluoxetine by AEDs has been reported.

Some antidepressants, such as paroxetine, venlafaxine, and duloxetine, have the unique characteristic of being both substrates and inhibitors of the same

CYP450 isoenzymes. Paroxetine is predominantly metabolized by CYP2D6. Since CYP2D6 is not considered to be an inducible enzyme, a clinically significant induction of paroxetine by carbamazepine, phenytoin, or phenobarbital would not be expected. However, administration of phenytoin has been shown to decrease the plasma levels of paroxetine by approximately 50%, whereas phenobarbital decreases the paroxetine concentration by 25% (GlaxoSmithKline, 2006).

Citalopram is a racemic mixture of R-citalopram and S-citalopram. Polymorphisms of CYP2C19 and CYP2D6 may alter citalopram levels (Forest Pharmaceuticals, Inc., 2005a; Grasmader et al., 2004). Carbamazepine administration in depressed patients not responding to citalopram decreased the plasma concentrations of S-citalopram and R-citalopram by 27% and 31%, respectively (Steinacher et al., 2002). Phenytoin and phenobarbital administration would be expected to have a similar effect on citalopram concentrations, but no reports have been published. Oxcarbazepine would be expected to have a less significant effect than carbamazepine on citalopram levels, and this is supported by case reports of increased citalopram levels in two patients after switching from carbamazepine to oxcarbazepine (Leinonen et al., 1996). Escitalopram is the S-enantiomer of citalopram, and it appears to be more affected by CYP2C19 metabolism than the racemic mixture (von Moltke et al., 2001). Since CYP2C19 is an inducible enzyme, metabolism-inducing AEDs would be expected to decrease the level of escitalopram.

Sertraline is metabolized by multiple CYP450 isoenzymes, including CYP3A4 (Kobayashi et al., 1999). Administration of carbamazepine and phenytoin significantly reduced sertraline levels. In one report, the ratio of plasma concentration to daily dose of sertraline was 1.06 nM/mg/day in the control group of 54 patients treated with sertraline alone; this ratio was 0.34 nM/mg/day among the 9 patients treated with carbamazepine or phenytoin as a comedication (Pihlsgard and Eliasson, 2002). Phenobarbital would be expected to have a similar effect on sertraline levels, but no reports have been published.

Venlafaxine metabolism to its major active metabolite is mediated by CYP2D6 (Lessard et al., 1999). Administration of AEDs would not be expected to alter venlafaxine concentrations, and in fact, no reports were found (effects on the active metabolite O-desmethyl-venlafaxine are unknown). Duloxetine is primarily metabolized by isoforms CYP1A2 and CYP2D6 (Caccia, 2004). The potential for induction of duloxetine metabolism by inducing AEDs is not known.

Mirtazapine is metabolized primarily by CYP2D6 and CYP3A4, with lesser contributions by CYP1A2 (Stormer et al., 2000). Coadministration of phenytoin decreased the plasma concentration of mirtazapine by 46% (Spaans et al., 2002). Coadministration of carbamazepine with mirtazapine decreased the area under the plasma concentration versus time curve (AUC) of mirtazapine by 44% (Sitsen et al., 2001).

There are notable differences in the effects of antidepressants on the metabolism of other drugs. Among the antidepressants reviewed, fluoxetine has the most significant potential for drug interactions with AEDs since it inhibits CYP2D6, CYP2C9, CYP2C19, and CYP3A4 (Lane, 1996). Fluoxetine and its metabolite norfluoxetine (which appears to also inhibit CYP3A4) (Graff et al., 2001) have long half-lives (Aronoff et al., 1984), and CYP2D6 metabolism can remain inhibited up to 5 weeks after fluoxetine discontinuation (Eli Lilly and Company, 2005). There are case reports of increased carbamazepine and phenytoin levels leading to adverse effects with fluoxetine administration (Darley, 1994; Gidal et al., 1993; Grimsley et al., 1991; Jalil, 1992). Fluoxetine also interacts with numerous other drugs, including lipophilic beta-blockers (Graff et al., 2001), antiarrhythmics (Cai et al., 1999), and warfarin (Dent and Orrock, 1997).

Paroxetine is the most potent CYP2D6 inhibitor in this therapeutic class (Ereshefsky et al., 1995). Although it has been reported not to change the plasma concentrations of carbamazepine, valproic acid, or phenytoin in a crossover clinical trial (Andersen et al., 1991), it would be expected to inhibit the metabolism of other drugs covered in this section that are substrates of CYP2D6. This assumption is substantiated by reports of clinically significant increases in metoprolol plasma levels in a study of paroxetine in healthy subjects (Hemeryck et al., 2000).

As a CYP2D6 inhibitor, sertraline is less potent than paroxetine. Although case reports have reported toxicity with the concomitant administration of sertraline with carbamazepine, phenytoin, and lamotrigine (Haselberger et al., 1997; Joblin, 1994; Kaufman and Gerner, 1998), clinical studies have not demonstrated inhibition of carbamazepine or phenytoin metabolism (Rapeport et al., 1996a,b). In a study comparing prothrombin time following dosing with warfarin before and after 21 days of sertraline administration, there was an 8% increase in prothrombin time; in addition, a delay in the normalization of prothrombin time to the value of the placebo group was reported (Pfizer Inc., 2006). The manufacturer suggests that patients be carefully monitored, especially during the addition or withdrawal of sertraline when patients are concomitantly receiving warfarin, propafenone, and flecainide (Pfizer Inc., 2006).

Citalopram has no effects on CYP3A4, CYP2C9, and CYP2E1 substrates, but it is a weak inhibitor of CYP1A2, CYP2C19, and CYP2D6 (Forest Pharmaceuticals, Inc., 2005a; von Moltke et al., 1999). Citalopram would not be expected to alter the pharmacokinetics of any AED, and this assumption was confirmed by a report that steady-state plasma levels of carbamazepine were not changed by the addition of citalopram (Moller et al., 2001). In vitro studies of escitalopram did not reveal an inhibitory effect on CYP1A2, CYP2C9, CYP2C19, or CYP3A4, but the drug weakly inhibited CYP2D6 (von Moltke et al., 2001). In addition, there are in vivo data that suggest that escitalopram is an inhibitor of CYP2D6 (Forest Pharmaceuticals, Inc., 2005b). There are no reports of escitalopram inhibiting AED metabolism. Administration of escitalopram resulted in a significant increase in plasma concentrations of metoprolol (Forest Pharmaceuticals, Inc., 2005b).

Venlafaxine is a less potent inhibitor of CYP2D6 than is fluoxetine (Albers *et al.*, 2000; Amchin *et al.*, 2001). It is also a weak inhibitor of CYP3A4 (Amchin *et al.*, 1998; von Moltke *et al.*, 1997). There are no reports of venlafaxine inhibiting the metabolism of any AEDs, and there is little potential for such inhibition. The inhibitory effect of duloxetine on CYP2D6 was compared to paroxetine in healthy men and women, and duloxetine was found to be a less potent inhibitor than paroxetine (Skinner *et al.*, 2003).

In vitro study results have not indicated that mirtazapine inhibits any CYP450 isoenzymes (Organon USA Inc., 2005). This finding is further supported by clinical trials in healthy subjects in which mirtazapine had no effect on the steady-state pharmacokinetics of phenytoin or carbamazepine (Sitsen *et al.*, 2001; Spaans *et al.*, 2002).

B. Drugs Used to Treat Dementia

The drugs used in the treatment of dementia include cholinesterase inhibitors (Ellis, 2005) and memantine (Ringman and Cummings, 2006). Donepezil, a cholinesterase inhibitor, is metabolized by CYP2D6, CYP3A4, and glucuronidation. Inducers of CYP3A4 (e.g., carbamazepine, phenytoin, and phenobarbital) theoretically could increase the rate of elimination of donepezil, but formal pharmacokinetic studies to evaluate the potential for enzyme induction of donepezil have not been conducted to date (Eisai Inc., 2005). Furthermore, there are no case reports of induction of donepezil metabolism. *In vitro* study results show that, at therapeutic plasma concentrations of donepezil, there is little likelihood of inhibition of CYP2D6 or CYP3A4 (Eisai Inc., 2005). There are no reports of donepezil affecting the metabolism of AEDs.

Galantamine, another cholinesterase inhibitor, is metabolized primarily by CYP2D6 and CYP3A4. There is no information on the possible effect of carbamazepine, phenytoin, or phenobarbital on galantamine metabolism. Paroxetine increases the AUC of galantamine by 40% (Ortho-McNeil Neurologies, Inc., 2005). On the basis of *in vitro* data, galantamine is not expected to alter CYP450 metabolism (Ortho-McNeil Neurologies, Inc., 2005). There are no case reports or clinical trials to evaluate the effects of galantamine on AEDs. Multiple doses of galantamine had no effect on the pharmacokinetics of warfarin or on prothrombin time (Ortho-McNeil Neurologies, Inc., 2005).

Rivastigmine is primarily metabolized by cholinesterase-mediated hydrolysis with minimal metabolism by CYP450 enzymes (Novartis Pharmaceuticals Corp., 2006a). No drug interactions related to CYP450 enzymes have been reported for rivastigmine.

The *N*-methyl-D-aspartate receptor antagonist memantine is predominantly eliminated without being metabolized. *In vitro* study results have shown that memantine produces minimal inhibition of CYP450 enzymes, and no pharmacokinetic interactions with drugs metabolized by CYP450 enzymes have been reported (Forest Pharmaceuticals, Inc., 2005c).

V. Cardiovascular Agents

A. ANTICOAGULANT AND ANTIPLATELET AGENTS

Warfarin has a narrow therapeutic index, and use of this drug is a significant concern in the elderly population. Warfarin metabolism declines significantly with advancing age, and the elderly have a higher risk of bleeding. The more potent *S*-warfarin enantiomer is metabolized primarily by CYP2C9, whereas *R*-warfarin is a substrate of CYP1A2, CYP2C19, and CYP3A4 (Kaminsky and Zhang, 1997; Wienkers *et al.*, 1996).

The propensity of enzyme-inducing AEDs to reduce warfarin levels and decrease prothrombin time is well documented. There are several case reports of carbamazepine and barbiturates reducing warfarin levels and decreasing prothrombin time, and there is also an associated risk of a rebound increase in prothrombin time after the enzyme-inducing AED is discontinued (Hansen *et al.*, 1971; Kendall and Boivin, 1981; Massey, 1983). Phenytoin has unpredictable effects on warfarin metabolism, and it may decrease or increase prothrombin time (Nappi, 1979). The propensity for oxcarbazepine to induce warfarin metabolism was investigated in a clinical trial in which 1 week of concomitant oxcarbazepine (900 mg/day) and warfarin administration in healthy subjects did not alter prothrombin time (Kramer *et al.*, 1992). A clinical study using therapeutic levels of levetiracetam concomitantly with warfarin did not demonstrate any pharmacokinetic or pharmacodynamic interaction between warfarin and levetiracetam (Ragueneau-Majlessi *et al.*, 2001). Lamotrigine, gabapentin, or topiramate would not be expected to alter the anticoagulant response to warfarin.

Clinically significant interactions with warfarin leading to prolongation of prothrombin time have been reported for other drugs covered in this chapter, including fluoxetine (Dent and Orrock, 1997), gemfibrozil (Ahmad, 1990a), fluvastatin (Trilli *et al.*, 1996), and lovastatin (Ahmad, 1990b).

There are no published reports on the effects of AEDs on ticlopidine metabolism. However, ticlopidine is a potent inhibitor of CYP2C19, and there are case reports and *in vivo* study reports that suggest that inhibition of both phenytoin and carbamazepine clearance occurs with concomitant ticlopidine administration (Brown and Cooper, 1997; Donahue *et al.*, 1999; Klaassen, 1998; Riva *et al.*, 1996).

Valproic acid is not an enzyme inducer, but it has the potential to inhibit drug metabolism and to adversely affect platelet function and platelet counts, as well as coagulation processes (Perucca *et al.*, 2006). Therefore, there are concerns with the use of valproic acid in patients taking anticoagulants and inhibitors of platelet aggregation, including aspirin; coagulation parameters and platelet counts should be monitored carefully if valproic acid is used in combination with these drugs.

B. Beta-Blockers

The hydrophilic beta-blockers (i.e., atenolol, nadolol, and sotalol) are not metabolized and are excreted in the urine; therefore, there is little potential for drug interactions. Many lipophilic beta-blockers (i.e., propranolol, metoprolol, and timolol) have been considered to be predominantly metabolized by CYP2D6 (Bertilsson *et al.*, 2002). *In vitro* studies in human liver microsomes found that propranolol is metabolized by CYP1A2 and CYP2D6, with large racial differences in the proportions of propranolol metabolized by these isoenzymes (Johnson *et al.*, 2000). There are no reports of induction of propranolol metabolism by AEDs, but the potential for such interaction does exist. There are no reports of metoprolol interactions with AEDs. Timolol metabolism was modestly induced (24% decrease in AUC) by a 7-day course of phenobarbital (Mantyla *et al.*, 1983). Carteolol is metabolized (at least partly) by CYP2D6 (Kudo *et al.*, 1997) and carvedilol by multiple isoenzymes, including CYP1A2, CYP2C9, and CYP3A4 (Oldham and Clarke, 1997). There are no reports of interactions between carteolol or carvedilol and AEDs, and there are no *in vivo* reports of drug interactions for either of these drugs.

C. Angiotensin-Converting Enzyme Inhibitors and Thiazide Diuretics

Many angiotensin-converting enzyme (ACE) inhibitors are prodrugs that are not metabolized by CYP450 enzymes; all other ACE inhibitors are excreted renally (Hoyer *et al.*, 1993). Angiotensin-converting enzyme inhibitors are not involved in CYP450-mediated drug interactions with other drugs, including AEDs. Similarly, thiazide diuretics are not metabolized, and there are no reported interactions between these drugs and AEDs (Flockhart and Tanus-Santos, 2002).

D. Angiotensin Receptor Blockers

The angiotensin receptor blocker (ARB) losartan is a CYP2C9 substrate with a small contribution to its own metabolism by CYP3A4 (Stearns *et al.*, 1995). The administration of phenytoin did not have an effect on losartan pharmacokinetics, but it did significantly reduce the concentration of the active metabolite of losartan, which is a more potent antihypertensive than the parent compound (Fischer *et al.*, 2002). Irbesartan is also a CYP2C9 substrate (Bourrie *et al.*, 1999), and there are no reports of interactions between irbesartan and AEDs.

E. CALCIUM CHANNEL BLOCKERS

Calcium channel blockers (CCBs) carry a significant risk of interaction with AEDs. Many of the CCBs are CYP3A4 substrates, including nimodipine, nilvadipine, nisoldipine, and felodipine (Flockhart and Tanus-Santos, 2002). Several studies have shown that the concurrent use of carbamazepine, phenytoin, and/or phenobarbital results in significant decreases in the levels of nimodipine, nilvadipine, and nisoldipine (Michelucci *et al.*, 1996; Tartara *et al.*, 1991; Yasui-Furukori and Tateishi, 2002). Oxcarbazepine has less of an inducing effect on CCBs than older-generation enzyme-inducing AEDs (Zaccara *et al.*, 1993). Diltiazem and verapamil effects are more complicated because these drugs are both CYP3A4 substrates and inhibitors. Increased clearance of verapamil after induction by phenobarbital has been reported (Rutledge *et al.*, 1988). Increased carbamazepine levels with the administration of diltiazem (Eimer and Carter, 1987) and verapamil (Macphee *et al.*, 1986) have been reported as well. There are two case series reports of increased phenytoin levels when phenytoin was concomitantly administered with diltiazem (Bahls *et al.*, 1991; Clarke *et al.*, 1993). Verapamil decreased the plasma levels of the monohydroxy metabolite of oxcarbazepine by 20% (Novartis Pharmaceuticals Corp., 2006b).

F. STATINS

The 3-hydroxy-3-methylglutaryl coenzyme A (HMG-CoA) reductase inhibitor lovastatin is metabolized predominantly by CYP3A4 and glucuronidation. Similarly, the HMG-CoA reductase inhibitors atorvastatin and simvastatin are metabolized predominantly by CYP3A4 and UGT1A1/1A3 (Prueksaritanont *et al.*, 2002). One case report showed that there was a loss of efficacy when simvastatin and atorvastatin were prescribed concomitantly with phenytoin (Murphy and Dominiczak, 1999). Rosuvastatin and pravastatin are predominantly excreted unchanged (Shitara and Sugiyama, 2006); therefore, there is less potential for interaction with AEDs.

Atorvastatin, lovastatin, and simvastatin are weak inhibitors of CYP3A4 (Shitara and Sugiyama, 2006). There are no reports of inhibition of carbamazepine or zonisamide metabolism when these AEDs are used concomitantly with these three statins. Fluvastatin is a weak inhibitor of CYP2C9 (Shitara and Sugiyama, 2006). However, there are no reports of phenytoin toxicity when phenytoin is prescribed with fluvastatin.

G. FIBRATES

Gemfibrozil is an inhibitor of CYP2C9 and CYP2C19 (Wen *et al.*, 2001), and its major metabolite, an acyl glucuronide, inhibits CYP2C8 (Ogilvie *et al.*, 2006; Shitara *et al.*, 2004). Although inhibition of phenytoin metabolism by gemfibrozil has not been reported in the literature, the potential for this interaction exists.

VI. Conclusions

The antidepressants recommended for use in the elderly epileptic population have favorable safety profiles, and generally, the effects of AEDs on the metabolism of these compounds are less than with other therapeutic classes. The antidepressants with the least potential for altering AED metabolism are citalopram, escitalopram, venlafaxine, duloxetine, and mirtazapine.

The use of carbamazepine, phenytoin, and phenobarbital could theoretically reduce the levels of donepezil and galantamine. There is little potential for these cholinesterase inhibitors to alter AED metabolism. Memantine is eliminated almost entirely without metabolism; therefore, there is little risk of interactions with this drug.

There is a significant risk of drug interactions with warfarin; carbamazepine and phenobarbital have been reported to decrease prothrombin time in warfarin-treated patients, whereas the effects of phenytoin on warfarin response are not predictable. Other drugs covered in this chapter, including fluoxetine, gemfibrozil, fluvastatin, and lovastatin, have been associated with increased prothrombin time and risk of hemorrhage. There are no reports of AEDs altering ticlopidine metabolism; however, there are several reports of ticlopidine coadministration resulting in carbamazepine and phenytoin toxicity.

Among the therapeutic classes of antihypertensives, no metabolic drug interactions would be expected with ACE inhibitors, ARBs, hydrophilic beta-blockers, and thiazide diuretics. There is a moderate risk for AEDs to decrease levels of lipophilic beta-blockers. There is a significant risk that carbamazepine, phenobarbital, and phenytoin will decrease the concentrations of CCBs. Furthermore, there is a significant risk of elevated carbamazepine levels when diltiazem and verapamil are administered. In addition, there are two case series reports of increased phenytoin levels when phenytoin was given with diltiazem.

In summary, there is a significant potential for interactions with AEDs among the drugs commonly prescribed in geriatric patients with epilepsy. This is especially true with older-generation AEDs, particularly carbamazepine, phenytoin, and phenobarbital. These interactions should be taken into account when prescribing comedications. Drug levels and clinical response should be monitored closely.

References

Ahmad, S. (1990a). Gemfibrozil interaction with warfarin sodium (coumadin). *Chest* **98**, 1041–1042.

Ahmad, S. (1990b). Lovastatin. Warfarin interaction. *Arch. Intern. Med.* **150**, 2407.

Albers, L. J., Reist, C., Vu, R. L., Fujimoto, K., Ozdemir, V., Helmeste, D., Poland, R., and Tang, S. W. (2000). Effect of venlafaxine on imipramine metabolism. *Psychiatry Res.* **96**, 235–243.

Amchin, J., Zarycranski, W., Taylor, K. P., Albano, D., and Klockowski, P. M. (1998). Effect of venlafaxine on the pharmacokinetics of alprazolam. *Psychopharmacol. Bull.* **34**, 211–219.

Amchin, J., Ereshefsky, L., Zarycranski, W., Taylor, K., Albano, D., and Klockowski, P. M. (2001). Effect of venlafaxine on metabolism of dextromethorphan, a CYP2D6 probe. *J. Clin. Pharmacol.* **41**, 443–451.

Andersen, B. B., Mikkelsen, M., Vesterager, A., Dam, M., Kristensen, H. B., Pedersen, B., Lund, J., and Mengel, H. (1991). No influence of the antidepressant paroxetine on carbamazepine, valproate and phenytoin. *Epilepsy Res.* **10**, 201–204.

Anderson, G. D. (2004). Pharmacogenetics and enzyme induction/inhibition properties of antiepileptic drugs. *Neurology* **63**, S3–S8.

Anderson, G. D., Yau, M. K., Gidal, B. E., Harris, S. J., Levy, R. H., Lai, A. A., Wolf, K. B., Wargin, W. A., and Dren, A. T. (1996). Bidirectional interaction of valproate and lamotrigine in healthy subjects. *Clin. Pharmacol. Ther.* **60**, 145–156.

Aronoff, G. R., Bergstrom, R. F., Pottratz, S. T., Sloan, R. S., Wolen, R. L., and Lemberger, L. (1984). Fluoxetine kinetics and protein binding in normal and impaired renal function. *Clin. Pharmacol. Ther.* **36**, 138–144.

Bahls, F. H., Ozuna, J., and Ritchie, D. E. (1991). Interactions between calcium channel blockers and the anticonvulsants carbamazepine and phenytoin. *Neurology* **41**, 740–742.

Bertilsson, L., Dahl, M. L., Dalen, P., and Al-Shurbaji, A. (2002). Molecular genetics of CYP2D6: Clinical relevance with focus on psychotropic drugs. *Br. J. Clin. Pharmacol.* **53**, 111–122.

Bourrie, M., Meunier, V., Berger, Y., and Fabre, G. (1999). Role of cytochrome P-4502C9 in irbesartan oxidation by human liver microsomes. *Drug Metab. Dispos.* **27**, 288–296.

Brown, R. I., and Cooper, T. G. (1997). Ticlopidine-carbamazepine interaction in a coronary stent patient. *Can. J. Cardiol.* **13**, 853–854.

Caccia, S. (2004). Metabolism of the newest antidepressants: Comparisons with related predecessors. *IDrugs* **7**, 143–150.

Cai, W. M., Chen, B., Zhou, Y., and Zhang, Y. D. (1999). Fluoxetine impairs the CYP2D6-mediated metabolism of propafenone enantiomers in healthy Chinese volunteers. *Clin. Pharmacol. Ther.* **66**, 516–521.

Clarke, W. R., Horn, J. R., Kawabori, I., and Gurtel, S. (1993). Potentially serious drug interactions secondary to high-dose diltiazem used in the treatment of pulmonary hypertension. *Pharmacotherapy* **13,** 402–405.

Darley, J. (1994). Interaction between phenytoin and fluoxetine. *Seizure* **3,** 151–152.

Dent, L. A., and Orrock, M. W. (1997). Warfarin-fluoxetine and diazepam-fluoxetine interaction. *Pharmacotherapy* **17,** 170–172.

DeToledo, J. C. (1999). Changing presentation of seizures with aging: Clinical and etiological factors. *Gerontology* **45,** 329–335.

Donahue, S., Flockhart, D. A., and Abernethy, D. R. (1999). Ticlopidine inhibits phenytoin clearance. *Clin. Pharmacol. Ther.* **66,** 563–568.

Eimer, M., and Carter, B. L. (1987). Elevated serum carbamazepine concentrations following diltiazem initiation. *Drug Intell. Clin. Pharm.* **21,** 340–342.

Eisai Inc. (2005). Aricept (donepezil hydrochloride) package insert. Eisai Inc., Teaneck, NJ.

Eli Lilly and Company (2005). Prozac (fluoxetine hydrochloride) package insert. Eli Lilly and Company, Indianapolis, IN.

Ellis, J. M. (2005). Cholinesterase inhibitors in the treatment of dementia. *J. Am. Osteopath. Assoc.* **105,** 145–158.

Ereshefsky, L., Riesenman, C., and Lam, Y. W. (1995). Antidepressant drug interactions and the cytochrome P450 system. The role of cytochrome P450 2D6. *Clin. Pharmacokinet.* **29**(Suppl. 1), 10–18; discussion 18–19.

Fischer, T. L., Pieper, J. A., Graff, D. W., Rodgers, J. E., Fischer, J. D., Parnell, K. J., Goldstein, J. A., Greenwood, R., and Patterson, J. H. (2002). Evaluation of potential losartan-phenytoin drug interactions in healthy volunteers. *Clin. Pharmacol. Ther.* **72,** 238–246.

Flockhart, D. A., and Tanus-Santos, J. E. (2002). Implications of cytochrome P450 interactions when prescribing medication for hypertension. *Arch. Intern. Med.* **162,** 405–412.

Forest Pharmaceuticals, Inc. (2005a). Celexa (citalopram hydrobromide) package insert. Forest Laboratories, Inc., St. Louis, MO.

Forest Pharmaceuticals, Inc. (2005b). Lexapro (escitalopram oxalate) package insert. Forest Laboratories, Inc., St. Louis, MO.

Forest Pharmaceuticals, Inc. (2005c). Namenda (memantine hydrochloride) package insert. Forest Laboratories, Inc., St. Louis, MO.

Gidal, B. E., Anderson, G. D., Seaton, T. L., Miyoshi, H. R., and Wilenksy, A. J. (1993). Evaluation of the effect of fluoxetine on the formation of carbamazepine epoxide. *Ther. Drug Monit.* **15,** 405–409.

GlaxoSmithKline (2006). Paxil (paroxetine hydrochloride) package insert. GlaxoSmithKline, Research Triangle Park, NC.

Gorelick, P. B., Sacco, R. L., Smith, D. B., Alberts, M., Mustone-Alexander, L., Rader, D., Ross, J. L., Raps, E., Ozer, M. N., Brass, L. M., Malone, M. E., Goldberg, S., *et al.* (1999). Prevention of a first stroke: A review of guidelines and a multidisciplinary consensus statement from the National Stroke Association. *J. Am. Med. Assoc.* **281,** 1112–1120.

Graff, D. W., Williamson, K. M., Pieper, J. A., Carson, S. W., Adams, K. F., Jr., Cascio, W. E., and Patterson, J. H. (2001). Effect of fluoxetine on carvedilol pharmacokinetics, CYP2D6 activity, and autonomic balance in heart failure patients. *J. Clin. Pharmacol.* **41,** 97–106.

Grasmader, K., Verwohlt, P. L., Rietschel, M., Dragicevic, A., Muller, M., Hiemke, C., Freymann, N., Zobel, A., Maier, W., and Rao, M. L. (2004). Impact of polymorphisms of cytochrome-P450 isoenzymes 2C9, 2C19 and 2D6 on plasma concentrations and clinical effects of antidepressants in a naturalistic clinical setting. *Eur. J. Clin. Pharmacol.* **60,** 329–336.

Grimsley, S. R., Jann, M. W., Carter, J. G., D'Mello, A. P., and D'Souza, M. J. (1991). Increased carbamazepine plasma concentrations after fluoxetine coadministration. *Clin. Pharmacol. Ther.* **50,** 10–15.

Hachad, H., Ragueneau-Majlessi, I., and Levy, R. H. (2002). New antiepileptic drugs: Review on drug interactions. *Ther. Drug Monit.* **24,** 91–103.

Hansen, J. M., Siersboek-Nielsen, K., and Skovsted, L. (1971). Carbamazepine-induced acceleration of diphenylhydantoin and warfarin metabolism in man. *Clin. Pharmacol. Ther.* **12,** 539–543.

Haselberger, M. B., Freedman, L. S., and Tolbert, S. (1997). Elevated serum phenytoin concentrations associated with coadministration of sertraline. *J. Clin. Psychopharmacol.* **17,** 107–109.

Hauser, W. A. (1992). Seizure disorders: The changes with age. *Epilepsia* **33**(Suppl. 4), S6–S14.

Hemeryck, A., Lefebvre, R. A., De Vriendt, C., and Belpaire, F. M. (2000). Paroxetine affects metoprolol pharmacokinetics and pharmacodynamics in healthy volunteers. *Clin. Pharmacol. Ther.* **67,** 283–291.

Honig, L. S., Tang, M. X., Albert, S., Costa, R., Luchsinger, J., Manly, J., Stern, Y., and Mayeux, R. (2003). Stroke and the risk of Alzheimer disease. *Arch. Neurol.* **60,** 1707–1712.

Hoyer, J., Schulte, K. L., and Lenz, T. (1993). Clinical pharmacokinetics of angiotensin converting enzyme (ACE) inhibitors in renal failure. *Clin. Pharmacokinet.* **24,** 230–254.

Jalil, P. (1992). Toxic reaction following the combined administration of fluoxetine and phenytoin: Two case reports. *J. Neurol. Neurosurg. Psychiatry* **55,** 412–413.

Joblin, M. (1994). Possible interaction of sertraline with carbamazepine. *N. Z. Med. J.* **107,** 43.

Johnson, J. A., Herring, V. L., Wolfe, M. S., and Relling, M. V. (2000). CYP1A2 and CYP2D6 4-hydroxylate propranolol and both reactions exhibit racial differences. *J. Pharmacol. Exp. Ther.* **294,** 1099–1105.

Kaminsky, L. S., and Zhang, Z. Y. (1997). Human P450 metabolism of warfarin. *Pharmacol. Ther.* **73,** 67–74.

Kanner, A. M. (2003). Depression in epilepsy: Prevalence, clinical semiology, pathogenic mechanisms, and treatment. *Biol. Psychiatry* **54,** 388–398.

Kaufman, K. R., and Gerner, R. (1998). Lamotrigine toxicity secondary to sertraline. *Seizure* **7,** 163–165.

Kendall, A. G., and Boivin, M. (1981). Warfarin-carbamazepine interaction. *Ann. Intern. Med.* **94,** 280.

Klaassen, S. L. (1998). Ticlopidine-induced phenytoin toxicity. *Ann. Pharmacother.* **32,** 1295–1298.

Kobayashi, K., Ishizuka, T., Shimada, N., Yoshimura, Y., Kamijima, K., and Chiba, K. (1999). Sertraline N-demethylation is catalyzed by multiple isoforms of human cytochrome P-450 in vitro. *Drug Metab. Dispos.* **27,** 763–766.

Kramer, G. (2001). Epilepsy in the elderly: Some clinical and pharmacotherapeutic aspects. *Epilepsia* **42**(Suppl. 3), 55–59.

Kramer, G., Tettenborn, B., Klosterskov Jensen, P., Menge, G. P., and Stoll, K. D. (1992). Oxcarbazepine does not affect the anticoagulant activity of warfarin. *Epilepsia* **33,** 1145–1148.

Kudo, S., Uchida, M., and Odomi, M. (1997). Metabolism of carteolol by cDNA-expressed human cytochrome P450. *Eur. J. Clin. Pharmacol.* **52,** 479–485.

Lackner, T. E., Cloyd, J. C., Thomas, L. W., and Leppik, I. E. (1998). Antiepileptic drug use in nursing home residents: Effect of age, gender, and comedication on patterns of use. *Epilepsia* **39,** 1083–1087.

Lane, R. M. (1996). Pharmacokinetic drug interaction potential of selective serotonin reuptake inhibitors. *Int. Clin. Psychopharmacol.* **11**(Suppl. 5), 31–61.

Larson, S. L., Owens, P. L., Ford, D., and Eaton, W. (2001). Depressive disorder, dysthymia, and risk of stroke: Thirteen-year follow-up from the Baltimore epidemiologic catchment area study. *Stroke* **32,** 1979–1983.

Leinonen, E., Lepola, U., and Koponen, H. (1996). Substituting carbamazepine with oxcarbazepine increases citalopram levels. A report on two cases. *Pharmacopsychiatry* **29,** 156–158.

Lessard, E., Yessine, M. A., Hamelin, B. A., O'Hara, G., LeBlanc, J., and Turgeon, J. (1999). Influence of CYP2D6 activity on the disposition and cardiovascular toxicity of the antidepressant agent venlafaxine in humans. *Pharmacogenetics* **9,** 435–443.

Lossius, M. I., Ronning, O. M., Mowinckel, P., and Gjerstad, L. (2002). Incidence and predictors for post-stroke epilepsy. A prospective controlled trial. The Akershus stroke study. *Eur. J. Neurol.* **9,** 365–368.

Macphee, G. J., McInnes, G. T., Thompson, G. G., and Brodie, M. J. (1986). Verapamil potentiates carbamazepine neurotoxicity: A clinically important inhibitory interaction. *Lancet* **1**(8483), 700–703.

Mantyla, R., Mannisto, P., Nykanen, S., Koponen, A., and Lamminsivu, U. (1983). Pharmacokinetic interactions of timolol with vasodilating drugs, food and phenobarbitone in healthy human volunteers. *Eur. J. Clin. Pharmacol.* **24,** 227–230.

Margolis, J. M., O'Donnell, J. P., Mankowski, D. C., Ekins, S., and Obach, R. S. (2000). (R)-, (S)-, and racemic fluoxetine N-demethylation by human cytochrome P450 enzymes. *Drug Metab. Dispos.* **28,** 1187–1191.

Massey, E. W. (1983). Effect of carbamazepine on Coumadin metabolism. *Ann. Neurol.* **13,** 691–692.

Michelucci, R., Cipolla, G., Passarelli, D., Gatti, G., Ochan, M., Heinig, R., Tassinari, C. A., and Perucca, E. (1996). Reduced plasma nisoldipine concentrations in phenytoin-treated patients with epilepsy. *Epilepsia* **37,** 1107–1110.

Moller, S. E., Larsen, F., Khant, A. Z., and Rolan, P. E. (2001). Lack of effect of citalopram on the steady-state pharmacokinetics of carbamazepine in healthy male subjects. *J. Clin. Psychopharmacol.* **21,** 493–499.

Murphy, M. J., and Dominiczak, M. H. (1999). Efficacy of statin therapy: Possible effect of phenytoin. *Postgrad. Med. J.* **75,** 359–360.

Nappi, J. M. (1979). Warfarin and phenytoin interaction. *Ann. Intern. Med.* **90,** 852.

Novartis Pharmaceuticals Corp. (2006a). Exelon (rivastigmine tartrate) package insert. Novartis Pharmaceuticals Corp., East Hanover, NJ.

Novartis Pharmaceuticals Corp. (2006b). Trileptal (oxcarbazepine) package insert. Novartis Pharmaceuticals Corp., East Hanover, NJ.

Ogilvie, B. W., Zhang, D., Li, W., Rodrigues, A. D., Gipson, A. E., Holsapple, J., Toren, P., and Parkinson, A. (2006). Glucuronidation converts gemfibrozil to a potent, metabolism-dependent inhibitor of CYP2C8: Implications for drug-drug interactions. *Drug Metab. Dispos.* **34,** 191–197.

Oldham, H. G., and Clarke, S. E. (1997). *In vitro* identification of the human cytochrome P450 enzymes involved in the metabolism of R(+)- and S(−)-carvedilol. *Drug Metab. Dispos.* **25,** 970–977.

Organon, USA Inc. (2005). Remeron (mirtazapine) package insert. Organon USA Inc., West Orange, NJ.

Ortho-McNeil Neurologies Inc. (2005). Razadyne ER (galantamine HBr) package insert. Ortho-McNeil Neurologies, Inc., Titusville, NJ.

Perucca, E., Berlowitz, D., Birnbaum, A., Cloyd, J. C., Garrard, J., Hanlon, J. T., Levy, R. H., and Pugh, M. J. (2006). Pharmacological and clinical aspects of antiepileptic drug use in the elderly. *Epilepsy Res.* **68**(Suppl. 1), S49–S63.

Pfizer Inc. (2006). Zoloft (sertraline hydrochloride) package insert. Pfizer Inc., New York, NY.

Pihlsgard, M., and Eliasson, E. (2002). Significant reduction of sertraline plasma levels by carbamazepine and phenytoin. *Eur. J. Clin. Pharmacol.* **57,** 915–916.

Prueksaritanont, T., Zhao, J. J., Ma, B., Roadcap, B. A., Tang, C., Qiu, Y., Liu, L., Lin, J. H., Pearson, P. G., and Baillie, T. A. (2002). Mechanistic studies on metabolic interactions between gemfibrozil and statins. *J. Pharmacol. Exp. Ther.* **301,** 1042–1051.

Ragueneau-Majlessi, I., Levy, R. H., and Meyerhoff, C. (2001). Lack of effect of repeated administration of levetiracetam on the pharmacodynamic and pharmacokinetic profiles of warfarin. *Epilepsy Res.* **47,** 55–63.

Rapeport, W. G., Muirhead, D. C., Williams, S. A., Cross, M., and Wesnes, K. (1996a). Absence of effect of sertraline on the pharmacokinetics and pharmacodynamics of phenytoin. *J. Clin. Psychiatry* **57**(Suppl. 1), 24–28.

Rapeport, W. G., Williams, S. A., Muirhead, D. C., Dewland, P. M., Tanner, T., and Wesnes, K. (1996b). Absence of a sertraline-mediated effect on the pharmacokinetics and pharmaco-dynamics of carbamazepine. *J. Clin. Psychiatry* **57**(Suppl. 1), 20–23.

Ringman, J. M., and Cummings, J. L. (2006). Current and emerging pharmacological treatment options for dementia. *Behav. Neurol.* **17**, 5–16.

Riva, R., Cerullo, A., Albani, F., and Baruzzi, A. (1996). Ticlopidine impairs phenytoin clearance: A case report. *Neurology* **46**, 1172–1173.

Robinson, R. G. (2003). Poststroke depression: Prevalence, diagnosis, treatment, and disease progression. *Biol. Psychiatry* **54**, 376–387.

Rutledge, D. R., Pieper, J. A., and Mirvis, D. M. (1988). Effects of chronic phenobarbital on verapamil disposition in humans. *J. Pharmacol. Exp. Ther.* **246**, 7–13.

Scripture, C. D., and Pieper, J. A. (2001). Clinical pharmacokinetics of fluvastatin. *Clin. Pharmacokinet.* **40**, 263–281.

Shitara, Y., and Sugiyama, Y. (2006). Pharmacokinetic and pharmacodynamic alterations of 3-hydroxy-3-methylglutaryl coenzyme A (HMG-CoA) reductase inhibitors: Drug-drug interac-tions and interindividual differences in transporter and metabolic enzyme functions. *Pharmacol. Ther.* May 19, 2006 [Epub ahead of print].

Shitara, Y., Hirano, M., Sato, H., and Sugiyama, Y. (2004). Gemfibrozil and its glucuronide inhibit the organic anion transporting polypeptide 2 (OATP2/OATP1B1:SLC21A6)-mediated hepatic uptake and CYP2C8-mediated metabolism of cerivastatin: Analysis of the mechanism of the clinically relevant drug-drug interaction between cerivastatin and gemfibrozil. *J. Pharmacol. Exp. Ther.* **311**, 228–236.

Sidhu, J., Job, S., Singh, S., and Philipson, R. (2006). The pharmacokinetic and pharmaco-dynamic consequences of the co-administration of lamotrigine and a combined oral contraceptive in healthy female subjects. *Br. J. Clin. Pharmacol.* **61**, 191–199.

Sitsen, J., Maris, F., and Timmer, C. (2001). Drug-drug interaction studies with mirtazapine and carbamazepine in healthy male subjects. *Eur. J. Drug Metab. Pharmacokinet.* **26**, 109–121.

Skinner, M. H., Kuan, H. Y., Pan, A., Sathirakul, K., Knadler, M. P., Gonzales, C. R., Yeo, K. P., Reddy, S., Lim, M., Ayan-Oshodi, M., and Wise, S. D. (2003). Duloxetine is both an inhibitor and a substrate of cytochrome P4502D6 in healthy volunteers. *Clin. Pharmacol. Ther.* **73**, 170–177.

Spaans, E., van den Heuvel, M. W., Schnabel, P. G., Peeters, P. A., Chin-Kon-Sung, U. G., Colbers, E. P., and Sitsen, J. M. (2002). Concomitant use of mirtazapine and phenytoin: A drug-drug interaction study in healthy male subjects. *Eur. J. Clin. Pharmacol.* **58**, 423–429.

Stearns, R. A., Chakravarty, P. K., Chen, R., and Chiu, S. H. (1995). Biotransformation of losartan to its active carboxylic acid metabolite in human liver microsomes. Role of cytochrome P4502C and 3A subfamily members. *Drug Metab. Dispos.* **23**, 207–215.

Steinacher, L., Vandel, P., Zullino, D. F., Eap, C. B., Brawand-Amey, M., and Baumann, P. (2002). Carbamazepine augmentation in depressive patients non-responding to citalopram: A pharma-cokinetic and clinical pilot study. *Eur. Neuropsychopharmacol.* **12**, 255–260.

Stephen, L. J., and Brodie, M. J. (2000). Epilepsy in elderly people. *Lancet* **355**, 1441–1446.

Stormer, E., von Moltke, L. L., Shader, R. I., and Greenblatt, D. J. (2000). Metabolism of the antidepressant mirtazapine *in vitro*: Contribution of cytochromes P-450 1A2, 2D6, and 3A4. *Drug Metab. Dispos.* **28**, 1168–1175.

Tartara, A., Galimberti, C. A., Manni, R., Parietti, L., Zucca, C., Baasch, H., Caresia, L., Muck, W., Barzaghi, N., Gatti, G., and Perucca, E. (1991). Differential effects of valproic acid and enzyme-inducing anticonvulsants on nimodipine pharmacokinetics in epileptic patients. *Br. J. Clin. Pharmacol.* **32**, 335–340.

Trilli, L. E., Kelley, C. L., Aspinall, S. L., and Kroner, B. A. (1996). Potential interaction between warfarin and fluvastatin. *Ann. Pharmacother.* **30**, 1399–1402.

Turnheim, K. (2003). When drug therapy gets old: Pharmacokinetics and pharmacodynamics in the elderly. *Exp. Gerontol.* **38**, 843–853.

von Moltke, L. L., Duan, S. X., Greenblatt, D. J., Fogelman, S. M., Schmider, J., Harmatz, J. S., and Shader, R. I. (1997). Venlafaxine and metabolites are very weak inhibitors of human cytochrome P450–3A isoforms. *Biol. Psychiatry* **41**, 377–380.

von Moltke, L. L., Greenblatt, D. J., Grassi, J. M., Granda, B. W., Venkatakrishnan, K., Duan, S. X., Fogelman, S. M., Harmatz, J. S., and Shader, R. I. (1999). Citalopram and desmethylcitalopram *in vitro*: Human cytochromes mediating transformation, and cytochrome inhibitory effects. *Biol. Psychiatry* **46**, 839–849.

von Moltke, L. L., Greenblatt, D. J., Giancarlo, G. M., Granda, B. W., Harmatz, J. S., and Shader, R. I. (2001). Escitalopram (S-citalopram) and its metabolites *in vitro*: Cytochromes mediating biotransformation, inhibitory effects, and comparison to R-citalopram. *Drug Metab. Dispos.* **29**, 1102–1109.

Wen, X., Wang, J. S., Backman, J. T., Kivisto, K. T., and Neuvonen, P. J. (2001). Gemfibrozil is a potent inhibitor of human cytochrome P450 2C9. *Drug Metab. Dispos.* **29**, 1359–1361.

Wienkers, L. C., Wurden, C. J., Storch, E., Kunze, K. L., Rettie, A. E., and Trager, W. F. (1996). Formation of (R)-8-hydroxywarfarin in human liver microsomes. A new metabolic marker for the (S)-mephenytoin hydroxylase, P4502C19. *Drug Metab. Dispos.* **24**, 610–614.

Yasui-Furukori, N., and Tateishi, T. (2002). Carbamazepine decreases antihypertensive effect of nilvadipine. *J. Clin. Pharmacol.* **42**, 100–103.

Zaccara, G., Gangemi, P. F., Bendoni, L., Menge, G. P., Schwabe, S., and Monza, G. C. (1993). Influence of single and repeated doses of oxcarbazepine on the pharmacokinetic profile of felodipine. *Ther. Drug Monit.* **15**, 39–42.

OUTCOMES IN ELDERLY PATIENTS WITH NEWLY DIAGNOSED AND TREATED EPILEPSY

Martin J. Brodie and Linda J. Stephen

Patient Services and Clinical Research Studies
Epilepsy Unit, Division of Cardiovascular and Medical Sciences
Western Infirmary, Glasgow, G11 6NT Scotland, United Kingdom

Epilepsy develops most commonly in the elderly. Seizures can severely affect a senior citizen's quality of life, and despite a growing elderly population with epilepsy, there is a paucity of good clinical data in this age group. To address some of the issues encountered by elderly patients with epilepsy, prospective information from elderly patients attending the Epilepsy Unit at the Western Infirmary in Glasgow, Scotland, was analyzed.

Ninety patients, aged 65–93 years, were diagnosed with epilepsy and started on antiepileptic drug (AED) treatment. Neuroimaging was performed in 84 patients (93%), with 69 evaluated via computerized tomography and 15 via magnetic resonance imaging; abnormalities were found in 45 patients (54%). Sixty-eight patients underwent interictal electroencephalography, which revealed epileptiform discharges in 18 patients (26%).

Fifty-eight of 90 patients (64%) became seizure free for at least 12 months on modest doses of the first prescribed AED. Seizures remained uncontrolled in 21 patients (23%), and the first AED was withdrawn in 11 patients (12%) because of adverse events. Following pharmacological manipulation, a total of 76 patients (84%) achieved seizure freedom. Patients starting treatment ≥ 2 years after their first seizure were less likely to achieve seizure control than patients who initiated treatment earlier. Newly diagnosed elderly patients were more likely to remain seizure free on AED treatment than newly diagnosed younger populations

($p < 0.001$). The majority of patients evaluated had partial-onset seizures, and underlying cerebral atrophy and infarcts were common.

Treating an older person with initial AED therapy can be complicated; taking adequate time and communicating clearly are paramount. Although most of the patients evaluated had a positive outcome, all AEDs have some disadvantages in this population. Choice of drug may depend on comorbidity and comedication, among other factors. Initial dosing should be low with a slow titration schedule. A holistic approach to care helps optimize the outcome for elderly people with epilepsy.

I. Introduction

Seizures most commonly develop in old age (Hauser, 1992), and the number of senior citizens with epilepsy will increase year by year as the global elderly population grows. In this population, it has been estimated that the annual prevalence of epilepsy is 1%, with an incidence of 134 per 100,000 (Stephen and Brodie, 2000). About 10–30% of these patients will present with tonic-clonic status epilepticus (Sung and Chu, 1989). Older people who present with a single seizure are also more likely to experience additional seizures than are younger adults (Hopkins *et al.*, 1988). Access to epilepsy services decreases with increasing age, and underdiagnosis and misdiagnosis of the condition are common (Stephen and Brodie, 2000). The situation is further compounded by a paucity of good clinical studies in elderly people with epilepsy. Health care services will become increasingly burdened by these patients.

Epilepsy can have profound physical and psychological consequences in the elderly (Tallis *et al.*, 2002), including stress resulting from the stigma of the diagnosis and physical injury sustained as a result of seizures. A range of neurodegenerative, cerebrovascular, and neoplastic comorbidities can complicate the situation, and problems with concomitant medications are common. The unpredictable nature of these epileptic events can lead to social withdrawal, and premature admission to nursing homes and residential care facilities can result from loss of confidence and reduced independence. Quality of life may also be adversely affected by the loss of a driving license, which may never be recovered. Mortality and rates of sudden unexpected death are high in elderly people with epilepsy (Luhdorf *et al.*, 1987).

The ability to study the outcomes of epilepsy in newly diagnosed elderly patients is severely limited by the lack of academic centers to which newly diagnosed individuals can be referred for evaluation and treatment. To better understand and address the issues encountered by elderly patients with epilepsy, the Epilepsy Unit at the Western Infirmary in Glasgow, Scotland, has for many years run a clinical program for elderly people diagnosed with epilepsy. This chapter will present the results of an analysis performed for the 2003 International Geriatric Epilepsy Symposium and discuss new information obtained from other analyses.

II. The Glasgow Registry

Data and clinical information from this cohort of elderly patients with epilepsy have been entered prospectively into a database. To examine outcomes, these data have been analyzed from patients aged 65 years and older with a diagnosis of epilepsy who were first treated with an antiepileptic drug (AED) between July 1, 1982, and May 1, 2001. A supplemental analysis also included patients who entered the clinic from May 2001 through June 2003.

Each patient was seen within 2 weeks of referral, preferably with a family member or caregiver. Diagnoses were made by obtaining witness accounts of epileptic events whenever possible. Most patients underwent routine brain neuroimaging; electroencephalography (EEG) was undertaken to aid seizure and syndrome classification in some cases. Additionally, prolonged electro-cardiographic recording, carotid and basilar ultrasound, orthostatic blood pressure measurements, tilt testing, and hematological, biochemical, and thyroid profiles were ordered, as necessary. A structured case sheet was developed for each individual, allowing longitudinal assessment over a period of years.

Patients received counseling about the implications of their diagnoses. After counseling, AED treatment was considered for any patient who had experienced two or more seizures; ideally, one seizure event had been witnessed. Patients who presented with a single seizure and had underlying neuropathology were offered medication. Those who reported a single seizure only and were not started on treatment were invited to contact the Epilepsy Unit via a dedicated telephone number should they experience another seizure. All treated patients were reviewed every 6–8 weeks until optimal seizure control or seizure freedom, defined as the absence of seizures or auras for at least 12 months on unchanged AED doses, was achieved. If no further events were reported after the initial 1 year of seizure freedom, patients were assumed to be in remission. If additional seizures were reported, patients were considered to have refractory epilepsy. The general extent of control over seizures was assessed at the time of the patient's last hospital visit.

III. Analysis of the 1982–2001 Cohort

A total of 90 elderly patients (53 men, 37 women) between the ages of 65 and 93 years (median, 73 years) were diagnosed with epilepsy and started on AED therapy at the Epilepsy Unit during the study period. The median time elapsed since first seizure occurrence was 4 months (range, 1–780 months). Eighty-eight patients (98%) had partial or secondary generalized seizures, and two patients (2%) had alcohol withdrawal seizures.

Eighty-four patients (93%) underwent brain neuroimaging. Computed tomography (CT) was performed in 69 patients, and magnetic resonance imaging (MRI) was performed in the remaining 15 patients. Scans were abnormal in 45 patients (54%); 24 patients (53%) had cerebral atrophy, 17 patients (38%) had infarcts, 2 patients (4%) had primary neoplasia, 1 patient (2%) had gliosis, and 1 patient (2%) had an arteriovenous malformation. The results of CT and MRI brain imaging were normal in 35 and 4 patients (42% and 5%), respectively. Interictal EEG was performed in 68 patients. Eighteen of these patients (26%) had epileptiform discharges, and 23 (34%) showed nonspecific changes; EEG results were normal in 27 patients (40%).

Fifty-eight of 90 patients (64%) became seizure free for at least 12 months while taking their first AED (Table I). Seizures remained uncontrolled in 21 patients (23%), and the drug had to be withdrawn in 11 patients (12%) because of adverse events (AEs). Following pharmacological manipulation, a total of 76 patients (84%) had achieved freedom from seizures for at least 12 months. Seventy-three patients experienced seizure control on monotherapy, whereas three patients required two AEDs (valproate and lamotrigine, $n = 2$; valproate and gabapentin, $n = 1$) to achieve seizure control. No individual monotherapy was statistically more likely than any other to confer seizure freedom.

Daily AED doses tended to be modest among the patients receiving monotherapy who became seizure free (Table II). No correlation was found between

TABLE I

Response to First Antiepileptic Drug in Elderly Patients With Newly Diagnosed Epilepsy

Antiepileptic drug	n	Seizure free, n (%)	Seizures uncontrolled, n (%)	Drug not tolerated, n (%)
Carbamazepine	35	22 (63)	7 (20)	6 (17)
Lamotrigine	26	17 (65)	6 (23)	3 (12)
Valproate	17	10 (59)	5 (29)	2 (12)
Other	12	9 (75)	3 (25)	0 (0)
Total	90	58 (64)	21 (23)	11 (12)

TABLE II

DAILY MONOTHERAPY DOSES IN SEIZURE-FREE ELDERLY PATIENTS WITH NEWLY DIAGNOSED EPILEPSY

Antiepileptic drug	n	Median dose (mg)	Range of dose (mg)
Carbamazepine	24	400	200–700
Lamotrigine	20	100	50–400
Valproate	20	800	300–1500
Oxcarbazepine	5	300	100–600
Phenytoin	3	200	100–600
Gabapentin	1	1800	NA

NA indicates not applicable.
Reprinted from Leppik *et al.* (2006), with permission from Elsevier. Copyright (2006).

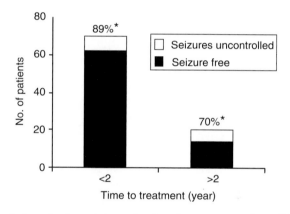

FIG. 1. Rates of seizure freedom relative to duration of pretreatment epilepsy. Percentages on top of bars correspond to rates of seizure freedom (*$p < 0.05$).

seizure control and abnormal imaging results or the presence of neurological deficits, and no statistical association was found between seizure freedom and the number of pretreatment seizures. However, patients who began treatment 2 or more years after experiencing their first seizure were less likely to gain full seizure control than those patients who began treatment closer to the onset of epilepsy (Fig. 1). In addition, elderly patients with newly diagnosed epilepsy were more likely to remain seizure free on AED therapy than younger populations with newly diagnosed epilepsy ($p < 0.001$; Table III).

TABLE III

PHARMACOLOGICAL OUTCOMES BY AGE AT STARTING TREATMENT IN NEWLY DIAGNOSED EPILEPSY

Patient groups	Age (years)	n	Remission (%)	Relapse (%)	Seizures uncontrolled (%)
Adolescent	<20	170	65	12	23
Adult	20–64	520	53	4	43
Elderly	>64	90	85[a]	1	14

[a]Remission rates in elderly patients versus adolescents and adults $p < 0.001$.

IV. Supplemental Analysis

The inclusion of patients who entered the clinic from May 2001 through June 2003 increased the number of patients to 117 (67 men and 50 women). The additional patients were similar to the initial population in age and had similar diagnostic test results. Seventy-three of the 117 patients (62%) became seizure free for at least 12 months on the first AED prescribed, and after further manipulation of the AED regimen, a total of 93 patients (79%) became seizure free (Stephen et al., 2006).

V. Difficulties in Diagnosing Epilepsy in the Elderly

Diagnosing epilepsy in elderly people can be difficult, and piecing together an accurate picture of events may take additional time. A history of unexplained trauma, such as cuts, bruising, and fractures, can be helpful, as can a description of postictal events. Other significant factors include a bitten tongue, unexplained urinary incontinence, and drowsiness. A reliable account of events from a witness is a valuable aid to diagnosis but is not always available, especially when an elderly person lives alone. Witnesses may report cyanosis, abnormal pallor, or unusual movements.

Conditions such as syncope, hypoglycemia, cardiac arrhythmias, drop attacks, transient ischemic attacks, transient global amnesia, and psychogenic episodes can mimic seizure activity at this time of life, complicating diagnoses (Stephen and Brodie, 2000). Investigative techniques that have proven useful in diagnosing elderly patients include ambulatory electrocardiography, orthostatic blood pressure measurement, tilt table testing, hematological and biochemical profiles, and evaluation of thyroid function.

The majority of patients in the 1982–2001 cohort had partial-onset seizures with or without secondary generalization. This is to be expected in an elderly population, as idiopathic epilepsy does not often occur de novo in this age group

(Grunewald and Panayiotopoulos, 1994). Half of the cohort had abnormal neuroimaging results, and this figure may have been higher had more patients undergone brain MRI, which is known to be a more sensitive tool than CT for detection of intracerebral lesions in people with epilepsy (Scottish Intercollegiate Guidelines Network, 2003).

The variety of identified pathologies reflects the diversity of seizure etiology in old age. The most common etiology for the elderly is cerebrovascular disease, which has accounted for 30–50% of seizures in some series (Stephen and Brodie, 2000). Cerebrovascular pathology is also the most frequently reported reason for status epilepticus in older people (Stephen and Brodie, 2000); it is the underlying pathology in over one third of cases (Sung and Chu, 1989). Although seizures are often provoked by larger areas of infarction, even a relatively small degree of disease can give rise to ictal activity.

Two patients in the 1982–2001 cohort were found to have primary neoplasia on cerebral imaging. Tumors are recognized as the cause of seizures in a small number of elderly people with epilepsy (Luhdorf et al., 1986; Sander et al., 1990). Tumors can range from primary lesions, such as gliomas and meningiomas, to metastases. Nonvascular cerebral degenerative diseases, such as Alzheimer dementia, are also linked to seizure activity (McAreavey et al., 1992). Head trauma may give rise to ictal activity, and underlying cerebral atrophy makes the development of a subdural hematoma more likely. It is also important to consider toxic and metabolic etiologies; alcohol withdrawal precipitated seizure activity in two of our patients. A wide variety of drugs (Tallis et al., 2002), cerebral infections or infestations, and metabolic abnormalities, such as uremia, thyroid disease, hypoglycemia, electrolyte disturbance, and hepatic impairment, should also be considered (Tallis, 1990).

One-quarter of patients in the 1982–2001 cohort showed epileptiform discharges on interictal EEG recordings. Although EEG can be useful in classifying seizure type in some elderly people with epilepsy, it should be used with caution in this age group because it is a less sensitive investigative tool in older patients (Drury and Beydoun, 1998). Results are best used only as supporting evidence. Prolonged EEG recording and recording during suspected seizure activity may be more valuable in establishing the diagnosis of epilepsy.

VI. Studies Comparing Newer and Older AEDs

Now that an array of AEDs is available, clinicians starting treatment in elderly patients with epilepsy are frequently faced with the dilemma of how to select the best "weapon" from this therapeutic armamentarium (Brodie and French, 2000). Because of a dearth of randomized monotherapy studies in this patient population, choice of AED is often based on evidence derived from studies in younger adults. Two exceptions are a multicenter, double-blind, randomized comparison between lamotrigine and carbamazepine in elderly patients (Brodie *et al.*, 1999) and a US Veterans Administration study that compared carbamazepine, lamotrigine, and gabapentin (Rowan *et al.*, 2005).

In the first study (Brodie *et al.*, 1999), 150 patients aged 65–94 years (mean, 77 years) with newly diagnosed epilepsy were randomized in a 2:1 ratio to receive lamotrigine or carbamazepine. Following a short titration period, the dosage was individualized for each patient while maintaining the blind over the next 24 weeks. The median dose of lamotrigine was 100 mg (range, 75–300 mg), and the median dose of carbamazepine was 400 mg (range, 200–800 mg). The greatest difference between the groups was seen in the rates of withdrawal due to AEs (lamotrigine 18%, carbamazepine 42%). A smaller percentage of patients taking lamotrigine complained of somnolence (lamotrigine 12%, carbamazepine 29%) and skin rash resulting in treatment discontinuation (lamotrigine 3%, carbamazepine 19%). Overall, a greater percentage of patients continued on treatment with lamotrigine than carbamazepine (lamotrigine 71%, carbamazepine 42%, $p < 0.001$; Fig. 2), and a greater percentage of lamotrigine-treated patients remained seizure free (lamotrigine 39%, carbamazepine 21%, $p < 0.05$) over the last 16 weeks of treatment; there was no difference between the drugs in time to first seizure. These data demonstrated a clear difference in effectiveness between the two drugs in favor of lamotrigine.

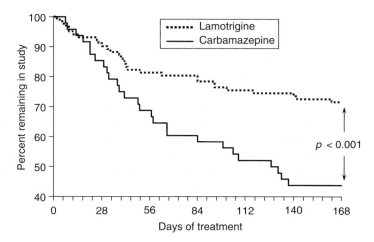

FIG. 2. Kaplan-Meier distribution curve for time to withdrawal from carbamazepine versus lamotrigine monotherapy. Adapted from Brodie *et al.* (1999), with permission.

The US Veterans Administration study was an 18-center, randomized, double-blind, double-dummy, parallel investigation of 593 elderly subjects with newly diagnosed seizures (Rowan *et al.*, 2005). The primary outcome measure was retention in the trial for 12 months. Early termination was reported for 44% of the lamotrigine-treated patients, 51% of patients treated with gabapentin, and 64% of patients prescribed carbamazepine ($p = 0.0002$). Withdrawals were primarily due to adverse effects. There was no significant difference in seizure control between drug treatment groups.

Although there are only two major comparative studies, it appears that the major difference between newer and older AEDs may be their tolerability, which is an important aspect to quality of life.

VII. Initiating AED Treatment

Starting AED treatment in elderly patients can be complicated. The support of family members and caregivers is paramount to a successful outcome. Time is needed to explain the etiology, diagnosis, and necessity for treatment. Written dosing instructions, careful explanation of the regimen, and provision of dosing trays can aid compliance; discussion of possible side effects is also an essential component in this process. Cohesive management of AED treatment in elderly patients can be greatly enhanced by the involvement of a multidisciplinary health care team including the epilepsy clinician, nurse specialist, general practitioner, and pharmacist. Good communication between these individuals will optimize treatment. Appropriate management of elderly patients will minimize socioeconomic costs, often allowing the individual to remain in his or her own home. The clinician should aim for seizure freedom with no medication-related side effects and an unchanged or enhanced quality of life for the patient (Tallis *et al.*, 2002).

The ideal "elderly-friendly" AED should be completely absorbed, undergo linear pharmacokinetics, and have its clearance unaffected by renal impairment. In addition, the ideal treatment would not induce or inhibit hepatic enzymes nor produce neurotoxic side effects, would achieve its target dose without titration, and would come in a range of formulations (Stephen and Brodie, 2000). Selecting an AED for an elderly patient requires some consideration. Pharmacokinetics in old age differ from those in younger patients; elderly patients exhibit altered volumes of distribution, reduced serum albumin concentrations, lower protein binding, slower hepatic metabolism, less hepatic enzyme inducibility, and reduced renal elimination (O'Mahony and Woodhouse, 1994). Older patients respond to lower AED doses and plasma concentrations than do younger individuals. There is a greater susceptibility to side effects among the elderly, and idiosyncratic reactions are more common (Willmore, 1996). Long-term AED treatment may be a risk factor for the development of osteoporosis (Stephen *et al.*, 1999), and interactions with comedications can be problematic as well.

All AEDs have some disadvantages in this patient population. Carbamazepine, phenytoin, phenobarbital, and primidone are all broad-spectrum enzyme inducers that will accelerate the turnover of a range of other lipid-soluble drugs, including cardiac antiarrhythmics, analgesics, antidepressants, neuroleptics, and anticoagulants (Patsalos *et al.*, 2002). These drugs also have sedative effects, and in addition, carbamazepine can produce hyponatremia and should therefore be avoided in patients taking diuretics (Brodie and Dichter, 1996). Rash and other idiosyncratic reactions have been associated with all four of these older agents. Sodium valproate can produce dose-dependent tremor and occasionally worsen the symptoms of idiopathic parkinsonism (Stephen, 2003). Among the newer AEDs, lamotrigine can cause rash and insomnia; gabapentin, dizziness, and headache; oxcarbazepine, rash, and hyponatremia; and topiramate, weight loss, and word-finding difficulties (Brodie and French, 2000).

The starting dose of the chosen AED should be low and the incremental pace of titration slow. A modest maintenance dose should be targeted and potential seizure events anticipated. A holistic approach to patient care should be the paramount consideration in older people with epilepsy, particularly given the wide choice of available drugs. If an AED fails due to an idiosyncratic reaction, poor tolerability, or lack of efficacy, another AED should be substituted. If the patient tolerates the first AED with a useful but suboptimal response, combination therapy using low doses of two agents should be considered.

VIII. Conclusions

The development of epilepsy is common in elderly people. The majority of patients will become seizure free on AED monotherapy, often at modest dosage. Prognosis may be better at this time of life than for younger people. Optimal management requires rapid investigation, accurate diagnosis, effective therapy, patient/caregiver education, and sympathetic support.

References

Brodie, M. J., and Dichter, M. A. (1996). Antiepileptic drugs. *N. Engl. J. Med.* **334,** 168–175. Erratum *N. Engl. J. Med.* **334,** 479.

Brodie, M. J., and French, J. A. (2000). Management of epilepsy in adolescents and adults. *Lancet* **356,** 323–329.

Brodie, M. J., Overstall, P. W., and Giorgi, L. (1999). Multicentre, double-blind, randomised comparison between lamotrigine and carbamazepine in elderly patients with newly diagnosed epilepsy. The UK Lamotrigine Elderly Study Group. *Epilepsy Res.* **37,** 81–87.

Drury, I., and Beydoun, A. (1998). Seizures and epilepsy in the elderly revisited. *Arch. Intern. Med.* **158,** 99–100.

Grunewald, R. A., and Panayiotopoulos, C. P. (1994). Diagnosing juvenile myoclonic epilepsy in an elderly patient. *Seizure* **3,** 239–241.

Hauser, W. A. (1992). Seizure disorders: The changes with age. *Epilepsia* **33**(Suppl. 4), S6–S14.

Hopkins, A., Garman, A., and Clarke, C. (1988). The first seizure in adult life. Value of clinical features, electroencephalography, and computerised tomographic scanning in prediction of seizure recurrence. *Lancet* **1,** 721–726.

Leppik, I. E., Brodie, M. J., Saetre, E. R., Rowan, A. J., Ramsay, R. E., and Jacobs, M. P. (2006). Outcomes research: Clinical trials in the elderly. *Epilepsy Res.* **68**(Suppl. 1), S71–S76.

Luhdorf, K., Jensen, L. K., and Plesner, A. M. (1986). Etiology of seizures in the elderly. *Epilepsia* **27,** 458–463.

Luhdorf, K., Jensen, L. K., and Plesner, A. M. (1987). Epilepsy in the elderly: Life expectancy and causes of death. *Acta Neurol. Scand.* **76,** 183–190.

McAreavey, M. J., Ballinger, B. R., and Fenton, G. W. (1992). Epileptic seizures in elderly patients with dementia. *Epilepsia* **33,** 657–660.

O'Mahony, M. S., and Woodhouse, K. W. (1994). Age, environmental factors and drug metabolism. *Pharmacol. Ther.* **61,** 279–287.

Patsalos, P. N., Froscher, W., Pisani, F., and van Rijn, C. M. (2002). The importance of drug interactions in epilepsy therapy. *Epilepsia* **43,** 365–385.

Rowan, A. J., Ramsay, R. E., Collins, J. F., Pryor, F., Boardman, K. D., Uthman, B. M., Spitz, M., Frederick, T., Towne, A., Carter, G. S., Marks, W., Felicetta, J., *et al.* and VA Cooperative Study 428 Group (2005). New onset geriatric epilepsy: A randomized study of gabapentin, lamotrigine, and carbamazepine. *Neurology* **64,** 1868–1873.

Sander, J. W., Hart, Y. M., Johnson, A. L., and Shorvon, S. D. (1990). National General Practice Study of Epilepsy: Newly diagnosed epileptic seizures in a general population. *Lancet* **336,** 1267–1271.

Scottish Intercollegiate Guidelines Network (2003). SIGN 70. "Diagnosis and Management of Epilepsy in Adults" (Scottish Intercollegiate Guidelines Network, Ed.). Royal College of Physicians, Edinburgh.

Stephen, L. J. (2003). Drug treatment of epilepsy in elderly people: Focus on valproic acid. *Drugs Aging* **20,** 141–152.

Stephen, L. J., and Brodie, M. J. (2000). Epilepsy in elderly people. *Lancet* **355,** 1441–1446.

Stephen, L. J., McLellan, A. R., Harrison, J. H., Shapiro, D., Dominiczak, M. H., Sills, G. J., and Brodie, M. J. (1999). Bone density and antiepileptic drugs: A case-controlled study. *Seizure* **8,** 339–342.

Stephen, L. J., Kelly, K., Mohanraj, R., and Brodie, M. J. (2006). Pharmacological outcomes in older people with newly diagnosed epilepsy. *Epilepsy. Behav.* **8,** 434–437.

Sung, C. Y., and Chu, N. S. (1989). Status epilepticus in the elderly: Etiology, seizure type and outcome. *Acta Neurol. Scand.* **80,** 51–56.

Tallis, R. (1990). Epilepsy in old age. *Lancet* **336,** 295–296.

Tallis, R., Boon, P., Perucca, E., and Stephen, L. (2002). Epilepsy in elderly people: Management issues. *Epileptic Disord.* **4**(Suppl. 2), S33–S39.

Willmore, L. J. (1996). Management of epilepsy in the elderly. *Epilepsia* **37**(Suppl. 6), S23–S33.

RECRUITMENT AND RETENTION IN
CLINICAL TRIALS OF THE ELDERLY

Flavia M. Macias,* R. Eugene Ramsay,*,† and A. James Rowan[‡,§]

*Department of Neurology, Miami VA Medical Center, Miami, Florida 33125, USA
†International Center for Epilepsy, Department of Neurology
University of Miami School of Medicine, Miami, Florida 33136, USA
‡Department of Neurology, Bronx VA Medical Center, Bronx, New York 10468, USA
§Mount Sinai School of Medicine, New York, New York, 10029, USA

The recruitment and retention of elderly patients in clinical trials provide many challenges. Factors affecting recruitment, retention, and cost of recruitment are discussed in this chapter. Various methods are described that were used in recruiting and retaining elderly patients in a Veterans Affairs (VA) Administration clinical trial that compared two newer antiepileptic drugs (AEDs), gabapentin and lamotrigine, to the established standard AED, carbamazepine. Various strategies were utilized in the VA study to improve recruitment, and each strategy's overall effectiveness was monitored. Modification of the patient inclusion criteria, by lowering the age of eligibility from 65 to 60 years, added approximately 100 patients to the study. Replacing five trial sites that had poor recruiting records, extending the patient recruitment period by 3 months, and conducting site visits also improved patient recruitment rates, such that 82.4% of target enrollment (720 patients) was achieved. The main reasons that screened patients were excluded from the study included: lack of seizures during the prior 3 months, unstable medical condition, adequate treatment with an AED, satisfaction with current treatment, and the inability to give informed consent. Retaining patients for 1 year was the primary outcome measure of this trial, with 46.8% of patients completing the year. The most common reasons for early termination were study drug–related adverse events (43.0%) and lack of seizure control (10.8%). Comorbidities and polypharmacy occurred more frequently in the elderly, and both had a negative influence on recruitment and retention.

I. Introduction

In the elderly, rates of participation in clinical trials have been shown to be particularly low (Hall, 1999). An analysis by Hutchins *et al.* (1999) of data from research participants in trials conducted by the Southwest Oncology Group revealed that persons over 65 years of age were significantly underrepresented. Given the rising number of older persons in our society and the associated increase in rates of neurodegenerative, musculoskeletal, and other medical disorders in this population, it is imperative that these individuals be included in research endeavors.

An effective recruitment plan is essential for the successful execution of any clinical trial. Recruiting for trials involving the elderly presents unique challenges for investigators, and the specific characteristics of the study often dictate the most effective recruitment strategy. For example, in a randomized trial of osteoporosis in older women that had narrow inclusion criteria, the optimal recruitment strategy was a mass media campaign (Flicker and Wark, 1997); in contrast, personal presentations or general mailouts were more effective for a health promotion study on home safety for the elderly (Thompson *et al.*, 1997).

Retention is as important as recruitment to the success of clinical trials in the elderly. Traditional retention strategies include, but are not limited to: provision of transportation, monetary incentives, continuity of care by the same health care provider(s), and frequent, regular contacts with research staff (Areán *et al.*, 2003).

II. Recruitment Outcomes of Veterans Affairs Cooperative Study 428

The multicenter clinical trial, "Treatment of Seizures in the Elderly Population," was headed by R. Eugene Ramsay, MD, and A. James Rowan, MD, and was conducted in 18 Veterans Affairs (VA) medical centers. Recruitment presented some unique challenges and was monitored carefully throughout the enrollment period. Recruitment strategies employed during the trial were evaluated for their overall effectiveness. Patient demographics, as well as other factors believed to influence recruitment, were compared in the enrolled and excluded groups to ascertain any significant differences between the groups.

This was a three-arm, double-blind, randomized clinical trial comparing two newer antiepileptic drugs (AEDs), gabapentin and lamotrigine, to the established

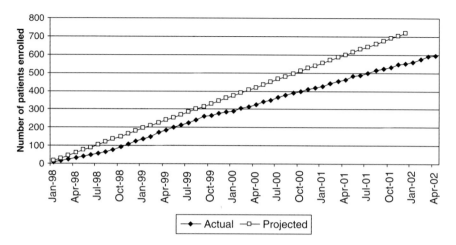

FIG. 1. Comparison of expected versus actual cumulative enrollment for VA Cooperative Study 428 (1998–2002). Adapted from Leppik *et al.* (2006), with permission.

standard AED, carbamazepine, in patients aged 65 years or older with new-onset seizures. In this study, unlike previous VA epilepsy trials, nonveterans were absolutely prohibited. There was a 4-year enrollment period with a projected total enrollment of 720 patients. The expected enrollment rate at each of the participating VA medical centers was 10 patients per year. Fifteen patients were expected to enroll each month. Recruitment began in January 1998 and was completed in April 2002.

Projected and actual cumulative enrollments are compared in Fig. 1; overall, 82.4% of the target enrollment was achieved. In Fig. 2, the difference between actual and projected enrollment is depicted by month. The study was placed on probation from March 1, 1999 (falling below the 90% target enrollment mark was grounds for being placed on probationary status) until August 31, 1999. During this 6-month period, actual enrollment reached 106.7% of projected enrollment, and the study was allowed to continue. The poorest recruitment months were consistently December and January. The holiday season, during which study coordinators and investigators often take vacation and clinics are canceled, may account for this finding.

The top five strategies for improving recruitment in clinical trials, as suggested by Collins *et al.* (1980), are outlined in Table I. Three of the five strategies were employed in this study:

1. *Modify the patient inclusion criteria.* After 1 year, the inclusion criteria were modified by lowering the age of eligibility from 65 to 60 years. As a result of lowering the age criterion, approximately 100 patients aged 60–64 were

enrolled in the study. The definition of "inadequately treated with an AED" was also clarified, which resulted in improved recruitment.

2. *Replace trial sites that have poor recruiting records.* Between October 1998 and May 1999, a total of five trial sites with the poorest recruiting records were replaced.

3. *Extend the patient recruitment period.* A 3-month extension of the recruitment period was approved. A total enrollment of approximately 600 (200 patients per treatment arm) was achieved.

Site visits were also an effective mechanism for improving recruitment. A single site visit increased enrollment by 365%, from two patients per year to an average of 9.3 patients per year.

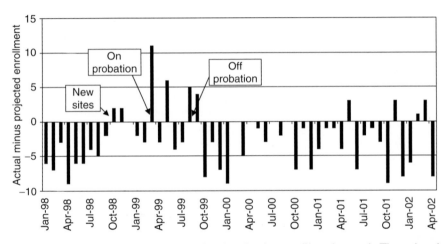

FIG. 2. The difference between actual and projected patient enrollment by month. The projected patient enrollment was 15 patients per month. Adapted from Leppik *et al.* (2006), with permission.

TABLE I

TOP FIVE STRATEGIES FOR IMPROVING RECRUITMENT IN CLINICAL TRIALS[a]

- Reevaluate the required sample size
- Add new trial sites to the study
- Replace trial sites that have poor recruiting records
- Extend the patient recruitment period
- Modify the patient inclusion criteria

[a]Collins *et al.* (1980).
Adapted from Leppik *et al.* (2006), with permission.

TABLE II

FACTORS THAT COULD HAVE INFLUENCED RECRUITMENT IN VA
COOPERATIVE STUDY 428

- Approach used by study team to introduce study to patient and family
- Seasonal residency
- Interpretation of subjective inclusion/exclusion criteria
- Admission to extended care facility
- Demographics of the recruiting center
- Family member participation
- Denial of disease by patient or family member
- Transportation
- Frequent hospital admissions

Adapted from Leppik *et al.* (2006), with permission.

Annual investigator meetings held in November did not improve recruitment. The ineffectiveness of these meetings may have been related to the time of year when the meetings were held. Other factors that were not measured, but that could have influenced recruitment, are listed in Table II.

Of 1358 patients screened, 593 were enrolled and 765 were excluded. The top five reasons for exclusion were: no seizures in the previous 3 months (24%), unstable medical condition (21%), adequately treated with an AED (21%), satisfied with current treatment (19%), and unable to give informed consent (19%). An age-specific analysis by decade revealed that only two reasons for exclusion increased in frequency with advancing age: unstable medical condition and inability to give informed consent. This result was to be expected, since the frequency of comorbid conditions is known to increase with aging. Black patients were excluded significantly more often than white patients for unstable medical condition ($p = 0.05$). The higher incidence in the black population of common chronic illnesses such as cardiac disease, stroke, and diabetes may have contributed to this difference.

Advancing age was not predictive of study entry. Level of education, on the other hand, played an important role in enrollment. Patients with less than 12 years of education (i.e., lacking a high school diploma) were excluded from the study at significantly higher levels than they were enrolled ($p < 0.00001$). When the level of education was between 12 and 14 years, or greater than 14 years, this relationship was reversed (i.e., proportionately more patients were enrolled than excluded). This may suggest that a higher educational level helps individuals better understand the purpose of the research and what is expected of them. Finally, a significantly higher proportion of blacks were excluded from the study than whites ($p = 0.011$). This may reflect a heightened sense of cautiousness among the black population, possibly related to the history of this subpopulation in research studies (e.g., Tuskegee Syphilis Study 1932–1972; http://ublib.buffalo.edu/libraries/units/hsl/history/tuskegee).

It is common for elderly patients and/or their respective family members to refuse participation in a research study. In this study, 30 eligible patients (17%) refused to participate. In over one-third of these cases, it was a family member (usually the spouse or adult child) who did not allow the patient to participate. Other common reasons reported for refusal to participate included denial of seizure diagnosis, the belief that treatment involved taking too many pills, travel difficulties, and refusal by the nursing home or assisted living facility to give patient study medication.

III. Factors Influencing Recruitment

The factors that influence recruitment are largely determined by the type of research study being done. In an analysis of health services utilization among older African Americans and whites, Brown and Topcu (2003) found no difference in willingness to participate between the two groups in a clinical cancer treatment trial. Gender and age, however, were statistically significant predictors of willingness to participate, with males more willing to participate than females ($p < 0.001$), and younger persons more willing to participate than older persons ($p < 0.001$). There was also a "race × age" interaction suggesting that older whites were more willing to participate than older African Americans.

Allsup and Gosney (2002) examined the difficulties in recruitment for a randomized, controlled trial involving influenza vaccination in healthy older adults. Of 6058 patients identified as potential candidates, only 729 (12%) were randomized. The more common reasons given for noninvolvement were reluctance to participate in a research study (53%), concerns about side effects (34%), patient's belief about not requiring an influenza vaccination (31.7%), and preference for patient's own doctor to give the vaccination (29.1%).

An attempt to better identify and understand recruitment challenges in the elderly was part of a primary prevention study conducted by Boles *et al.* (2000), testing the effect of aspirin in an HMO population. A total of 47,453 eligible patients over the age of 65 were identified by random sampling. Although 44% responded to recruitment efforts, only 3% were enrolled. A subset of "eligible refusers" was randomly selected, and these patients were asked their reason(s) for nonparticipation. The most common reasons provided were hesitation to give up choice of aspirin (21.2%), unwillingness to travel to the research center (33.6%), and the patient's belief that, because of potential risks, his or her good health should not be jeopardized by participation in primary prevention (10%).

IV. Cost of Recruitment

The cost of recruitment varies widely based on the methods used. Garrett *et al.* (2000) examined the cost-effectiveness of four recruitment strategies used in the Vitamin E, Cataract, and Age-Related Maculopathy (VECAT) Study, a 4-year interventional study examining the effects of daily vitamin E supplementation on the progression of cataract and age-related maculopathy in patients aged 55–80 years. A personalized letter mailout and newspaper advertising were the most cost-effective methods at AUS$71.53 and AUS$74.61 per participant, respectively. Radio advertising was less cost-effective at AUS$121.09 per participant, and enlisting the assistance of general practitioners was the least cost-effective method at US$241.29 per participant. Another randomized clinical trial on frailty and injury prevention for older adults had costs of recruitment estimated at AUS$300 per participant (Ory *et al.*, 2002). Recruitment costs were influenced by the additional time spent by staff to address concerns about patients' health, mediate essential family interactions, and arrange for patient transportation.

In the Diabetes Prevention Program, a multicenter, randomized controlled trial of the effect of diet and exercise or medication on the delay of onset of type 2 diabetes, the estimated cost of recruitment was US$1075.00 per participant (Rubin *et al.*, 2002). This number stemmed from the large amount of staff time devoted to recruitment, which reached 86.8 h per week in each clinic.

V. Retention Results From VA Cooperative Study 428

Patient retention at Week 52 was the primary outcome measure for this trial; 276 patients (46.8%) were completers and 314 patients (53.2%) were early terminators. Of the early terminators, 131 (41.7%) withdrew due to study drug–related adverse reactions, and 33 (10.5%) withdrew due to lack of seizure control. The most commonly cited reasons for voluntary withdrawal from the study were adverse effects, family members insisting that the patient withdraw, failure of the patient to take part in follow-up, transfer to an extended care facility, patient perception of taking too many pills, and no recurrence of seizures. Other reasons for early termination were death (not related to study treatment) and noncompliance. Retention was affected by other important factors, including increased frequency of hospitalizations (leading to changes in medication regimens), population migration (refers to seasonal residency), and variability among investigators in the interpretation of what constituted a "treatment failure."

This study demonstrated that an elderly population may present unique recruitment and retention challenges for researchers. Both recruitment and retention were affected by the greater prevalence of comorbid conditions and polypharmacy in the elderly. Retention was primarily influenced by side effects, frequent hospitalizations, and admission to nursing home and assisted living facilities. Investigators should pay special attention to the approach used during recruitment while introducing the study to patients and their families. Additionally, consideration should be given to potential changes in living conditions such as prolonged hospitalizations, admissions to extended care facilities, or seasonal residency variations.

References

Allsup, S. J., and Gosney, M. A. (2002). Difficulties of recruitment for a randomized controlled trial involving influenza vaccination in healthy older people. *Gerontology* **48,** 170–173.

Areán, P. A., Alvidrez, J., Nery, R., Estes, C., and Linkins, K. (2003). Recruitment and retention of older minorities in mental health services research. *Gerontologist* **43,** 36–44.

Boles, M., Getchell, W. S., Feldman, G., McBride, R., and Hart, R. G. (2000). Primary prevention studies and the healthy elderly: Evaluating barriers to recruitment. *J. Community Health* **25,** 279–292.

Brown, D. R., and Topcu, M. (2003). Willingness to participate in clinical treatment research among older African Americans and Whites. *Gerontologist* **43,** 62–72.

Collins, J. F., Bingham, S. F., Weiss, D. G., Williford, W. O., and Kuhn, R. M. (1980). Some adaptive strategies for inadequate sample acquisition in veterans administration cooperative clinical trials. *Control Clin. Trials* **1,** 227–248.

Flicker, L., and Wark, J. D. (1997). Recruitment strategies for randomised clinical trials in elderly Australians. *Med. J. Aust.* **167,** 438–439.

Garrett, S. K., Thomas, A. P., Cicuttini, F., Silagy, C., Taylor, H. R., and McNeil, J. J. (2000). Community-based recruitment strategies for a longitudinal interventional study: The VECAT experience. *J. Clin. Epidemiol.* **53,** 541–548.

Hall, W. D. (1999). Representation of blacks, women, and the very elderly (aged > or = 80) in 28 major randomized clinical trials. *Ethn. Dis.* **9,** 333–340.

Hutchins, L. F., Unger, J. M., Crowley, J. J., Coltman, C. A., Jr., and Albain, K. S. (1999). Underrepresentation of patients 65 years of age or older in cancer-treatment trials. *N. Engl. J. Med.* **341,** 2061–2067.

Leppik, I. E., Brodie, M. J., Saetre, E. R., Rowan, A. J., Ramsay, R. E., and Jacobs, M. P. (2006). Outcomes research: Clinical trials in the elderly. *Epilepsy Res.* **68**(Suppl. 1), S71–S76.

Ory, M. G., Lipman, P. D., Karlen, P. L., Gerety, M. B., Stevens, V. J., Singh, M. A., Buchner, D. M., and Schechtman, K. B. (2002). Recruitment of older participants in frailty/injury prevention studies. *Prev. Sci.* **3,** 1–22.

Rubin, R. R., Fujimoto, W. Y., Marrero, D. G., Brenneman, T., Charleston, J. B., Edelstein, S. L., Fisher, E. B., Jordan, R., Knowler, W. C., Lichterman, L. C., Prince, M., and Rowe, P. M. (2002). The Diabetes Prevention Program: Recruitment methods and results. *Control. Clin. Trials* **23,** 157–171.

Thompson, P. G., Somers, R. L., and Wilson, R. (1997). Recruiting older people to a home safety program. *Med. J. Aust.* **167,** 439–440.

TREATMENT OF CONVULSIVE STATUS EPILEPTICUS

David M. Treiman*

Barrow Neurological Institute, Phoenix, Arizona 85013, USA

Status epilepticus (SE) is a medical and neurological emergency requiring prompt and aggressive treatment, particularly for elderly individuals in whom comorbid conditions may increase the severity of consequences in SE. Generalized convulsive status epilepticus (GCSE) is the most common and life-threatening type of SE. It may be overt or subtle in its presentation. Most cases are overt, but as the duration of GCSE increases, its presentation may become more subtle. Progressive electroencephalographic changes also occur during GCSE. A predictable sequence of five electroencephalographic patterns has been identified: (1) discrete seizures with interictal slowing, (2) merging seizures with waxing and waning ictal discharges, (3) continuous ictal sharp or spike-wave discharges, (4) continuous ictal discharges with episodes of generalized flattening, and (5) periodic epileptiform discharges superimposed on a relatively flat background.

Several factors affect the prognosis of GCSE, including etiology, age, seizure type, gender, and duration. GCSE may lead to systemic complications and neuronal damage and is often fatal if untreated or inadequately treated. Treatment of GCSE should begin with basic life support measures and monitoring. Ideally, pharmacological treatment should be easy to administer and fast acting. Analysis of data on elderly patients with overt GCSE from a Veterans Affairs cooperative study revealed that success rates of first-line treatment were 71.4% for phenobarbital, 63.0% for lorazepam, 53.3% for diazepam followed by phenytoin, and 41.5% for phenytoin alone. In elderly patients with subtle GCSE, success rates

*Co-author: Veterans Affairs Status Epilepticus Cooperative Study Group.

INTERNATIONAL REVIEW OF
NEUROBIOLOGY, VOL. 81
DOI: 10.1016/S0074-7742(06)81018-4

for first-line treatment were 30.8% for phenobarbital, 14.3% for lorazepam, 11.8% for phenytoin, and 5.6% for diazepam followed by phenytoin. Because each drug has advantages and disadvantages, the choice of which agent to use as first-line treatment depends on individual patient characteristics.

I. Introduction

Status epilepticus (SE) is a neurological emergency that requires immediate aggressive intervention to prevent neuronal damage or death (Treiman, 1983). This is especially true for the geriatric population, in which comorbid conditions may make the consequences of repeated uncontrolled seizures much more severe and life threatening (Claassen et al., 2002; Cloyd et al., 2006; Hui et al., 2005; Rossetti et al., 2006).

The annual incidence of SE has been reported to range from 2.9 to 57 per 100,000 individuals, depending on the age and population studied (Table I). It has been estimated that there are at least 2.5 million cases of SE per year worldwide (Treiman, 2001). A number of study reports have suggested an increased annual incidence of SE in elderly patients, with estimates ranging from 14.6 to 86 per 100,000 persons (Chin et al., 2004; DeLorenzo et al., 1995;

TABLE I
ANNUAL INCIDENCE OF STATUS EPILEPTICUS

Population	Incidence (cases per 100,000)	References
Richmond, VA	41	DeLorenzo et al. (1996)
Caucasian	20	
African-American	57	
Rochester, MN	18.3	Hesdorffer et al. (1998)
French Switzerland	10.3	Coeytaux et al. (2000)
Germany	17.1	Knake et al. (2001)
<60 years	4.2	
≥60 years	54.5	
California	6.2	Wu et al. (2002)
<5 years	7.5	
>75 years	22.3	
Bologna, Italy		Vignatelli et al. (2003)
<60 years	5.2	
≥60 years	26.2	
Rural North Italy		Vignatelli et al. (2005)
<60 years	2.9	
≥60 years	38.6	

Hesdorffer *et al.*, 1998; Knake *et al.*, 2001; Vignatelli *et al.*, 2003, 2005; Wu *et al.*, 2002). Among elderly individuals, the annual incidence of SE continues to increase with advancing age; annual incidence estimates of 35.8–98.9 per 100,000 persons have been reported for patients aged 80 years and older (Hesdorffer *et al.*, 1998; Vignatelli *et al.*, 2003).

Gastaut (1983) suggested that any type of seizure that lasts long enough or recurs frequently enough can be considered a type of SE. Traditionally, SE is defined as two or more epileptic seizures without full recovery of neurological function between seizures, or as continuous clinical and/or electrical seizure activity lasting 30 min or longer (Treiman, 1993).

Several types of SE exist. Clinically, perhaps the most important distinction is between convulsive and nonconvulsive SE. SE can be further classified as generalized from onset or partial at onset, a distinction usually made via electro-encephalogram (EEG) readings. Generalized (from onset) SE may be classified as tonic-clonic, absence, myoclonic, tonic, or clonic, depending on the phenotype of the seizure, and partial onset SE may be simple or complex, depending on whether consciousness is maintained during the seizure. Generalized convulsive status epilepticus (GCSE), which may begin as a generalized convulsion or as a simple or complex partial seizure that secondarily generalizes, is the most common and life-threatening type of SE (Gaitanis and Drislane, 2003; Treiman, 1993). Annual incidence estimates of GCSE range from 6.2 to 20 per 100,000 persons (Kalviainen *et al.*, 2005; Wu *et al.*, 2002), and it appears to be even more common among elderly individuals (estimated incidence, 22.3 per 100,000 individuals) (Wu *et al.*, 2002). This chapter will review the presentation, diagnosis, and treatment of GCSE in elderly patients.

II. Presentation, Progression, and Diagnosis

Characteristics of GCSE include paroxysmal or continuous clonic and/or tonic motor activity associated with a marked impairment of consciousness and bilateral ictal discharge patterns on the EEG (Treiman, 1993). Most GCSE cases are overt in their presentation and consist of recurrent generalized convulsions without complete recovery between the convulsions (Treiman *et al.*, 1998). However, if GCSE is untreated, inadequately treated, or sufficiently encephalopatho-genic, the overt motor activity becomes progressively more subtle (Treiman, 2001). Approximately 75% of GCSE is overt and 25% is subtle in presentation (Treiman *et al.*, 1998), and the longer that GCSE lasts, the more likely it is that the clinical motor manifestations will be subtle. Subtle GCSE is characterized by continuous rhythmic subtle motor phenomena, bilateral ictal discharges, and coma (Treiman, 1993). Subtle GCSE is a reflection of a profound encephalopathy and is not always preceded by overt GCSE. In some cases, an encephalopathic insult may cause the episode of SE (Treiman and Walker, 2006). This is especially true in elderly patients, who have a higher incidence of serious encephalopathic insults than do younger adults. The most common cause of SE in elderly patients is cerebro-vascular disease (Vignatelli *et al.*, 2005; Waterhouse *et al.*, 1998). Estimates of the mortality of GCSE range from 3% to 35% (Hauser, 1990), and mortality may be higher when certain associated conditions are present. Acute cerebrovascular accidents, in combination with SE, have a very high mortality compared with the mortality of either disorder alone (Waterhouse *et al.*, 1998).

Just as there may be a clinical progression from overt to subtle GCSE, progressive electrographic changes occur during episodes of GCSE. Five EEG patterns that occur in a predictable sequence during GCSE have been identified (Treiman *et al.*, 1990). The first pattern consists of discrete seizures and interictal slowing, which is often accompanied by overt generalized, frequently asym-metric, tonic-clonic seizures. The second EEG pattern consists of merging sei-zures with waxing and waning ictal discharges. Overt generalized seizures may be associated with this pattern, but focal intermittent tonic and/or clonic con-vulsions are more commonly observed. The third pattern is characterized by continuous ictal sharp or spike-wave discharges. The patient may experience continuous clonic jerks or subtle clonic movements during this time. Continuous ictal discharges with episodes of generalized flattening compose the fourth EEG pattern. Overt or subtle focal clonic movements may be seen, or the patient may have no motor symptoms. Finally, the fifth EEG pattern consists of periodic epileptiform discharges superimposed on a relatively flat background.

Usually, GCSE is not difficult to diagnose, but certain conditions may mimic GCSE. Postanoxic myoclonus following cardiac arrest, as well as emergence from phenobarbital-induced coma or general anesthesia, may be confused with GCSE (Cascino, 1996). However, it is important to note that because GCSE is a medical and neurological emergency, treatment should not be delayed pending the results of laboratory tests. Psychogenic nonepileptic SE (PNESE) may also be mistaken for GCSE. A recent study identified certain characteristics of patients with PNESE that may aid clinicians in making the correct diagnosis: younger age, presence of a port system (likely implanted because of repeated PNESE presentation), persis-tence of seizures without respiratory failure despite high-dose benzodiazepine administration, and normal serum creatine kinase levels (Holtkamp *et al.*, 2006).

III. Consequences and Prognosis

Several factors help to determine the prognosis of patients with SE. Etiology is an important determinant of outcome; higher mortality rates appear to be associated with etiologies such as hypoxia, anoxia, cerebrovascular disease, hemorrhage, and metabolic abnormalities (Claassen *et al.*, 2002; Gaitanis and Drislane, 2003; Logroscino *et al.*, 1997; Towne *et al.*, 1994), whereas mortality rates are lower with antiepileptic drug discontinuation and alcohol-related etiologies (Gaitanis and Drislane, 2003; Towne *et al.*, 1994). Other factors associated with increased mortality in SE include increased age, acute symptomatic seizures, and male gender (Claassen *et al.*, 2002; Logroscino *et al.*, 1997; Rossetti *et al.*, 2006; Towne *et al.*, 1994).

Duration of GCSE is also an important determinant of its prognosis. Prompt treatment of prolonged or repetitive seizures may prevent the eventual development of SE. With longer SE duration, meeting the increased metabolic demands of the brain becomes more difficult, and several systemic and neurological complications may result. Loss of systemic autoregulation may lead to complications such as hypotension, hypoxia, hypoglycemia, metabolic acidosis, hyperpyrexia, and respiratory failure (Gaitanis and Drislane, 2003; Treiman, 1993, 2001). If untreated or inadequately treated, SE is often fatal. Systemic complications, as well as prolonged seizure activity, may lead to neuronal damage (Meldrum, 1991; Meldrum *et al.*, 1973). Patients with prolonged SE have an increased risk of cognitive decline, epilepsy, and other neurological abnormalities (Dodson *et al.*, 1993). Significantly higher mortality rates have been reported for patients with prolonged SE (>1 h) than for patients with nonprolonged SE (from 30 min to 1 h, $p < 0.001$) (DeLorenzo *et al.*, 1992). Therefore, prompt and aggressive treatment of SE is important to reduce the risk of death and adverse neurological consequences. In fact, although seizure episodes lasting under 30 min are not technically considered SE, the Working Group on Status Epilepticus (Dodson *et al.*, 1993) recommended that treatment with antiepileptic drugs be considered for any seizure lasting longer than 10 min.

IV. Treatment

A. LIFE SUPPORT AND MONITORING

Treatment of GCSE should begin with basic life support measures to stabilize the patient. Body temperature, blood pressure, electrocardiogram readings, and respiratory function should be monitored (Dodson *et al.*, 1993). Pulse oximetry or arterial blood gas studies should be performed for assessment of oxygenation (Cascino, 1996; Gaitanis and Drislane, 2003). Respiratory support should be provided as needed, and intravenous (IV) access should be established (Dodson *et al.*, 1993; Gaitanis and Drislane, 2003; Kalviainen *et al.*, 2005). Hydration and the use of vasopressors may be necessary if the patient is hypotensive, and cooling may be required if the patient's body temperature exceeds 40°C (Cascino, 1996). Blood samples should be drawn for determination of hematologic and serum chemistry values, as well as serum antiepileptic drug concentrations (Treiman, 2001). If the patient is hypoglycemic, IV glucose should be administered (Kalviainen *et al.*, 2005; Treiman, 2001). IV thiamine should also be administered before or with glucose if thiamine deficiency is a concern (Treiman, 2001). Bicarbonate may be administered to treat acidosis, but this is advisable only if the patient's blood pH is life-threateningly low (Treiman, 2001). Whenever possible, the underlying cause of the SE should be identified and treated. Electroencephalographic monitoring of patients in GCSE is also important during treatment because if overt seizures stop, but the patient remains unconscious, he or she may still be experiencing continuing electrographic seizure activity (Treiman *et al.*, 1990).

B. PHARMACOLOGICAL TREATMENT: THE VETERANS AFFAIRS COOPERATIVE STUDY

Ideally, pharmacological treatment for GCSE should be easy to administer rapidly, stop seizure activity quickly, and have minimal adverse effects. There have been no prospective clinical trials designed to compare treatment options for GCSE in elderly patients. However, because nearly half of the patients enrolled in a large U.S. Department of Veterans Affairs (VA) cooperative study (Treiman *et al.*, 1998) to compare treatments of GCSE were ≥65 years of age, it was possible to study the characteristics and treatment responses in this group of 236 elderly patients with GCSE (Treiman and Walker, 2006).

1. *Methods*

Patients presenting with GCSE were classified as having overt or subtle presentations. Overt GCSE was defined as at least two generalized convulsions without full recovery of consciousness between the episodes, or continuous convulsion for more than 30 min. Patients were classified as having subtle GCSE if coma and ictal discharge patterns were present on the EEG, with or without the presence of subtle convulsive movements. Within the overt and subtle GCSE groups, patients were randomized into the following four IV treatment groups:

- Phenobarbital (15 mg/kg)
- Lorazepam (0.1 mg/kg)
- Phenytoin alone (18 mg/kg)
- Diazepam (0.15 mg/kg) followed by phenytoin (18 mg/kg)

The study was conducted using a double-blind design. Chi-square techniques were used to analyze rates of treatment success, recurrence, and adverse events. The alpha level was set at 0.05 for analyses of all four treatments. Electroencephalographic recording began as soon as possible after initiation of the treatment protocol. Seizure activity, blood pressure, heart and respiratory rates, and level of consciousness were recorded every 5 min during the first 20 min of drug infusion and every 10 min for the following 40 min. Thereafter, seizure activity and level of consciousness were recorded hourly until the end of the 12-h study period. Adverse effects occurring during the study were documented. The primary outcome measure of the study was success of the first treatment. Treatment success was defined as cessation of all clinical and electrical seizure activity occurring 20–60 min after the start of the drug infusion.

2. Results

Table II provides information regarding the elderly cohort. Of the 236 elderly patients, GCSE was overt in 167 patients (70.8%) and subtle in 69 patients (29.2%). Mean age was 72.1 years (±SD, 6.6 years) in the overt GCSE group and 73.6 years (±SD, 5.5 years) in the subtle GCSE group. More than 80% of patients were veterans and more than 85% were male. Remote neurological insult (past history of stroke, traumatic brain injury, and so on) accounted for

TABLE II

DEMOGRAPHIC INFORMATION FOR OVERT AND SUBTLE GENERALIZED
CONVULSIVE STATUS EPILEPTICUS, ELDERLY COHORT

	Overt GCSE	Subtle GCSE
Number of persons enrolled	167	69
Mean age (±SD) (year)	72.1 (6.6)	73.6 (5.5)
Gender (male, %)	86.2	85.5
Race (%)		
Caucasian	41.3	65.2
African-American	49.7	33.3
Other	9.0	1.4
Veteran (%)	82.6	85.5
Prior history of seizures (%)	46.1	26.1
Prior history of status epilepticus (%)	10.2	5.8
Etiology (%)		
Remote neurological	69.5	42.0
Acute neurological	29.9	37.7
Life-threatening condition	32.9	46.4
Cardiopulmonary arrest	6.0	36.2
Drug toxicity	3.0	4.3
Alcohol withdrawal	6.6	0.0
Treatment location (%)		
Emergency room	40.1	1.4
Ward	22.8	14.5
Intensive care unit	32.3	78.3
Other	4.8	5.8

Adapted with permission from Treiman and Walker (2006).

approximately 70% of the overt GCSE cases, whereas a life-threatening condition (acute stroke, infection, head injury, hypoxia) was the most common etiological factor in the subtle GCSE patients (46.4%). Patients with overt GCSE were most commonly treated in the emergency department (40.1%) or the intensive care unit (32.3%), whereas patients with subtle GCSE were usually treated in the intensive care unit (78.3%) (Treiman and Walker, 2006).

In elderly patients with overt GCSE, first-line treatment with phenobarbital stopped GCSE in 71.4% of cases (Fig. 1). First-line treatment with diazepam and phenytoin or phenytoin alone was not as effective, with success rates of 53.3% and 41.5%, respectively. Lorazepam as first-line therapy was successful in 63.0% of cases. Overall, the χ^2 analysis was significant ($\chi^2 = 8.69$, $p = 0.03$). This allowed pairwise comparison between individual treatment groups. There was no significant difference between the treatment success of phenobarbital and lorazepam. However, in pairwise treatment comparisons, phenytoin alone was significantly worse than lorazepam ($p = 0.03$) or phenobarbital ($p = 0.02$). These results from elderly patients with GCSE differ only slightly from those of the entire original study population ($N = 518$, 384 of whom had overt GCSE). In the original study, the treatment success rate was highest with lorazepam (64.9%), followed by phenobarbital (58.2%), diazepam and phenytoin (55.8%), and phenytoin alone (43.6%) (Treiman et al., 1998).

In the elderly subtle GCSE group, phenobarbital was most effective at stopping seizures (30.8% of cases) (Fig. 2). Treatment with lorazepam, phenytoin alone, or diazepam and phenytoin was successful in 14.3%, 11.8%, and 5.6% of

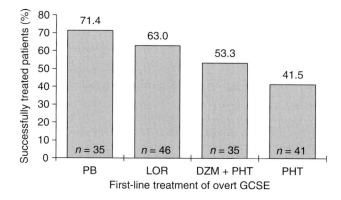

FIG. 1. Success of first-line treatment of overt GCSE in patients ≥65 years of age. Overall, there were statistically significant differences in rates of success ($\chi^2 = 8.69$, $p = 0.03$). There was no significant difference between PB and LOR in treatment success, but in pairwise treatment comparisons, PHT given alone was significantly worse than LOR ($p = 0.03$) or PB ($p = 0.02$). PB, phenobarbital; LOR, lorazepam; DZM, diazepam; PHT, phenytoin.

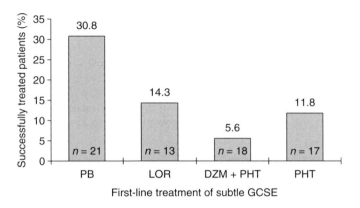

FIG. 2. Success of first-line treatment of subtle GCSE in patients ≥65 years of age. Chi-square analysis showed no significant differences among groups ($\chi^2 = 4.042$, $p = 0.257$). PB, phenobarbital; LOR, lorazepam; DZM, diazepam; PHT, phenytoin.

cases, respectively (Treiman and Walker, 2006). The chi-square analysis showed no significant differences between treatments. These results were similar to those of all subtle GCSE patients in the original study, independent of age: the treatment success rate was highest for phenobarbital (24.2%), followed by lorazepam (17.9%), diazepam and phenytoin (8.3%), and phenytoin alone (7.7%) (Treiman et al., 1998). This similarity in outcome between the entire study population and the elderly cohort is probably a reflection of the high proportion of older patients in the study; in the entire study population, the mean age was 58.6 years (±SD, 15.6 years) in the overt GCSE patients and 62.0 years (±SD, 15.1 years) in the subtle GCSE patients (Treiman et al., 1998).

As in the entire original study population, there were no significant differences in the elderly cohort between the subtle and overt GCSE groups in rate of SE recurrence, 30-day outcome, and adverse events. Overall success rates of first-line treatment in the elderly cohort were 56.9% for patients with overt GCSE and 14.5% for patients with subtle GCSE (Treiman and Walker, 2006). In this elderly cohort, the mean duration from onset of GCSE to enrollment in the study was 4.5 h (±SD, 17.9 h) for patients with overt GCSE, but 17.4 h (±SD, 21.5 h) for patients with subtle GCSE ($p < 0.001$). Again, these data are similar to those from the entire population of the original study (Treiman et al., 1998) and indicate that the longer SE persists, the more difficult it is to stop.

3. Discussion

The results of this analysis suggest that lorazepam, phenobarbital, and possibly diazepam followed by phenytoin are equally likely to be effective at stopping GCSE in elderly patients. Lorazepam has the advantages of ease of use (rapid administration) and of having a shorter duration of sedation than does pheno-barbital. Diazepam followed by phenytoin requires a prolonged period for administration, but recovery from sedation is likely to be quicker than with either lorazepam or phenobarbital. Use of diazepam/fosphenytoin rather than diazepam/phenytoin would substantially reduce the time required for administration, so this combination should be considered when avoidance of drug-induced sedation is important in the management of the individual patient. The potential role of valproate or levetiracetam as a second drug following treatment with a benzodiazepine is not yet clear. Table III outlines a treatment protocol for the management of GCSE, which is appropriate for elderly patients, and Table IV

TABLE III

TREATMENT PROTOCOL FOR GENERALIZED CONVULSIVE STATUS EPILEPTICUS IN THE ELDERLY

Time (min)	
0	Establish the diagnosis by observing one additional seizure in a patient with recent seizures or impaired consciousness, or by observing continuous behavioral and/or electrical seizure activity for >10 min.
	Start EEG as soon as possible, but do not delay treatment unless EEG verification of the diagnosis is necessary
5	Establish IV catheter with normal saline (dextrose solutions may precipitate phenytoin; with fosphenytoin either dextrose or saline is acceptable)
	Draw blood for serum chemistry, hematologic values, and antiepileptic drug concentrations. Test for hypoglycemia by finger stick. Administer 100 mg of thiamine (if indicated) followed by 50 ml of 50% glucose by direct push into the IV line
10	Administer lorazepam (0.1 mg/kg) by IV push (<2 mg/min)
25	If status epilepticus continues, start fosphenytoin (20 mg/kg PE) by fast IV push (up to 150 mg PE/min) directly into the IV port nearest to patient; if only phenytoin is available, give by slow IV push (<50 mg/min); with either preparation, monitor blood pressure and electrocardiogram during infusion
	If status epilepticus continues after 20 mg/kg PE of fosphenytoin (or 20 mg/kg of phenytoin), administer an additional 5 mg/kg PE and, if necessary, another 5 mg/kg PE, to a maximum dose of 30 mg/kg PE
60	If status epilepticus persists, give phenobarbital (20 mg/kg) by IV push (<100 mg/min) or start barbiturate coma; support respiration by endotracheal intubation; give pentobarbital (5–15 mg/kg) slowly as an initial IV dose to suppress all epileptiform activities and continue 0.5–5 mg/kg/h to maintain suppression; slow infusion rate periodically to determine cessation of seizure activity; monitor blood pressure, electrocardiogram, and respiratory function. If unable to suppress all epileptiform activity, change to continuous infusions of propofol (1 mg/kg given over 5 min, then 2–4 mg/kg/h; adjust to 1–15 mg/kg/h) or use midazolam (0.2 mg/kg bolus injection, followed by infusion of 0.05–0.5 mg/kg/h)

EEG, electroencephalogram; IV, intravenous; PE, phenytoin-equivalents.
Adapted with permission from Treiman, D. M. (2001).

TABLE IV

MANAGEMENT OF REFRACTORY GENERALIZED CONVULSIVE STATUS EPILEPTICUS

- After lorazepam followed by phenytoin, start IV general anesthesia with pentobarbital (5–15 mg/kg load, 0.5–5 mg/kg/h maintenance); titrate to suppress all epileptiform activity
- Monitor closely for hypotension and infection
- Search for etiology, correct if possible
- Adjust phenytoin concentration to 30 µg/ml, load phenobarbital to 100–150 µg/ml
- Slow infusion periodically (every 24–72 h) to see if epileptiform activity returns
- Do not give up; patients have recovered consciousness after >2 months of coma

outlines a useful approach to the management of GCSE when it is refractory to initial treatment.

V. Conclusions

GCSE occurs more frequently in elderly individuals than in younger adults and is associated with increased morbidity and mortality. Cerebrovascular disease is the most common cause of GCSE in the elderly. Approximately 75% of elderly patients present with overt GCSE and about 25% with subtle GCSE. Compared with phenytoin, lorazepam and phenobarbital are more effective at stopping overt GCSE when used as first-line treatment in elderly patients. Each drug has advantages and disadvantages, so choice of drug treatment in the elderly should to be based on individual patient characteristics.

References

Cascino, G. D. (1996). Generalized convulsive status epilepticus. *Mayo Clin. Proc.* **71,** 787–792.

Chin, R. F. M., Nevill, B. G. R., and Scott, R. C. (2004). A systematic review of the epidemiology of status epilepticus. *Eur. J. Neurol.* **11,** 800–810.

Claassen, J., Lokin, J. K., Fitzsimmons, B. F., Mendelsohn, F. A., and Mayer, S. A. (2002). Predictors of functional disability and mortality after status epilepticus. *Neurology* **58,** 139–142.

Cloyd, J., Hauser, W., Towne, A., Ramsay, R., Mattson, R., Gilliam, F., and Walczak, T. (2006). Epidemiological and medical aspects of epilepsy in the elderly. *Epilepsy Res.* **68**(Suppl. 1), S39–S48.

Coeytaux, A., Jallon, P., Galobardes, B., and Morabia, A. (2000). Incidence of status epilepticus in French-speaking Switzerland: (EPISTAR). *Neurology* **55,** 693–697.

DeLorenzo, R. J., Towne, A. R., Pellock, J. M., and Ko, D. (1992). Status epilepticus in children, adults, and the elderly. *Epilepsia* **33**(Suppl. 4), S15–S25.

DeLorenzo, R. J., Pellock, J. M., Towne, A. R., and Boggs, J. G. (1995). Epidemiology of status epilepticus. *J. Clin. Neurophysiol.* **12,** 316–325.

DeLorenzo, R. J., Hauser, W. A., Towne, A. R., Boggs, J. G., Pellock, J. M., Penberthy, L., Garnett, L., Fortner, C. A., and Ko, D. (1996). A prospective, population-based epidemiologic study of status epilepticus in Richmond, Virginia. *Neurology* **46,** 1029–1035.

Dodson, W. E., DeLorenzo, R. J., Pedley, T. A., Shinnar, S., Treiman, D. M., and Wannamaker, B. B. (1993). Treatment of convulsive status epilepticus: Recommendations of the Epilepsy Foundation of America's Working Group on Status Epilepticus. *JAMA* **270,** 854–859.

Gaitanis, J. N., and Drislane, F. W. (2003). Status epilepticus: A review of different syndromes, their current evaluation, and treatment. *Neurologist* **9,** 61–76.

Gastaut, H. (1983). Classification of status epilepticus. *In* "Status Epilepticus: Mechanisms of Brain Damage and Treatment. Advances in Neurology" (A. V. Delgado-Escueta, C. G. Wasterlain, D. M. Treiman, and R. J. Porter, Eds.), Vol. 34, pp. 15–35. Raven Press, New York.

Hauser, W. A. (1990). Status epilepticus: Epidemiologic considerations. *Neurology* **40**(5 Suppl. 2), 9–13.

Hesdorffer, D., Logroscino, G., Cascino, G., Annegers, J. F., and Hauser, W. A. (1998). Incidence of status epilepticus in Rochester, Minnesota, 1965–1984. *Neurology* **50,** 735–741.

Holtkamp, M., Othman, J., Buchheim, K., and Meierkord, H. (2006). Diagnosis of psychogenic nonepileptic status epilepticus in the emergency setting. *Neurology* **66,** 1727–1729.

Hui, A. C. F., Lam, A. K., Wong, A., Chow, K.-M., Chan, E. L. Y., Choi, S. L., and Wong, K.-S. (2005). Generalized tonic-clonic status epilepticus in the elderly in China. *Epileptic Disord.* **7,** 27–31.

Kalviainen, R., Eriksson, K., and Parviainen, I. (2005). Refractory generalised convulsive status epilepticus: A guide to treatment. *CNS Drugs* **19,** 759–768.

Knake, S., Rosenow, F., Vescovi, M., Oertel, W. H., Mueller, H. H., Wirbatz, A., Katsarou, N., and Hamer, H. M., and Status Epilepticus Study Group Hessen (SESGH) (2001). Incidence of status epilepticus in adults in Germany: A prospective, population-based study. *Epilepsia* **42,** 714–718.

Logroscino, G., Hesdorffer, D., Cascino, G. D., Annegers, J. F., and Hauser, W. A. (1997). Short-term mortality after a first episode of status epilepticus. *Epilepsia* **38,** 1344–1349.

Meldrum, B. (1991). Excitotoxicity and epileptic brain damage. *Epilepsy Res.* **10,** 55–61.

Meldrum, B. S., Vigouroux, R. A., and Brierley, J. B. (1973). Systemic factors and epileptic brain damage. Prolonged seizures in paralyzed, artificially ventilated baboons. *Arch. Neurol.* **29,** 82–87.

Rossetti, A. O., Hurwitz, S., Logroscino, G., and Bromfield, E. B. (2006). Prognosis of status epilepticus: Role of aetiology, age, and consciousness impairment at presentation. *J. Neurol. Neurosurg. Psychiatry* **77,** 611–615.

Towne, A. R., Pellock, J. M., Ko, D., and DeLorenzo, R. J. (1994). Determinants of mortality in status epilepticus. *Epilepsia* **35,** 27–34.

Treiman, D. M. (1983). General principles of treatment: Responsive and intractable status epilepticus in adults. *In* "Status Epilepticus (Advances in Neurology, Vol. 34)" (A. V. Delgado-Escueta, C. G. Wasterlain, D. M. Treiman, and R. J. Porter, Eds.), pp. 377–384. Raven Press, New York.

Treiman, D. M. (1993). Generalized convulsive status epilepticus in the adult. *Epilepsia* **34**(Suppl. 1), S2–S11.

Treiman, D. M. (2001). Status epilepticus. *In* "Treatment of Epilepsy: Principles and Practice" (E. Wyllie, Ed.), pp. 681–697. Lippincott Williams & Wilkins, Philadelphia.

Treiman, D. M., and Walker, M. C. (2006). Treatment of seizure emergencies: Convulsive and non-convulsive status epilepticus. *Epilepsy Res.* **68**(Suppl. 1), S77–S82.

Treiman, D. M., Walton, N. Y., and Kendrick, C. (1990). A progressive sequence of electroencephalographic changes during generalized convulsive status epilepticus. *Epilepsy Res.* **5,** 49–60.

Treiman, D. M., Meyers, P. D., Walton, N. Y., Collins, J. F., Colling, C., Rowan, A. J., Handforth, A., Faught, E., Calabrese, V. P., Uthman, B. M., Ramsay, R. E., and Mamdani, M. B. (1998).

A comparison of four treatments for generalized convulsive status epilepticus. Veterans Affairs Status Epilepticus Cooperative Study Group. *N. Engl. J. Med.* **339,** 792–798.

Vignatelli, L., Tonon, C., and D'Alessandro, R. (2003). Incidence and short-term prognosis of status epilepticus in adults in Bologna, Italy. *Epilepsia* **44,** 964–968.

Vignatelli, L., Rinaldi, R., Galeotti, M., de Carolis, P., and D'Alessandro, R. (2005). Epidemiology of status epilepticus in a rural area of northern Italy: A 2-year population-based study. *Eur. J. Neurol.* **12,** 897–902.

Waterhouse, E. J., Vaughan, J. K., Barnes, T. Y., Boggs, J. G., Towne, A. R., Kopec-Garnett, L., and DeLorenzo, R. J. (1998). Synergistic effect of status epilepticus and ischemic brain injury on mortality. *Epilepsy Res.* **29,** 175–183.

Wu, Y. W., Shek, D. W., Garcia, P. A., Zhao, S., and Johnston, S. C. (2002). Incidence and mortality of generalized convulsive status epilepticus in California. *Neurology* **58,** 1070–1076.

TREATMENT OF NONCONVULSIVE STATUS EPILEPTICUS

Matthew C. Walker

Department of Clinical and Experimental Epilepsy, Institute of Neurology
University College London, London WC1N 3BG, United Kingdom

Nonconvulsive status epilepticus (NCSE) is relatively common; it comprises at least one third of all cases of status epilepticus. NCSE may be an even more common, yet more elusive, condition in the elderly population. NCSE can be divided into complex partial status epilepticus (CPSE), NCSE in coma, and typical absence status epilepticus (TAS). The clinical manifestations may be subtle, and thus the diagnosis of these conditions is critically dependent on electroencephalography (EEG). When EEG demonstrates typical ictal patterns, the diagnosis is usually straightforward. However, in many circumstances the EEG pattern has to be differentiated from other encephalopathic patterns, and this differentiation can prove troublesome; clinical and electrographic response to treatment can prove helpful in these situations.

The prognosis for NCSE in the elderly is generally poor due to the underlying etiology rather than the persistence of electrographic discharges. Whether the neuronal damage that occurs in convulsive status epilepticus and in animal models of limbic status epilepticus also occurs in NCSE in humans is still a matter of debate. Intravenous treatment is not benign, especially in the elderly, who may be at greater risk of systemic complications from hypotensive and sedative agents. Therefore, a more conservative approach to the treatment of NCSE in the elderly is warranted.

INTERNATIONAL REVIEW OF
NEUROBIOLOGY, VOL. 81
DOI: 10.1016/S0074-7742(06)81019-6

287

Oral benzodiazepines should be used for the treatment of TAS and CPSE in noncomatose patients with a prior history of epilepsy, and in some circumstances, intravenous medication may be necessary. Generally, anesthetic coma should not be advised in either of these conditions. A more aggressive approach may be required with NCSE in coma, in the hope of improving a very poor prognosis. Treatment regimens will remain largely speculative until there are more relevant animal models and controlled trials of conservative versus aggressive treatment.

I. Introduction

The term status epilepticus was originally reserved for continuous convulsive seizure activity but is now applied to all seizure types. In certain forms, there is minimal or no convulsive activity, and these forms are collectively termed non-convulsive status epilepticus (NCSE), which comprises at least one-third of all cases of status epilepticus (Shorvon, 1994). This term is a descriptive classification and it is of most use because of the difficulty in making a diagnosis in these cases. Undoubtedly, it is an underdiagnosed and a misdiagnosed condition, as other encephalopathies share characteristics with NCSE.

Hughlings Jackson described prolonged "fugue" states associated with epilepsy (Jackson, 1931). However, it was the advent of electroencephalography (EEG) that enabled the diagnosis of these "fugue" states as epileptic, termed by Gastaut as "état de mal temporal" (Gastaut et al., 1956). NCSE has been transformed from a rare to a common condition by the introduction of more widely accessible EEG, particularly on intensive care units. The increasing use of EEG has raised new questions, such as what are the EEG criteria needed to differentiate status epilepticus from other encephalopathic conditions, and once diagnosed, how aggressively should we treat? This last point depends on the degree to which NCSE contributes to morbidity and mortality and whether treatment prevents these. These questions remain a matter of considerable debate.

NCSE can be divided into complex partial status epilepticus (CPSE), NCSE in coma, and typical absence status epilepticus (TAS). This chapter will concentrate on the diagnosis and treatment of these forms of NCSE in the elderly.

II. Diagnosis

Considerable debate exists concerning the diagnosis of NCSE conditions, with very few conclusions. NCSE can be classified as follows (Drislane, 1999; Krumholz et al., 1995; Walker, 2001).

- CPSE, which can be subdivided into:
  - CPSE in patients with epilepsy
  - CPSE *de novo*
- NCSE in coma, which can be further broken down into (Lowenstein and Aminoff, 1992):
  - Patients with a history of status epilepticus
  - Patients with subtle signs of seizure activity
  - Patients without clinical signs
- TAS, which can be subdivided into:
  - TAS in patients with idiopathic generalized epilepsy
  - TAS *de novo*

It is important that CPSE be differentiated not only from TAS, but also from postictal states and other encephalopathies, in order to determine appropriate management strategies.

Electroencephalogram interpretation for the diagnosis of NCSE is not always straightforward. Difficulties can occur in differentiating CPSE from encephalopathies of other causes (Husain *et al.*, 1999). Triphasic waves due to metabolic encephalopathies (particularly hepatic or hyperammonemic) can be frequent and are often mistakenly attributed to CPSE. Strict criteria for electrographic diagnosis of NCSE must therefore be used (Kaplan, 1996), such as (Husain *et al.*, 1999; Kaplan, 2000):

- Unequivocal electrographic seizure activity in which there is a typical evolution of changes in the EEG recording with a buildup of rhythmic activity
- Periodic epileptiform discharges or rhythmic discharge with clinical seizure activity
- Rhythmic discharge with either clinical or electrographic response to treatment

Despite these criteria, difficulties can still arise. Response to treatment is not a definitive indication of an epileptic cause, as triphasic waves can respond to treatment with benzodiazepines (Kaplan, 2000). Uncertainty also exists about the relevance of periodic lateralized epileptiform discharges (PLEDs) in diagnosing and predicting seizures (Pohlmann-Eden *et al.*, 1996). This EEG pattern is most notable following severe encephalitis or hypoxic injury, where discharges can occur with such periodicity that they can be confused with the periodic discharges seen following prolonged status epilepticus. Some have argued that such discharges represent ongoing seizure activity and should be treated as such. The general consensus, however, is that a multitude of etiologies can underlie PLEDs and that they should be treated as epileptic only if there is other evidence of ictal activity (Pohlmann-Eden *et al.*, 1996). In most instances, it is probably best to assume that such discharges represent cortical damage or metabolic derange-ment, regardless of cause, rather than ongoing seizure activity. Finally, there can also be difficulty in distinguishing between CPSE and TAS by EEG (Treiman and Delgado-Escueta, 1983), as rapid generalization can occur despite an initial focus that may only become apparent after treatment (Tomson *et al.*, 1992).

The clinical presentation of NCSE is highly variable from patient to patient. Prolonged change in personality, confusion, prolonged postictal state (greater than 30 min), or recent-onset psychosis can all be presentations of NCSE in patients with a previous diagnosis of epilepsy and should be investigated with EEG (Fagan and Lee, 1990; Kaplan, 1996; Tomson *et al.*, 1992). However, NCSE can present as confusion or personality change in noncomatose patients without a history of epilepsy. Because of the nonspecific signs and symptoms of NCSE, it is underdiagnosed in the confused elderly, in whom the confusion associated with NCSE is frequently attributed to other underlying conditions (Labar *et al.*, 1998; Litt *et al.*, 1998; Martin *et al.*, 2004). Specifically, the diagnosis of CPSE has caused much debate. The proposed definition of CPSE is "a prolonged epileptic episode in which fluctuating or frequently recurring electro-graphic epileptic discharges, arising in temporal or extratemporal regions, result in a confusional state with variable clinical symptoms" (Shorvon, 1994). This definition of CPSE allows emphasis to be placed on the fact that CPSE can originate in any cortical region and can fluctuate in a cyclic fashion in the absence of coma.

NCSE is an important, treatable cause of persistent coma following con-vulsive status epilepticus (DeLorenzo *et al.*, 1998). Indeed, convulsive status epilepticus can evolve into NCSE in which there is minimal or no motor activity but ongoing electrical activity. The subtle motor manifestations often consist of twitching of the limbs or facial muscles, or nystagmoid eye jerking; this activity can also result from brain damage exacerbated by hypoxia (Treiman, 1993).

NCSE in coma may occur in up to 8% of patients who have had no prior seizures or outward signs of seizure activity; this emphasizes the value of EEG in the evaluation of all comatose patients (Towne *et al.*, 2000). However, the interpretation of the EEG recording in coma can be difficult. In many instances, burst suppression patterns, periodic discharges, and encephalopathic triphasic patterns have been proposed to represent electrographic status epilepti-cus, but these most probably indicate underlying widespread cortical damage or dysfunction (see above).

Another form of NCSE diagnosed in the elderly is absence status epilepticus. Considered by most to be a childhood condition, absence status epilepticus in the elderly occurs in two important circumstances: TAS with a history of idiopathic generalized epilepsy (Agathonikou *et al.*, 1998) and late-onset absence status epilepticus developing *de novo* (usually following drug or alcohol withdrawal)

(Agathonikou *et al.*, 1998; Thomas *et al.*, 1993). Most absence epilepsies remit by adulthood, but certain syndromes, in particular juvenile myoclonic epilepsy, absence epilepsy with eyelid myoclonia, and generalized tonic-clonic seizures with phantom absences, are lifelong conditions with frequent occurrence of absence status epilepticus (Agathonikou *et al.*, 1998). Perhaps the specific diagnosis of TAS should be reserved for prolonged absence attacks with continuous or discontinuous 3-Hz spike-and-wave activity occurring in patients with generalized epilepsy (Cockerell *et al.*, 1994; Shorvon, 1994). However, the EEG recording may include irregular spike-and-wave activity, prolonged bursts of spike activity, sharp wave activity, or polyspike-and-wave activity. Whether to include such cases as TAS is uncertain.

III. Neuronal Damage Models

The risk of neuronal damage is another concern with continued seizure activity. Animal experiments helped to elucidate the mechanisms underlying this damage (Meldrum, 1991). Initial experiments demonstrated that neuronal damage was caused by convulsive status epilepticus even if the systemic compromise that occurs in convulsive status epilepticus was controlled, leading to the concept of excitotoxic neuronal damage; the presence of continuous electrographic seizure activity eventually results in neuronal damage (Olney *et al.*, 1986). Because animal models of NCSE have been proposed to replicate human CPSE, it has been assumed that neuronal damage unavoidably results from CPSE in humans. However, the methods of seizure generation do not extrapolate well to humans; powerful chemoconvulsants or prolonged high-frequency repetitive stimulation were used to induce status epilepticus in naïve animals. Comparable precipitants in humans, such as domoic acid poisoning from mussels, where pathological changes occur that are similar to the animal models, are rare (Cendes *et al.*, 1995). NCSE in humans is very different from these animal models. In humans, NCSE tends to have lower-frequency discharges (Granner and Lee, 1994), which if reproduced in animal models, produce substantially less neuronal damage than high-frequency discharges (Drislane, 1999; Krsek *et al.*, 2001). Furthermore, epileptic animals, and animals pretreated with antiepileptic drugs (AEDs), are resistant to chemoconvulsant-induced neuronal damage (Kelly and McIntyre, 1994; Najm *et al.*, 1998; Pitkanen *et al.*, 1999). Thus, prior AED use and history of epilepsy may confer neuroprotection. Finally, CPSE in humans often results from an acute precipitant; any further pathology caused by the status epilepticus may be negligible. Reports of human data of NCSE causing neuronal damage are confounded by etiology, concomitant illness, and treatment. Importantly, large case series of prolonged CPSE with no neurological sequelae have been published (Cockerell *et al.*, 1994; Williamson *et al.*, 1985).

Despite the absence of an acute neurological insult, CPSE resulted in significant rises in serum neuron-specific enolase (NSE), a marker for acute neuronal injury (DeGiorgio *et al.*, 1996, 1999). However, the rises in serum NSE could partially be the result of a breakdown in the blood–brain barrier rather than an increase in neuronal death; thus, cerebrospinal fluid NSE would be a more accurate predictor of neuronal damage (Correale *et al.*, 1998). The degree to which serum NSE correlates with neurological and cognitive disability in CPSE is unknown, and it is unknown whether the results of these studies hold for the majority of patients. Neuroimaging has also been inconclusive; reversible changes do occur, and in some selected patients, mild atrophy can be associated with CPSE (Lansberg *et al.*, 1999).

IV. Complex Partial Status Epilepticus

A. MORBIDITY AND MORTALITY

Acute neurological deficit and poor outcome have been reported in patients with CPSE (Engel *et al.*, 1978; Krumholz *et al.*, 1995; Treiman and Delgado-Escueta, 1983). The true incidence of morbidity and mortality following CPSE remains largely unknown because these have been selected case reports. Results from a study of 100 randomly selected patients with NCSE suggest that the mortality rate may be as high as 18% (Shneker and Fountain, 2003). Those with a prior diagnosis of epilepsy had a much lower mortality rate than those with NCSE in the setting of an acute medical illness. Indeed, considering just patients with epilepsy in randomly selected case series of CPSE, a low mortality rate of only 2 of 98 patients is revealed; one of these deaths was probably secondary to treatment (Cockerell *et al.*, 1994; Scholtes *et al.*, 1996; Shneker and Fountain, 2003; Tomson *et al.*, 1992; Williamson *et al.*, 1985). Furthermore, none of these study reports contained findings of serious morbidity in patients with epilepsy who developed NCSE. Thus, acutely precipitated CPSE and CPSE in the setting of a person with epilepsy should be considered to be two separate conditions. CPSE in epilepsy patients is likely a benign condition; patients commonly have repeated episodes, which may respond to oral benzodiazepines (Cockerell *et al.*, 1994).

Nonconvulsive CPSE in the setting of an acute medical illness has a high mortality and morbidity (Shneker and Fountain, 2003). The mortality usually relates to the underlying condition and medical complications. Reports of NCSE in elderly patients, who have a high incidence of concomitant medical illness and in whom NCSE is often the result of an ischemic/hypoxic event, have revealed a very high mortality and morbidity; in one case series only 2 of 10 patients returned to baseline function, and in another small case series only 3 of 15 patients were alive at 6 months following the status epilepticus episode (Labar *et al.*, 1998; Martin *et al.*, 2004). In both case series, concomitant complications and underlying etiology appeared to be the main determinants of the poor prognosis. Indeed, results from these series (Labar *et al.*, 1998; Martin *et al.*, 2004) and another study (Shneker and Fountain, 2003) emphasize the importance of treating underlying medical complications. Since aggressive AED treatment in such patients can increase mortality (Litt *et al.*, 1998), treatment decisions have to be based on a risk/benefit analysis. Randomized trials are needed to explore treatment options for CPSE in patients with acute illness or comorbidities.

B. Treatment

There is considerable evidence that CPSE in epilepsy patients is probably a benign condition and should be treated as such. In epilepsy patients, CPSE resolves with oral benzodiazepines (such as clobazam) (Corman *et al.*, 1998; Tinuper *et al.*, 1986). Many patients have repetitive attacks of CPSE (Cockerell *et al.*, 1994), and a specific treatment plan with oral benzodiazepines over a period of 2–3 days administered at home with the first symptoms can usually abort the status epilepticus; this strategy and others should be discussed with patients and caregivers.

In patients presenting to the hospital with CPSE, early recognition is a critical goal, as the delay in treatment results from failure to diagnose the condition rather than choice of therapeutic strategy (intravenous vs oral drug loading). Even when there is a resolution of the electrographic status epilepticus, the clinical response to benzodiazepines can be disappointing, possibly due to postictal effects (Cockerell *et al.*, 1994). Thus, initial clinical response may be a poor indicator of drug efficacy.

The medications commonly used to treat status epilepticus are not without adverse effects and can result in hypotension, respiratory depression, and on occasion, cardiorespiratory arrest (Shorvon, 1994). This effect is more pronounced with intravenous administration, due to rapid high serum levels. In an uncontrolled, nonrandomized study, intravenous benzodiazepines worsened the prognosis in the critically ill elderly, and aggressive intensive care unit treatment prolonged hospitalization (Litt *et al.*, 1998). Further study is warranted to determine if aggressive AED treatment for CPSE in the elderly is justified. The potential side effects of high-dose AEDs on the patient have to be considered. Intravenous valproate has been shown to avoid negative cardiovascular effects in some patients, and it thus may offer an advantage over intravenous benzodiazepines (Sinha and Naritoku, 2000). The use of general anesthesia to treat CPSE remains a matter for speculation. Of course, treatment of the underlying condition (e.g., encephalitis or metabolic derangement) is paramount and often leads to resolution of the status epilepticus. The best approach to treatment will be determined only in randomized studies of aggressive versus more conservative management.

V. NCSE in Coma

A. Morbidity and Mortality

NCSE is common in critically ill and comatose patients. Morbidity and mortality in these patients are increased due to the underlying etiology of NCSE (see Section IV) and potential exposure to infection in the hospital setting (Waterhouse and DeLorenzo, 2001).

B. Treatment

Scant data exist to guide treatment of NCSE in coma. NCSE following convulsive status epilepticus in coma should perhaps be treated aggressively with deep anesthesia and concomitant AEDs (Walker, 2001). Similarly, the association of electrographic status epilepticus with subtle motor activity related to hypoxic brain activity has a poor prognosis, and aggressive therapy with intravenous medication is justified, since the little evidence available indicates that such treatment improves prognosis (Lowenstein and Aminoff, 1992). Electrographic status epilepticus with no overt clinical signs is difficult to interpret, as it may represent status epilepticus or widespread cortical damage. Since these patients have a poor prognosis, aggressive treatment is recommended in the hope of improving outcomes. Finally, there is a group of patients in whom there are clinical signs of repetitive movements without electrographic seizure activity. In these patients, antiepileptic treatment and aggressive sedation are not recommended (Lowenstein and Aminoff, 1992).

VI. Typical Absence Status Epilepticus

A. MORBIDITY AND MORTALITY

Patients with TAS have a better prognosis than do patients with CPSE. In one study of 11 patients with absence status epilepticus *de novo*, there was no reported morbidity or mortality (Thomas *et al.*, 1992). There was no evidence of pathological damage or long-term behavioral effects in an animal model of absence status epilepticus (Wong *et al.*, 2003). Thus, in clinical settings, it seems unlikely that TAS causes any cerebral damage; there are many patients who have had frequent or prolonged episodes without any clinical sequelae. Therefore, aggressive treatment is not warranted.

B. TREATMENT

Absence status epilepticus responds rapidly to intravenous or oral benzodiazepines; these are so effective that the response is almost a *sine qua non*. Lorazepam at 0.05–0.1 mg/kg is the benzodiazepine of choice; however, the effect may be transient, and a longer-acting AED may be required. In most cases, oral therapy is adequate, but if a more rapid response is desired, then intravenous medication can be used. In cases where intravenous benzodiazepines are ineffective or contraindicated, intravenous valproate (20–40 mg/kg) can be administered. If a precipitating factor can be identified in late-onset *de novo* cases, long-term therapy is not usually indicated.

VII. Conclusions

NCSE can be classified as CPSE, NCSE in coma, or TAS and should be treated accordingly. The prognosis largely depends on the degree of impairment of consciousness and the underlying etiology. Patients with epilepsy who have an episode of NCSE with preserved consciousness have a good prognosis and rarely require aggressive treatment. Elderly patients with NCSE in the setting of an acute illness have a very poor prognosis, usually related to the underlying etiology; caution has to be exercised because in some instances aggressive treatment can worsen the prognosis. Treating the underlying cause and medical complications are the most important aspects of treatment in these patients. NCSE in coma often has a dismal prognosis and aggressive treatment is warranted.

References

Agathonikou, A., Panayiotopoulos, C. P., Giannakodimos, S., and Koutroumanidis, M. (1998). Typical absence status in adults: Diagnostic and syndromic considerations. *Epilepsia* **39,** 1265–1276.

Cendes, F., Andermann, F., Carpenter, S., Zatorre, R. J., and Cashman, N. R. (1995). Temporal lobe epilepsy caused by domoic acid intoxication: Evidence for glutamate receptor-mediated excitotoxicity in humans. *Ann. Neurol.* **37,** 123–126.

Cockerell, O. C., Walker, M. C., Sander, J. W., and Shorvon, S. D. (1994). Complex partial status epilepticus: A recurrent problem. *J. Neurol. Neurosurg. Psychiatry.* **57,** 835–837.

Corman, C., Guberman, A., and Benavente, O. (1998). Clobazam in partial status epilepticus. *Seizure* **7,** 243–247.

Correale, J., Rabinowicz, A. L., Heck, C. N., Smith, T. D., Loskota, W. J., and DeGiorgio, C. M. (1998). Status epilepticus increases CSF levels of neuron-specific enolase and alters the blood-brain barrier. *Neurology* **50,** 1388–1391.

DeGiorgio, C. M., Gott, P. S., Rabinowicz, A. L., Heck, C. N., Smith, T. D., and Correale, J. D. (1996). Neuron-specific enolase, a marker of acute neuronal injury, is increased in complex partial status epilepticus. *Epilepsia* **37,** 606–609.

DeGiorgio, C. M., Heck, C. N., Rabinowicz, A. L., Gott, P. S., Smith, T., and Correale, J. (1999). Serum neuron-specific enolase in the major subtypes of status epilepticus. *Neurology* **52,** 746–749.

DeLorenzo, R. J., Waterhouse, E. J., Towne, A. R., Boggs, J. G., Ko, D., DeLorenzo, G. A., Brown, A., and Garnett, L. (1998). Persistent nonconvulsive status epilepticus after the control of convulsive status epilepticus. *Epilepsia* **39,** 833–840.

Drislane, F. W. (1999). Evidence against permanent neurologic damage from nonconvulsive status epilepticus. *J. Clin. Neurophysiol.* **16,** 323–331; discussion 353.

Engel, J., Jr., Ludwig, B. I., and Fetell, M. (1978). Prolonged partial complex status epilepticus: EEG and behavioral observations. *Neurology* **28,** 863–869.

Fagan, K. J., and Lee, S. I. (1990). Prolonged confusion following convulsions due to generalized nonconvulsive status epilepticus. *Neurology* **40,** 1689–1694.

Gastaut, H., Roger, J., and Roger, A. (1956). The significance of certain epileptic fugues; concerning a clinical and electrical observation of temporal status epilepticus. *Rev. Neurol. (Paris)* **94,** 298–301.

Granner, M. A., and Lee, S. I. (1994). Nonconvulsive status epilepticus: EEG analysis in a large series. *Epilepsia* **35,** 42–47.

Husain, A. M., Mebust, K. A., and Radtke, R. A. (1999). Generalized periodic epileptiform discharges: Etiologies, relationship to status epilepticus, and prognosis. *J. Clin. Neurophysiol.* **16,** 51–58.

Jackson, J. H. (1931). On epilepsy and epileptiform convulsions. *In* "Selected Writings of John Hughlings Jackson" (J. Taylor, Ed.), Vol. 1. Hodder and Stoughton, London.

Kaplan, P. W. (1996). Nonconvulsive status epilepticus in the emergency room. *Epilepsia* **37,** 643–650.

Kaplan, P. W. (2000). Prognosis in nonconvulsive status epilepticus. *Epileptic Disord.* **2,** 185–193.

Kelly, M. E., and McIntyre, D. C. (1994). Hippocampal kindling protects several structures from the neuronal damage resulting from kainic acid-induced status epilepticus. *Brain Res.* **634,** 245–256.

Krsek, P., Mikulecka, A., Druga, R., Hlinak, Z., Kubova, H., and Mares, P. (2001). An animal model of nonconvulsive status epilepticus: A contribution to clinical controversies. *Epilepsia* **42,** 171–180.

Krumholz, A., Sung, G. Y., Fisher, R. S., Barry, E., Bergey, G. K., and Grattan, L. M. (1995). Complex partial status epilepticus accompanied by serious morbidity and mortality. *Neurology* **45,** 1499–1504.

Labar, D., Barrera, J., Solomon, G., and Harden, C. (1998). Nonconvulsive status epilepticus in the elderly: A case series and a review of the literature. *J. Epilepsy* **11,** 74–78.

Lansberg, M. G., O'Brien, M. W., Norbash, A. M., Moseley, M. E., Morrell, M., and Albers, G. W. (1999). MRI abnormalities associated with partial status epilepticus. *Neurology* **52,** 1021–1027.

Litt, B., Wityk, R. J., Hertz, S. H., Mullen, P. D., Weiss, H., Ryan, D. D., and Henry, T. R. (1998). Nonconvulsive status epilepticus in the critically ill elderly. *Epilepsia* **39,** 1194–1202.

Lowenstein, D. H., and Aminoff, M. J. (1992). Clinical and EEG features of status epilepticus in comatose patients. *Neurology* **42,** 100–104.

Martin, Y., Artaz, M. A., and Bornand-Rousselot, A. (2004). Nonconvulsive status epilepticus in the elderly. *J. Am. Geriatr. Soc.* **52,** 476–477.

Meldrum, B. (1991). Excitotoxicity and epileptic brain damage. *Epilepsy Res.* **10,** 55–61.

Najm, I. M., Hadam, J., Ckakraverty, D., Mikuni, N., Penrod, C., Sopa, C., Markarian, G., Luders, H. O., Babb, T., and Baudry, M. (1998). A short episode of seizure activity protects from status epilepticus-induced neuronal damage in rat brain. *Brain Res.* **810,** 72–75.

Olney, J. W., Collins, R. C., and Sloviter, R. S. (1986). Excitotoxic mechanisms of epileptic brain damage. *Adv. Neurol.* **44,** 857–877.

Pitkanen, A., Nissinen, J., Jolkkonen, E., Tuunanen, J., and Halonen, T. (1999). Effects of vigabatrin treatment on status epilepticus-induced neuronal damage and mossy fiber sprouting in the rat hippocampus. *Epilepsy Res.* **33,** 67–85.

Pohlmann-Eden, B., Hoch, D. B., Cochius, J. I., and Chiappa, K. H. (1996). Periodic lateralized epileptiform discharges—a critical review. *J. Clin. Neurophysiol.* **13,** 519–530.

Scholtes, F. B., Renier, W. O., and Meinardi, H. (1996). Non-convulsive status epilepticus: Causes, treatment, and outcome in 65 patients. *J. Neurol. Neurosurg. Psychiatry* **61,** 93–95.

Shneker, B. F., and Fountain, N. B. (2003). Assessment of acute morbidity and mortality in nonconvulsive status epilepticus. *Neurology* **61,** 1066–1073.

Shorvon, S. (1994). "Status Epilepticus: Its Clinical Features and Treatment in Children and Adults." Cambridge University Press, Cambridge.

Sinha, S., and Naritoku, D. K. (2000). Intravenous valproate is well tolerated in unstable patients with status epilepticus. *Neurology* **55,** 722–724.

Thomas, P., Beaumanoir, A., Genton, P., Dolisi, C., and Chatel, M. (1992). '*De novo*' absence status of late onset: Report of 11 cases. *Neurology* **42,** 104–110.

Thomas, P., Lebrun, C., and Chatel, M. (1993). *De novo* absence status epilepticus as a benzodiazepine withdrawal syndrome. *Epilepsia* **34,** 355–358.

Tinuper, P., Aguglia, U., and Gastaut, H. (1986). Use of clobazam in certain forms of status epilepticus and in startle-induced epileptic seizures. *Epilepsia* **27**(Suppl. 1), S18–S26.

Tomson, T., Lindbom, U., and Nilsson, B. Y. (1992). Nonconvulsive status epilepticus in adults: Thirty-two consecutive patients from a general hospital population. *Epilepsia* **33,** 829–835.

Towne, A. R., Waterhouse, E. J., Boggs, J. G., Garnett, L. K., Brown, A. J., Smith, J. R., Jr., and DeLorenzo, R. J. (2000). Prevalence of nonconvulsive status epilepticus in comatose patients. *Neurology* **54,** 340–345.

Treiman, D. M. (1993). Generalized convulsive status epilepticus in the adult. *Epilepsia* **34**(Suppl. 1), S2–S11.

Treiman, D. M., and Delgado-Escueta, A. V. (1983). Complex partial status epilepticus. *Adv. Neurol.* **34,** 69–81.

Walker, M. C. (2001). Diagnosis and treatment of nonconvulsive status epilepticus. *CNS Drugs* **15,** 931–939.

Waterhouse, E. J., and DeLorenzo, R. J. (2001). Status epilepticus in older patients: Epidemiology and treatment options. *Drugs Aging* **18,** 133–142.

Williamson, P. D., Spencer, D. D., Spencer, S. S., Novelly, R. A., and Mattson, R. H. (1985). Complex partial status epilepticus: A depth-electrode study. *Ann. Neurol.* **18,** 647–654.

Wong, M., Wozniak, D. F., and Yamada, K. A. (2003). An animal model of generalized nonconvulsive status epilepticus: Immediate characteristics and long-term effects. *Exp. Neurol.* **183,** 87–99.

ANTIEPILEPTIC DRUG FORMULATION AND TREATMENT IN THE ELDERLY: BIOPHARMACEUTICAL CONSIDERATIONS

Barry E. Gidal

School of Pharmacy and Department of Neurology
University of Wisconsin, Madison, Wisconsin 53705, USA

The pharmacokinetics of antiepileptic drugs (AEDs) determine their effectiveness in the treatment of patients with epilepsy. Given the likelihood of comorbid medical conditions that require polytherapy, as well as the normal physiological changes associated with aging, an understanding of AED pharmacokinetics and pharmacodynamics in the elderly patient is critical. There is a relative sparsity of data regarding changes in the oral absorption patterns of AEDs that may accompany aging. Therefore, the objective of this chapter is to discuss fundamental principles related to oral drug absorption, and to discuss their potential impact on AED treatment in the older patient. Although most drugs are absorbed via the diffusion process, active transport also plays a role in absorption. While the gastrointestinal tract shows remarkable resilience during aging, physiological changes that influence oral and esophageal function, gastric pH, gastric emptying rates, and intestinal transit times do occur. Oral administration of AEDs may be affected by changes associated with aging, including altered oral protective reflexes, xerostomia, thickening of the esophageal smooth muscle layer, reduced contraction velocity and duration, altered esophageal emptying rates, and enteric plexus neuron reduction. Gastric acid secretion is similar between older and younger patients, but older patients require more time to return to baseline gastric pH values and have prolonged gastric emptying rates compared to younger patients. Elderly patients may similarly have reduced numbers of myenteric neurons, decreased postprandial contractions, reduced frequency of migrating motor complex, and diminished rectal compliance as well as reduced sphincter tones. All of these effects observed in the aging patient, in turn, produce numerous opportunities for changes in AED absorption, particularly for those agents demonstrating poor water solubility or variable absorption patterns.

I. Introduction

The management of antiepileptic drug (AED) pharmacokinetics presents a significant challenge in the treatment of elderly patients with epilepsy. Pharmacokinetic characteristics of AEDs, particularly mechanisms of elimination, protein binding, and the potential for pharmacokinetic or pharmacodynamic interactions, determine the effectiveness of epilepsy treatment and are important in drug selection. Elderly patients often require polytherapy for comorbid medical conditions, and this factor, along with the normal physiological changes associated with aging, may be of high importance in the treatment of elderly patients with epilepsy. While certain changes in drug pharmacokinetics, including declines in either hepatic oxidative metabolism or renal glomerular filtration rate (McLean and Le Couteur, 2004; Schwartz, 2006), have been frequently described in older patients, one of the most fundamental aspects of pharmacotherapy, oral absorption of a drug product, has been frequently overlooked or assumed to be unchanged in the older patient. However, specific changes may occur in the gastrointestinal (GI) tract, and these changes have the potential to introduce additional variability in drug pharmacokinetics.

Given the relatively complex pharmacokinetic profiles of several of the commonly used AEDs, one might assume that substantial clinical or experimental data exist to delineate these potential changes in oral drug absorption. Unfortunately, relatively few studies have been conducted in this particular area. A synopsis of these data was published in a review article in the journal *Epilepsy Research* (Gidal, 2006). The purpose of this chapter is to discuss the basic principles associated with the process of drug absorption, particularly biopharmaceutical characteristics of AEDs and the physiological changes expected to occur in the GI tract of the elderly patient.

II. Biopharmaceutical Considerations

The systemic exposure, or bioavailability, of a drug following oral administration is a multifactorial process. For a drug to ultimately reach the systemic circulation, a number of steps or processes must occur, and these include drug disintegration and dissolution, diffusion through GI fluids, and finally mucosal membrane permeation and uptake into blood or lymph. Important factors that can affect any of these steps include alterations in aqueous solubility, membrane permeability (including efflux transport), or presystemic metabolism (Burton *et al.*, 2002). Recognition of pharmaceutical characteristics, such as solubility and permeability of a specific compound, may help predict product- or patient-specific variables (e.g., comorbid diseases), either of which may influence drug bioavailability. The Biopharmaceutics Classification System (BCS) has been developed as a predictive tool for identifying drugs in which oral absorption patterns may be influenced by formulation or physiological variables (Amidon *et al.*, 1995). According to this system, pharmaceutical compounds may be grouped into categories based on solubility and permeability properties.

Class I BCS compounds exhibit high solubility and high permeability. These compounds are generally very well absorbed, although certain factors (either pathophysiological or drug related) that affect gut surface area or GI transit time may alter their bioavailability. Class II compounds display low aqueous solubility but high membrane permeability. The oral absorption of these compounds is usually rate-limited by dissolution. Anything that increases the rate of *in vivo* dissolution will tend to increase bioavailability. Class III compounds display high solubility but have low permeability characteristics. In general, the absorption of these drugs is limited by their membrane permeability rates. Finally, Class IV agents display low aqueous solubility and low permeability. These compounds typically have very poor oral bioavailability.

High-solubility compounds have dose to solubility volume ratios that are ≤ 250 ml. In this context, solubility is defined as the minimum solubility over a pH range of 1–8, at a temperature of $37\,^{\circ}\mathrm{C}$. Aqueous solubility of a compound can be estimated by determining the ability of a drug to partition between lipid (octanol) and aqueous milieus. Very soluble drugs need less than 1 part solvent to dissolve 1 part solute (e.g., drug compound), whereas sparingly soluble and slightly soluble drugs require 30–100 and 100–1000 parts to dissolve 1 part solute, respectively. Drugs tend to display increased aqueous solubility in the ionized state, as compared to the unionized state. Therefore, the pH of the GI fluid will in large part determine the solubility behavior of a given drug (Martinez and Amidon, 2002).

Permeability of the human jejunal wall is expressed in units of 10^{-4} cm/sec. Highly permeable drugs display oral absorption $\geq 90\%$ and do not appear to have GI tract instability. Clearly, the fraction of an oral dose of a drug that is absorbed is dependent on the drug's permeability across the GI mucosa. For drugs with permeability constants $<2 \times 10^{-4}$ cm/sec, one would expect incomplete absorption (Martinez and Amidon, 2002).

Intestinal permeability is a complex, multifactorial process that involves variables such as molecular size, lipophilicity, van der Waals surface area, and hydrogen bond formation (Palm *et al.*, 1996). In general, the degree of intestinal permeability is inversely correlated with van der Waals surface area (Palm et al., 1996) and molecular size, and it is positively correlated with lipophilicity (Camenisch et al., 1998). In addition to gut membrane barriers and drug molecular characteristics, specific components of gastric and mucosal layers, such as cholesterol, phosphatidylcholine, and linoleic acid, can influence the drug permeation of small lipophilic molecules (Larhed et al., 1998). Drugs are often absorbed through transcellular diffusion, but absorption can also occur via paracellular diffusion through tiny, water-filled channels with diameters of about 5–18 Å (Fanning et al., 1999). In humans, the intestinal surface area available for paracellular transport is limited to 0.01% of total membrane surface area. Therefore, paracellular transport plays a limited role in the absorption of most drugs. In general, passive diffusion of most drugs is not likely to be influenced significantly by normal aging processes (Saltzman et al., 1995).

Although most drugs are absorbed via the diffusion processes discussed above, active transport has been shown to be important in the absorption of several drugs, including levodopa (Morris, 1978) and gabapentin (Beydoun et al., 1995). Membrane transport proteins appear to be involved with both influx (absorption) and efflux functions. Intestinal transporter locations appear to be site specific and have been best described for vitamins and nutrients. Iron transport and absorption occur primarily in the upper duodenum (Lash and Saleem, 1995), whereas transport and absorption of vitamin B12 occur in the ileum (Kapadia, 1995). Examples of drug transporters include organic anionic transporters [e.g., salicylic acid, valproic acid, nonsteroidal anti-inflammatory drugs (NSAIDs)] (Van Aubel et al., 2000); organic cation transporters (e.g., dopamine, antiarrhythmics) (Siest et al., 2004); nucleoside transporters (e.g., nucleoside analogues) (Kong et al., 2004); and L-amino acid transporters (e.g., gabapentin, levodopa) (Tsuji, 2005). Important efflux transporters are the adenosine 5′-triphosphate (ATP)-binding cassette transporter proteins, including the multidrug resistance proteins (MRP), of which the MRP1 gene product, P-glycoprotein, has been shown to be important in modulating the extent of absorption of a number of drugs (Sakaeda et al., 2002). Several AEDs have been shown to be substrates for this transporter (Kwan and Brodie, 2005). It is unclear whether these transporters significantly influence oral bioavailability of the AEDs or whether age impacts the activity of this transporter system.

Some drugs may rely on all three processes of absorption (transcellular and paracellular diffusion as well as active transport); therefore, the bioavailability of a given drug may be influenced by physiological changes in the structure or function of the GI membrane. Given the importance of drug solubility in the absorption process, formulation changes in a given drug compound have the potential to introduce pharmacokinetic variability in patients in whom the GI milieu or structure has been altered, whether due to disease- or age-related alterations. These potential changes in GI function and their subsequent impact on drug absorption will be discussed next.

III. GI Function and Age

Although the GI tract shows remarkable resilience during aging, both healthy and nonhealthy aging processes lead to physiological changes that influence oral and esophageal function, gastric pH, gastric emptying rates, and intestinal transit times (Blechman and Gelb, 1999; Firth and Prather, 2002). During the normal aging process, moderate reductions in small intestine absorptive function are seen, as are slowed GI transit times and prolonged gastric emptying times, all of which may be due to regional losses of neurons (Orr and Chen, 2002). It has been generally accepted that in otherwise healthy individuals, sufficient functional reserve exists, so these modest changes are unlikely to substantially alter absorption and bioavailability for most drugs. However, several studies have suggested that advancing age may, in fact, alter drug absorption. For example, the bioavailability of drugs like propranolol, verapamil, and labetalol almost doubles in older patients (Wilkinson, 1997). Bioavailability of levodopa has also been shown to increase in older patients (Robertson et al., 1989). Although results of some experiments in animals have suggested no change in valproic acid absorption (Cato et al., 1995), others have suggested increased intestinal permeability with aging (Mullin et al., 2002). Active transport of several nutrients, including glucose, vitamin B12, and calcium, may, on the other hand, demonstrate reduced absorption (McLean and Le Couteur, 2004; Nordin et al., 2004). Whether these changes are primarily due to absorption properties or due to presystemic metabolism can be difficult to discern clinically.

In cases where the reduction in functional absorptive capacity suggests a reduction in intestinal absorptive cells, one would expect to see a reduction in both rate and extent of drug absorption. Conversely, depending on the physicochemical characteristics of the particular drug, alterations in gut physiology may actually facilitate drug absorption. In the case of a drug in which the absorption rate is limited by dissolution, a decrease in gastric emptying time may increase the overall extent of absorption, thereby increasing systemic drug exposure. For drug products that are acid labile, an increase in gastric pH may also result in enhanced drug absorption. Of all the potential transformations in physiological activity, changes in gastric function are the most likely to influence AED absorption kinetics.

A. GASTRIC FUNCTION

Although gastric acid secretion is generally similar in older and younger individuals, the incidence of achlorhydria is approximately 10–20% among elderly patients compared to <1% in younger subjects (Blechman and Gelb, 1999).

Similarly, while average fasting gastric pH is similar in older and younger individuals, the postprandial pH response may differ significantly (Russell *et al.*, 1993; Table I), with older individuals requiring significantly more time to return to baseline pH values. In addition, elderly persons may also display marked day-to-day, and even hour-to-hour, variability in fasting gastric pH (Fig. 1).

Altered gastric pH could modify drug absorption in at least two ways. Since drugs tend to be more soluble when ionized, increased gastric pH can decrease the dissolution of weak bases and increase the solubility of weak acids. Weakly acidic drugs dissolve more quickly in environments where the pH is higher and a greater fraction of the drug is in an ionized form. Conversely, weakly basic drugs display slower dissolution rates at higher gastric pH, since more of the drug would exist in its unionized form. Thus, fluctuations in gastric pH may distort the absorption and bioavailability profiles of certain drugs, such as the commonly used AED phenytoin, which is weakly acidic (Yamaoka *et al.*, 1983). Although specific absorption studies in elderly patients have not been published, one could speculate that in situations where gastric pH was elevated during drug administration, phenytoin bioavailability might be enhanced, whereas reductions in gastric pH might lead to reduced dissolution and absorption.

In aging patients, concomitant (and potentially inconsistent) use of nonprescription drugs, such as H2 antagonists and proton pump inhibitors, can increase gastric pH and perhaps subtly influence day-to-day absorption of drugs. In addition to drug influences, ingestion of meals that elevate gastric pH may influence drug product dissolution (Toothaker and Welling, 1980). For example, postprandial elevations in gastric pH would increase the proportion of a weakly basic drug existing in the unionized state, thereby decreasing its dissolution. This would

TABLE I
COMPARISON OF GASTRIC pH BETWEEN YOUNG AND ELDERLY PATIENTS

	Young ($n = 24$)	Elderly ($n = 79$)	p Value
Fasted			
Median pH (range)	1.7 (1.4–2.0)	1.3 (1.1–1.6)	0.014
During the meal			
Median pH (range)	5.0 (4.4–5.6)	4.9 (3.9–5.5)	0.006
Postprandial response median (range), min			
pH 5	8 (2–17)	23 (6–46)	0.015
pH 4	14 (8–40)	52 (27–115)	0.0002
pH 3	42 (26–83)	89 (44–167)	0.0026
pH 2	100 (44–143)	154 (82–210)	0.026

Reproduced from Russell *et al.* (1993), with kind permission of Springer Science and Business Media, Copyright 1993.

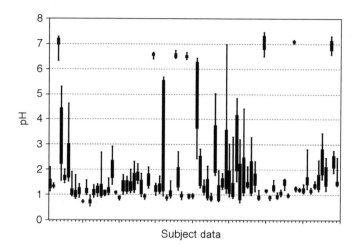

FIG. 1. Gastric pH in healthy elderly persons. Box-whisker plots of the distribution for fasted gastric pH within each of the 79 subjects. Each box represents the distribution of pH values over the entire 1-h fasted pH recording period (*n* = 240). Adapted from Russell *et al.* (1993), with kind permission of Springer Science and Business Media, Copyright 1993.

likely impact both maximal peak concentrations and overall systemic exposure (area under the plasma drug concentration vs time curve). The same meal might have the opposite effect for a weakly acidic drug such as phenytoin. Therefore, clinicians should consider meal effects and concomitant use of acid-modifying drugs in patients displaying fluctuating serum concentrations of phenytoin.

The rate and extent of absorption of acid-labile products, prodrugs that require an acidic medium for conversion, or other agents, such as ketoconazole, that require an acidic environment for optimal absorption (Carlson *et al.*, 1983) may be influenced by gastric pH changes. Clorazepate is a prodrug whose active moiety is desmethyldiazepam (Brooks *et al.*, 1977). The bioconversion of clorazepate takes place in the stomach and is pH dependent (Abruzzo *et al.*, 1977). Results from several studies have suggested that the impaired absorption of desmethyldiazepam following clorazepate administration seen in some older subjects (Abruzzo *et al.*, 1977; Ochs *et al.*, 1979) may occur, at least in part, because of elevation in gastric pH. Similarly, elevated gastric pH (greater than 5.0) resulted in impaired absorption of ketoconazole in elderly individuals (Hurwitz *et al.*, 2003).

Modified gastric emptying rates can also significantly influence the bioavailability of certain drugs, depending on their physicochemical properties. Gastric emptying time may be significantly prolonged in elderly individuals; this may, in part, be due to autonomic neuropathy in some patients (Altomare *et al.*, 1999;

Orr and Chen, 2002). In one study, elderly subjects (aged 70–84 years) had significantly slower gastric emptying for both solids and liquids as compared to younger subjects (aged 23–50 years) (Clarkston *et al.*, 1997). Alterations in gastric emptying would generally be expected to have an impact on drugs that were either very soluble or poorly soluble. Absorption of highly soluble drugs may be delayed by the reduced gastric emptying that presents in some elderly patients (Clarkston *et al.*, 1997), whereas absorption of poorly soluble compounds may actually be facilitated. Increased oral bioavailability of levodopa has been observed in older patients, and it has been suggested that this is due to delayed gastric emptying (Evans *et al.*, 1981). Solid dosages are more susceptible to these effects than liquid formulations (Christensen *et al.*, 1985).

Finally, gastric emptying rates can be influenced by the volume of fluid ingested. In general, increases in fluid volume lead to increases in gastric empty-ing rates (Moran *et al.*, 1999). Fluid volume may influence the bioavailability of drugs with rapid dissolution for which the rate-limiting step is movement of drug from the stomach to the intestinal site of absorption (BCS Class I and III drugs). In this setting, one might expect the time to maximal absorption to decrease as the amount of liquid taken with the medication increases.

B. Intestinal Changes

Age-related changes may also be seen in the intestine. Whereas enterocytes are essentially unchanged in the elderly individual, reduced numbers of myen-teric neurons have been noted in the small intestine (Shankle *et al.*, 1993). Changes in manometric patterns, including decreased postprandial contractions and reduced frequency of migrating motor complex, have been seen, prompting questions about altered GI transit times in some individuals (Firth and Prather, 2002). These changes are minor and would not be expected to significantly alter the absorption of most drugs. An exception is the absorption of some extended-release AED formulations, which is probably sensitive to GI transit times, although the clinical significance of this is still unclear. In one evaluation of an osmotically controlled release formulation of carbamazepine, systemic exposure varied substantially as a function of the transit time from stomach and small intestine (Wilding *et al.*, 1991). It is important to recognize that the GI tract is best viewed as a heterogeneous organ system, and that physiological properties such as gut motility and number of villi and microvilli may change across the length of the gut (Martinez and Amidon, 2002).

Elderly patients may have diminished rectal compliance as well as reduced sphincter tones. Delayed colonic transit times can also be seen, particularly in inactive elderly patients (Firth and Prather, 2002). Whether these changes, or the increased incidence of constipation in older patients, result in altered rectal drug absorption is not clear.

Transport proteins involving both drug influx and efflux are also likely to be factors in the overall extent of intestinal drug absorption. Protein transporters include organic anionic and cationic transporters, amino acid transporters, P-glycoprotein transporters, vitamin transporters, and intestinal dipeptide transporters (Martinez and Amidon, 2002). Gabapentin absorption, which utilizes the L-amino acid transporter (system L), shows considerable variation in absorption rates between young, healthy subjects (Fig. 2; Gidal *et al.*, 2000). However, there are no marked differences in gabapentin absorption rates between elderly patients and younger subjects (Boyd *et al.*, 1999). In addition, pregabalin, although structurally related to gabapentin, does not appear to display saturable absorption (via system L) from the GI tract; indeed, unlike gabapentin, pregabalin absorption appears to be dose proportional (Gilron and Flatters, 2006). One would therefore not expect to see substantial variations in oral absorption between older and younger subjects receiving this medication. The ways in which various transporters influence drug absorption, and how they respond to intestinal changes associated with aging, is an area of increasing research interest.

C. ORAL AND ESOPHAGEAL CHANGES

A great number of esophageal changes in response to aging have been documented; in particular, altered oral protective reflexes and xerostomia may complicate oral administration of certain medications (Firth and Prather, 2002).

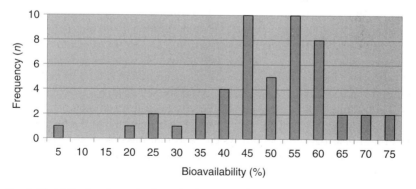

FIG. 2. Variability in gabapentin absorption in young, healthy subjects. Frequency distribution of bioavailability values following single-dose administration of gabapentin 600 mg to 50 healthy volunteers. Reprinted from Gidal *et al.* (2000), with permission from Elsevier, Copyright 2000.

Other esophageal changes include thickening of the smooth muscle layer, reduced contraction velocity and duration, and delayed esophageal emptying. Advancing age has been associated with an approximately 20–60% reduction in enteric plexus neurons (Blechman and Gelb, 1999; Dharmarajan et al., 2001). While these changes probably do not directly affect absorption, they may have an impact on the ability to administer certain dosage forms orally.

IV. Conclusions

The complex process of drug absorption involves the interaction of numerous drug-related and physiological variables. The impact of both healthy and unhealthy aging on various GI functions is incompletely characterized in many cases. At this time, the increased research focus on AED pharmacokinetics in the elderly often centers on changes in processes involved in drug elimination. However, the absorption kinetics of certain drug products may also be substantially affected by changes in GI function that are associated with aging.

When treating the elderly epilepsy patient, it is important to realize that the GI tract is a heterogeneous, dynamic organ and that certain physiological changes in GI function may be associated with aging. For example, although enterocytes are similar in older and younger individuals, slower intestinal transit may be observed in the elderly. Similarly, while fasting gastric pH is generally unchanged in elderly patients, postprandial return to basal levels slows with age. These effects and others could alter drug solubility, modify transport protein activity, and have other consequences that may reshape drug absorption profiles among the elderly. It is likely that AEDs with either low solubility, low permeability, or both are more likely to have absorption profiles that alter with age. Conversely, high-solubility, high-permeability agents would not be expected to display impaired or increased absorption profiles in the aging patient. This is particularly true of drugs that do not undergo presystemic oxidative metabolism.

The generalization that drug absorption is unaffected by aging is clearly simplistic, and clinicians must consider the biopharmaceutical properties of a given drug formulation in concert with individual patient physiological variables. Unfortunately, little information exists in the literature that specifically addresses the issue of variability in AED absorption in the older patient. In many cases, data are extrapolated from younger, healthier patient populations, or from healthy volunteers. On the basis of the biopharmaceutical principles discussed previously, we can, however, make some reasonable predictions. Among the currently available, commonly used AEDs, water solubility is likely to be one of the most important parameters in assessing oral absorption patterns. AEDs that display poor water solubility are likely to show increased variability in the aging patient. The use of drugs such as phenytoin and carbamazepine may therefore be particularly problematic. Newer-generation AEDs with improved solubility characteristics might therefore be expected to show less intra- and interpatient variability in oral absorption patterns.

References

Abruzzo, C. W., Macasieb, T., Weinfeld, R., Rider, J. A., and Kaplan, S. A. (1977). Changes in the oral absorption characteristics in man of dipotassium clorazepate at normal and elevated gastric pH. *J. Pharmacokinet. Biopharm.* **5,** 377–390.

Altomare, D. F., Portincasa, P., Rinaldi, M., Di Ciaula, A., Martinelli, E., Amoruso, A., Palasciano, G., and Memeo, V. (1999). Slow-transit constipation: Solitary symptom of a systemic gastrointestinal disease. *Dis. Colon Rectum* **42,** 231–240.

Amidon, G. L., Lennernas, H., Shah, V. P., and Crison, J. R. (1995). A theoretical basis for a biopharmaceutic drug classification: The correlation of *in vitro* drug product dissolution and *in vivo* bioavailability. *Pharm. Res.* **12,** 413–420.

Beydoun, A., Uthman, B. M., and Sackellares, J. C. (1995). Gabapentin: Pharmacokinetics, efficacy, and safety. *Clin. Neuropharmacol.* **18,** 469–481.

Blechman, M. B., and Gelb, A. M. (1999). Aging and gastrointestinal physiology. *Clin. Geriatr. Med.* **15,** 429–438.

Boyd, R. A., Turck, D., Abel, R. B., Sedman, A. J., and Bockbrader, H. N. (1999). Effects of age and gender on single-dose pharmacokinetics of gabapentin. *Epilepsia* **40,** 474–479.

Brooks, M. A., Hackman, M. R., Weinfeld, R. E., and Macasieb, T. (1977). Determination of clorazepate and its major metabolites in blood and urine by electron capture gas-liquid chromatography. *J. Chromatogr.* **135,** 123–131.

Burton, P. S., Goodwin, J. T., Vidmar, T. J., and Amore, B. M. (2002). Predicting drug absorption: How nature made it a difficult problem. *J. Pharmacol. Exp. Ther.* **303,** 889–895.

Camenisch, G., Alsenz, J., van de Waterbeemd, H., and Folkers, G. (1998). Estimation of permeability by passive diffusion through Caco-2 cell monolayers using the drugs' lipophilicity and molecular weight. *Eur. J. Pharm. Sci.* **6,** 317–324.

Carlson, J. A., Mann, H. J., and Canafax, D. M. (1983). Effect of pH on disintegration and dissolution of ketoconazole tablets. *Am. J. Hosp. Pharm.* **40,** 1334–1336.

Cato, A., III, Pollack, G. M., and Brouwer, K. L. (1995). Age-dependent intestinal absorption of valproic acid in the rat. *Pharm. Res.* **12,** 284–290.

Christensen, F. N., Davis, S. S., Hardy, J. G., Taylor, M. J., Whalley, D. R., and Wilson, C. G. (1985). The use of gamma scintigraphy to follow the gastrointestinal transit of pharmaceutical formulations. *J. Pharm. Pharmacol.* **37,** 91–95.

Clarkston, W. K., Pantano, M. M., Morley, J. E., Horowitz, M., Littlefield, J. M., and Burton, F. R. (1997). Evidence for the anorexia of aging: Gastrointestinal transit and hunger in healthy elderly vs. young adults. *Am. J. Physiol.* **272,** R243–R248.

Dharmarajan, T. S., Pitchumoni, C. S., and Kokkat, A. J. (2001). The aging gut. *Pract. Gastroenterol.* **25,** 15–27.

Evans, M. A., Broe, G. A., Triggs, E. J., Cheung, M., Creasey, H., and Paull, P. D. (1981). Gastric emptying rate and the systemic availability of levodopa in the elderly parkinsonian patient. *Neurology* **31,** 1288–1294.

Fanning, A. S., Mitic, L. L., and Anderson, J. M. (1999). Transmembrane proteins in the tight junction barrier. *J. Am. Soc. Nephrol.* **10,** 1337–1345.

Firth, M., and Prather, C. M. (2002). Gastrointestinal motility problems in the elderly patient. *Gastroenterology* **122,** 1688–1700.

Gidal, B. E. (2006). Drug absorption in the elderly: Biopharmaceutical considerations for the antiepileptic drugs. *Epilepsy Res.* **68**(Suppl. 1), S65–S69.

Gidal, B. E., Radulovic, L. L., Kruger, S., Rutecki, P., Pitterle, M., and Bockbrader, H. N. (2000). Inter- and intra-subject variability in gabapentin absorption and absolute bioavailability. *Epilepsy Res.* **40,** 123–127.

Gilron, I., and Flatters, S. J. (2006). Gabapentin and pregabalin for the treatment of neuropathic pain: A review of laboratory and clinical evidence. *Pain Res. Manag.* **11**(Suppl. A), 16A–29A.

Hurwitz, A., Ruhl, C. E., Kimler, B. F., Topp, E. M., and Mayo, M. S. (2003). Gastric function in the elderly: Effects on absorption of ketoconazole. *J. Clin. Pharmacol.* **43,** 996–1002.

Kapadia, C. R. (1995). Vitamin B12 in health and disease. Part I. Inherited disorders of function, absorption, and transport. *Gastroenterologist* **3,** 329–344.

Kong, W., Engel, K., and Wang, J. (2004). Mammalian nucleoside transporters. *Curr. Drug Metab.* **5,** 63–84.

Kwan, P., and Brodie, M. J. (2005). Potential role of drug transporters in the pathogenesis of medically intractable epilepsy. *Epilepsia* **46,** 224–235.

Larhed, A. W., Artursson, P., and Bjork, E. (1998). The influence of intestinal mucus components on the diffusion of drugs. *Pharm. Res.* **15,** 66–71.

Lash, A., and Saleem, A. (1995). Iron metabolism and its regulation. A review. *Ann. Clin. Lab. Sci.* **25,** 20–30.

Martinez, M. N., and Amidon, G. L. (2002). A mechanistic approach to understanding the factors affecting drug absorption: A review of fundamentals. *J. Clin. Pharmacol.* **42,** 620–643.

McLean, A. J., and Le Couteur, D. G. (2004). Aging biology and geriatric clinical pharmacology. *Pharmacol. Rev.* **56,** 163–184.

Moran, T. H., Wirth, J. B., Schwartz, G. J., and McHugh, P. R. (1999). Interactions between gastric volume and duodenal nutrients in the control of liquid gastric emptying. *Am. J. Physiol.* **276,** R997–R1002.

Morris, J. G. (1978). A review of some aspects of the pharmacology of levodopa. *Clin. Exp. Neurol.* **15,** 24–50.

Mullin, J. M., Valenzano, M. C., Verrecchio, J. J., and Kothari, R. (2002). Age- and diet-related increase in transepithelial colon permeability of Fischer 344 rats. *Dig. Dis. Sci.* **47,** 2262–2270.

Nordin, B. E., Need, A. G., Morris, H. A., O'Loughlin, P. D., and Horowitz, M. (2004). Effect of age on calcium absorption in postmenopausal women. *Am. J. Clin. Nutr.* **80,** 998–1002.

Ochs, H. R., Greenblatt, D. J., Allen, M. D., Harmatz, J. S., Shader, R. I., and Bodem, G. (1979). Effect of age and Billroth gastrectomy on absorption of desmethyldiazepam from clorazepate. *Clin. Pharmacol. Ther.* **26,** 449–456.

Orr, W. C., and Chen, C. L. (2002). Aging and neural control of the GI tract: IV. Clinical and physiological aspects of gastrointestinal motility and aging. *Am. J. Physiol. Gastrointest. Liver Physiol.* **283,** G1226–G1231.

Palm, K., Luthman, K., Ungell, A. L., Strandlund, G., and Artursson, P. (1996). Correlation of drug absorption with molecular surface properties. *J. Pharm. Sci.* **85,** 32–39.

Robertson, D. R., Wood, N. D., Everest, H., Monks, K., Waller, D. G., Renwick, A. G., and George, C. F. (1989). The effect of age on the pharmacokinetics of levodopa administered alone and in the presence of carbidopa. *Br. J. Clin. Pharmacol.* **28,** 61–69.

Russell, T. L., Berardi, R. R., Barnett, J. L., Dermentzoglou, L. C., Jarvenpaa, K. M., Schmaltz, S. P., and Dressman, J. B. (1993). Upper gastrointestinal pH in seventy-nine healthy, elderly, North American men and women. *Pharm. Res.* **10,** 187–196.

Sakaeda, T., Nakamura, T., and Okumura, K. (2002). MDR1 genotype-related pharmacokinetics and pharmacodynamics. *Biol. Pharm. Bull.* **25,** 1391–1400.

Saltzman, J. R., Kowdley, K. V., Perrone, G., and Russell, R. M. (1995). Changes in small-intestine permeability with aging. *J. Am. Geriatr. Soc.* **43,** 160–164.

Schwartz, J. B. (2006). Erythromycin breath test results in elderly, very elderly, and frail elderly persons. *Clin. Pharmacol. Ther.* **79,** 440–448.

Shankle, W. R., Landing, B. H., Ang, S. M., Chui, H., Villarreal-Engelhardt, G., and Zarow, C. (1993). Studies of the enteric nervous system in Alzheimer disease and other dementias of the elderly: Enteric neurons in Alzheimer disease. *Mod. Pathol.* **6,** 10–14.

Siest, G., Jeannesson, E., Berrahmoune, H., Maumus, S., Marteau, J. B., Mohr, S., and Visvikis, S. (2004). Pharmacogenomics and drug response in cardiovascular disorders. *Pharmacogenomics* **5,** 779–802.

Toothaker, R. D., and Welling, P. G. (1980). The effect of food on drug bioavailability. *Annu. Rev. Pharmacol. Toxicol.* **20,** 173–179.

Tsuji, A. (2005). Small molecular drug transfer across the blood-brain barrier via carrier-mediated transport systems. *NeuroRx* **2,** 54–62.

Van Aubel, R. A., Masereeuw, R., and Russel, F. G. (2000). Molecular pharmacology of renal organic anion transporters. *Am. J. Physiol. Renal Physiol.* **279,** F216–F232.

Wilding, I. R., Davis, S. S., Hardy, J. G., Robertson, C. S., John, V. A., Powell, M. L., Leal, M., Lloyd, P., and Walker, S. M. (1991). Relationship between systemic drug absorption and gastrointestinal transit after the simultaneous oral administration of carbamazepine as a controlled-release system and as a suspension of ^{15}N-labelled drug to healthy volunteers. *Br. J. Clin. Pharmacol.* **32,** 573–579.

Wilkinson, G. R. (1997). The effects of diet, aging and disease-states on presystemic elimination and oral drug bioavailability in humans. *Adv. Drug Deliv. Rev.* **27,** 129–159.

Yamaoka, Y., Roberts, R. D., and Stella, V. J. (1983). Low-melting phenytoin prodrugs as alternative oral delivery modes for phenytoin: A model for other high-melting sparingly water-soluble drugs. *J. Pharm. Sci.* **72,** 400–405.

INDEX

CONTENTS OF RECENT VOLUMES

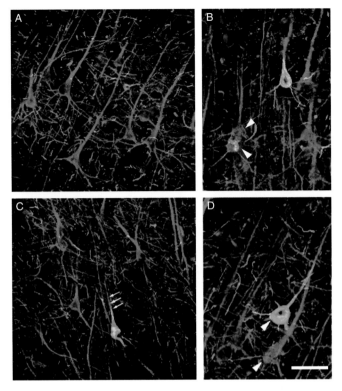

Morrison and Hof, Chapter 3, Fig. 1. (See Legend in Text.)

	Base-line	Treat-ment	30 min post	60 min post	90 min post	120 min post
A Control						
B GLU						
C GLU + Low $[Ca^{2+}]_e$						
D GLU + APV						
E GLU + CNQX						
F GLU + MCPG						

100 nM   1000 nM

DeLorenzo et al., Chapter 4, Fig. 6. (See Legend in Text.)

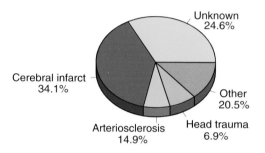

RAMSAY *ET AL.*, CHAPTER 7, FIG. 1. (See Legend in Text.)

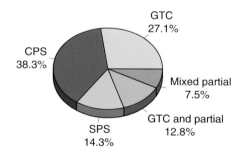

RAMSAY *ET AL.*, CHAPTER 7, FIG. 2. (See Legend in Text.)